D0806228

MECHANICAL BEHAVIOUR OF COMPOSITES AND LAMINATES

Proceedings of the European Mechanics Colloquium 214 'Mechanical Behaviour of Composites and Laminates' held in Kupari, Yugoslavia, 16–19 September 1986.

MECHANICAL BEHAVIOUR OF COMPOSITES AND LAMINATES

Edited by

W. A. GREEN

Department of Theoretical Mechanics,
University of Nottingham, UK

and

M. MIĆUNOVIĆ

Faculty of Mechanical Engineering,
University of Svetozar Marković, Yugoslavia

ELSEVIER APPLIED SCIENCE
LONDON and NEW YORK

ELSEVIER APPLIED SCIENCE PUBLISHERS LTD
Crown House, Linton Road, Barking, Essex IG11 8JU, England

Sole Distributor in the USA and Canada
ELSEVIER SCIENCE PUBLISHING CO., INC.
52 Vanderbilt Avenue, New York, NY 10017, USA

WITH 23 TABLES AND 126 ILLUSTRATIONS

British Library Cataloguing in Publication Data

European Colloquium 214 Mechanical Behaviour
of Composites and Laminates (*1986:*
Yugoslavia)
Mechanical behaviour of composites and
laminates
1. Composite materials
I. Title II. Green, W. A. III. Mićunović, M.
620′.1′18 TA418.9.C6

ISBN 1-85166-144-1

Library of Congress CIP data applied for

Printed in Great Britain by Galliard (Printers) Ltd, Great Yarmouth

PREFACE

The first announcement for Euromech Colloquium 214 on the Mechanical Behaviour of Composites and Laminates indicated the intention of the Chairmen to provide a forum for experimentalists and theoreticians to report on the current state of development in this rapidly expanding field. The invitation called for contributions concerned with four main topics. These were: the formulation of constitutive equations; the experimental determination of mechanical response; wave propagation and vibrations; and methods of solution of boundary value problems.

We believe that the papers contained in this report of the proceedings amply demonstrate the wide-ranging response to this call. The participants at the Colloquium included materials scientists, engineers, physicists, applied mathematicians and pure mathematicians and the papers in turn reflect this variety of disciplines which contribute to the study of composites and laminates.

In compiling the papers for this volume we have arranged them into five groups. The first group consists of nine contributions dealing with the topics of edge effects, impact damage and fracture criteria including both experimental and theoretical aspects of these topics. This is followed by a set of six papers devoted to the theoretical study of wave propagation and vibration. The third group consists of six articles concerned with homogenization theory applied to derive mathematical models of inhomogeneous media and structures. The four papers comprising the next group are devoted to the derivation of constitutive equations and the solution of boundary value problems for non-linear and inelastic composites. Finally there is a set of five contributions concerned with numerical methods and optimization. This grouping is by no means exclusive and many of the articles which we have assigned under one of these headings could equally well have been allocated under another. We would therefore encourage the reader whose main field of interest may be covered by one of these group headings to explore the possibilities in the remaining groups.

In the discussion session which formed the closing stage of the Colloquium a number of the participants laid emphasis on the interdisciplinary nature of the study of composites. The topic covers materials science, mathematical modelling and mechanics and structural analysis. There was a feeling that these aspects had in the past been considered in isolation but that some attempts at coordination were now under way. On the mechanics aspect, stress was laid on the interrelation between theory, experiment and analysis. In this respect it was felt that the meeting had been of value in bringing together practitioners from each of these three areas, in a relaxed, informal atmosphere which gave the opportunity for fruitful interaction. We believe that we speak for all the participants when we say that the presentations were both stimulating and informative. Each session of talks evoked a lively response in terms of questions and discussion and one of the exciting features of the meeting was the chance to see the wider aspects of the field in relation to one's own work. We are convinced that the papers contained in this compilation make a significant contribution to the research activity in this field and we welcome this opportunity to make them accessible to a wider audience. In doing so, we earnestly hope that the reader will experience the enthusiasm and enjoyment for their subjects displayed by the authors which was one of the features of the Colloquium.

It is a pleasure to thank Miss Gordana Avramović, Mr Nenad Grujović and Dr Dragan Milosavljević whose hard work and enthusiasm ensured the smooth running of the sessions and the welfare of the delegates. They were assisted in the preparatory work by Mr Miroslav Živković and to him also we express our thanks. We are grateful to the Serbian Scientific Council, the Regional Scientific Society of Sumadija and the University of Svetozar Marković, Kragujevac, for their financial assistance towards the costs of the meeting and to the management of the Hotel Kupari complex for affording us the use of their conference facilities. Finally we acknowledge with thanks the secretarial assistance of Mrs Dušanka Žugić and Mrs Anne Perkins.

W. A. GREEN
Nottingham

M. V. MIĆUNOVIĆ
Kragujevac

CONTENTS

II. Dynamics Theory

III. Homogenization

IV. Nonlinear, Inelastic and Thermal Behaviour

V. Numerical Methods and Optimization

1

EDGE EFFECTS IN FAILURE OF COMPRESSION PANELS

by

G.A.O. Davies, N. Buskell, K.A. Stevens

(Department of Aeronautics, Imperial College, London)

Summary

A theoretical and experimental investigation is conducted into the behaviour after initial buckling of a series of carbon-composite compression panels. The panels are loaded to failure which is found to be precipitated by interlaminar shear stresses near the edge due to both membrane forces and twisting moments. Failure is predictable using either a fine finite-element model or a simple boundary layer approximation. The effect of these edge stresses is to reduce the strength of the panels by about 40%.

Introduction

The Aircraft Industry has pioneered the use of high grade fibre-composite structures in the pursuit for optimistic performances. Carbon composites in particular should be attractive in view of their high specific stiffness, which is the governing material parameter for obtaining high buckling resistance in thin-walled or slender structures. However the development of carbon composite designs for structures in compression has been slow and their full potential has not been realised. 'Knock-down' factors of 2 have been used by industry on permissible strains simply to play safe on strength criterion, with the result that the new lithium-aluminium alloys look competitive even though their specific stiffness is a moderate 20% better than traditional alloys. The shortfall in compressive performance of carbon composites has been due largely to three factors:

(a) Commonly used epoxy resin matrix materials do degenerate when warm and moist and this leads to destabilisation of the fibres. This environmental degradation can be accelerated if the structure suffers moderate impact damage.

(b) Composite laminates do have in-built imperfections and this can lead to strength reductions in imperfection-sensitive structures like some shells or stiffened panels which are optimised for coincident buckling modes.

(c) Thin-walled compression panels are usually stabilised by deploying discretely spaced stiffeners of some variety to reduce the effective breadth/thickness (b/t) ratio. Such discontinuities inevitably raise the spectre of three-dimensional stress fields - the Achilles heel of all composite laminate constructions.

The first weakness is being overcome with newer matrix materials, and the last two should be solved by better detailed design and production technology, particularly since the new high-strain carbon fibres offer potentially 50% more strength. There is therefore every incentive to understand, and be able to predict, the mechanisms involved in the failure of stiffened compression panels.

Buckling of Stiffened Panels

The behaviour of compression panels stiffened, or supported along their longitudinal edges, is well understood for homogeneous isotropic metals. Thus as the load is increased the panel will buckle and lose stiffness at a critical stress σ_{cr} proportional to $E(t/b)^2$ where E is an effective modulus. If the b/t ratio is large, this initial buckling stress may be much below any potential material failure, and considerable **post-buckling** strength may be exploited. The immediate drop in stiffness after initial buckling is of order 1/3 (see Fig. 7) and as the panel is further loaded the central region remains dormant whilst the extra load is carried by the regions near the supports or stiffeners. Eventual failure may be a further instability mode of the support-stiffener, a pure material failure, or a combination of both. A characteristic buckling mode of a long panel is shown in Fig. 1, where the main feature to observe is the shedding of membrane load N_{xx} to the edge supports, particularly in the region of the buckle crests. Also shown is a typical distribution of twisting moment M_{xy} which is a maximum at the node lines. This feature will be shown to be important.

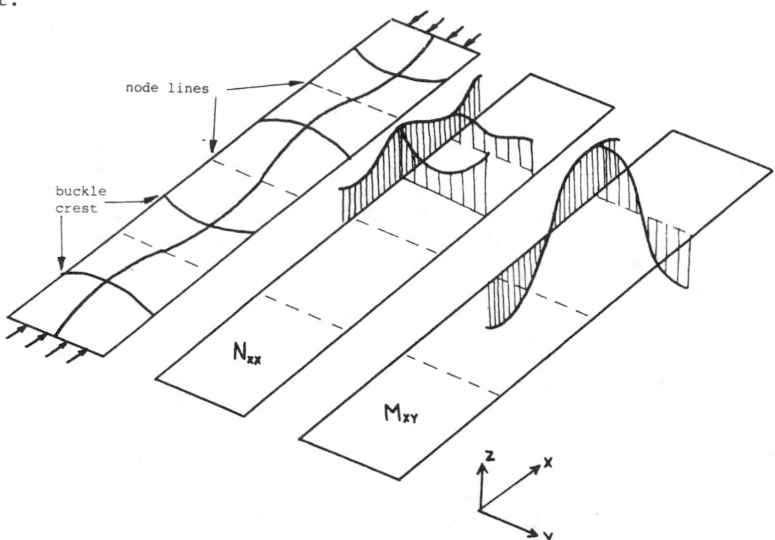

Fig. 1. Post-buckling Modes and Stress Distributions in a long panel

In carbon composite panels very little work has been reported but the behaviour after the initial bcukling load must be radically different from familiar metal behaviour which usually involves a gradual reduction in the panel stiffness, leading to ultimate failure. The reduced post-buckled stiffness will remain fairly constant. Failure is likely to be sudden, without warning, and will probably destroy the evidence of the failure mechanism.

The prediction of post-buckling stiffness is possible using classical laminate theory and the plate equations for moderately large displacements. However, to predict the ultimate load capacity we need to know the location and the mechanism of the failure process, and in the case of post-buckled panels the source may be near a supported edge where the stress concentrations of Fig. 1 are highest. In addition to the membrane action

N_{xx}, the post-buckling deformation will feed upon the initial imperfections and produce twisting moments M_{xy} and edge shears Q_y all of which are likely to interact with three-dimensional edge effects - the object of this study.

Experimental work by Starnes (1) has identified the source of failure along the edge at the node lines of Fig. 1. This partially contradicts the above assertion since the membrane compression is not a maximum here, and the Kirchoff shear ($Q_y - \partial M_{xy}/\partial x$) is actually zero! Industry has therefore been understandably cautious, and to the authors' knowledge no military aircraft wing-box has been designed with post-buckled strength. Other structures such as fuselage panels, control surfaces, and civil aircraft wings have b/t ratios in the range 20 to 50 where the ultimate strength may be up to 3 times the initial buckling value. Tests have been undertaken by the U.K. industry and failure originating at the node points has been confirmed (2). The source has been proposed as the twisting moments M_{xy} which produce high fibre strains in the 45° fibres near the surface. A finite element analyis was used to evaluate N_{xx} and M_{xy} and the consequent maximum strains were evaluated in the surface fibres. The ultimate strength was correctly predicted using a permissible fibre strain of 5,300 microstrain. However this value is unrealistically low for multiply laminates and no attempt was made to predict failure for stacking sequences with 0° surface fibres. A programme has therefore been completed by the authors to identify the source of postbuckling failure and validate some analytical predictive capability (3,4).

Experimental Tests

Panels have been tested having b/t ratios in the range 20 to 60, and having aspect ratios between 3 and 8; consequently their behaviour is close to that of infinitely long panels. The loaded ends were clamped and the edges were simply supported. Three layups were used of 16-ply, medium stiffness XAS 914 prepreg, arranged in 'quasi-isotropic' sequences as follows:-

$(0/90/\pm45)_{2S}$; $(0/\pm45/90)_{2S}$; $(45/0/-45/90)_{S2}$

These three layups produce quite different through-thickness variations in any edge boundary layer.

A 'hard' displacement-controlled testing machine was used to avoid destroying the panels once failure started. Nevertheless it was very necessary to use an acoustic emission system to provide a signal of imminent failure and so enable us to examine and monitor the failure process. This technique was extremely successful. Fig. 2a shows that ring-down counting is an accurate prophet of composite failure, even though the dramatic increase in noise may occur only 5% below the failure load. The plates were removed for ultrasonic scanning and then reloaded again until further emission indicated new damage. The C-scans were most revealing and completely repeatable for several tests. They showed progressive delamination sites at the edges near the node lines and occurring at the 45° interfaces near the plate neutral axis as shown in Fig. 2b. Fig. 3 shows the pattern of delamination sites and a typical C-scan. Propagation of the delamination is extremely rapid in compression, since the plate is effectively reduced to two separate plates having only 1/8 of the original rigidity and buckling resistance.

The delamination interfaces occurred at positions above and below the neutral surface in an alternating fashion along the edge sides, and this pattern clearly corresponded to the changing sign of the twisting moment M_{xy} along the edges. The delamination source was confirmed to be a shear, rather than tensile peeling, from a fractographic survey of the face

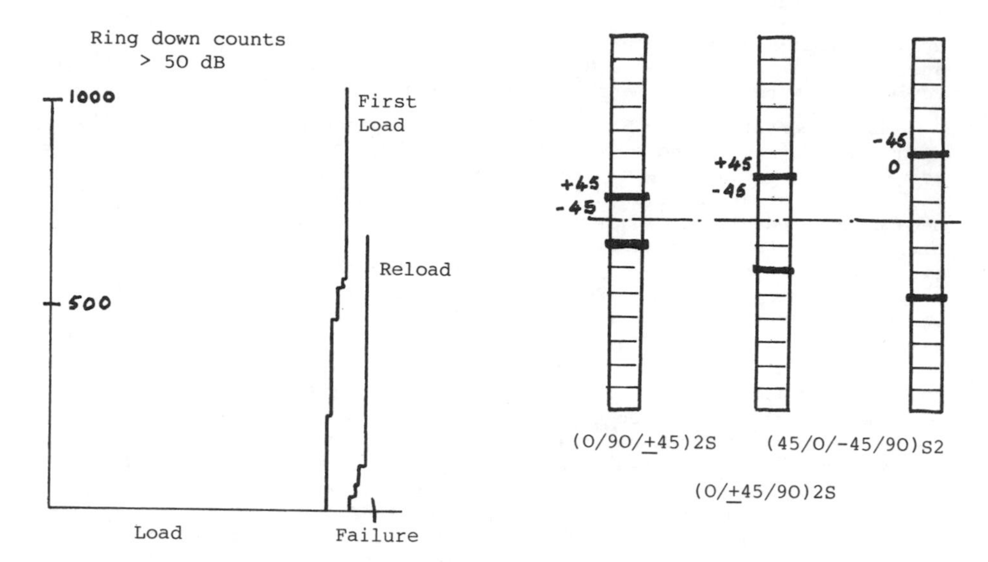

Fig. 2a. Acoustic Emission

Fig. 2b. Delamination Sites

performed by the Royal Aircraft Establishment. [See ref. 5 for details of this technique.] The fact that shear delamination takes place between +45° layers and -45° or 0° layers is not surprising since these are the weaker interfaces, but the fractographic evidence showed the component σ_{xz} to be the driving shear. It is usual to look for the normal component σ_{yz} as a source of edge delamination.

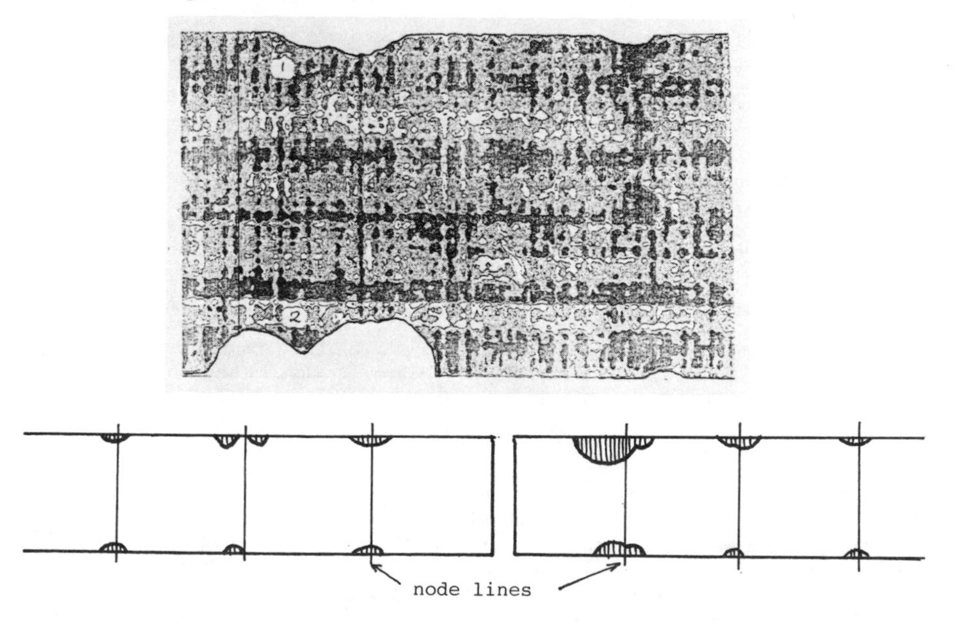

Fig. 3. Delamination Sites and Ultrasonic C-scan

The reasons for this behaviour, the dependence on M_{xy}, and an explanation for the sites near the neutral axis rather than the plate surface (where M_{xy} produces the largest fibre strains) needs an investigation into the nature of the edge effects and the three-dimensional stress field in a boundary layer at the plate edge.

Edge Effects in Composite Plates

The presence of a boundary layer zone, in which interlaminar stresses σ_z, τ_{xz}, τ_{yz} occur, is well known and has been covered by Pipes & Pagano amongst others (6,7). Full finite-element analyses have been used to substantiate the variation in these edge stresses and to confirm that the boundary layer extends an order of the plate thickness t from the edge. Previous arguments will not be repeated here but they have concentrated on the effects of longitudinal membrane loading, and in particular on the consequent peeling stress σ_z which is likely to cause internal delamination. A simplified analysis is possible using equilibrium arguments to evaluate the interlaminar stresses but some assumption has to be made on the profile of their variation away from the edge. We show later that this assumption is an important feature.

A similar edge effect due to the in-plane shears τ_{xy}, induced by M_{xy}, was suspected. Classical laminate and plate-bending theory will predict these as a linear variation with z. However, these shear stresses must decay to zero at the free edge, and it is well known (8) that they are turned through a right angle into τ_{xz} components in the "Kirchoff" Boundary Layer of thickness order t. The decay of τ_{xy} and the sympathetic rise of τ_{xz} was thought to be the mechanism for the onset of edge delamination.

The Kirchoff boundary layer in isotropic metals predicts both shear stresses to be of order M_{xy}/t^2 but this cannot be true in composites where the two components have differing order of stiffness. The in-plane shears τ_{xy} are resisted largely by the 45° angle-ply layers and have a respectable effective modulus G_{xy} compared with the transverse modulus G_{xz} which is largely dependent on the matrix stiffness. An approximate analysis in the Appendix shows this, and also demonstrates that the boundary layer now extends a distance of order $t(G_{xy}/G_{xz})^{\frac{1}{2}}$, which in the panels tested is in excess of 3t.

It was decided therefore to resort to a three-dimensional finite element analysis to confirm these speculations, but it was clearly necessary to control the number of elements used. The failure locations had been identified as the regions near the node lines where both M_{xy} and N_{xx} have local maxima. The variation in these two forces occurs over a distance of order panel-width b which is large compared to t, and consequently it is justifiable to isolate a small region where the variations in M_{xy} and N_{xx} are ignored. Fortunately there exists a program developed by Bartholomew and Mercer (9) for analysing this two-dimensional flexure-torsion problem using a bolt-on constraint to any standard finite-element program - in this case MSC-NASTRAN was used. The problem was idealised as a doubly-symmetric finite strip of width 4t in which effectively the boundary layers on two edges are brought together and it is hoped that diffusion to plate behaviour is complete at the centre line. Fig. 4 shows the scheme. The elements used were the 8-node rectangular CHEXA, using 3 elements through each lamina to capture through-thickness variations. The full mesh shown deploys 1600 elements which taxed the resources of a CRAY 1S, and explains why a rather risky diffusion length of 2t was assumed. The values of N_{xx} and M_{xy} were taken from a separate finite element analysis of the whole plate corrected for the small shortfall between these

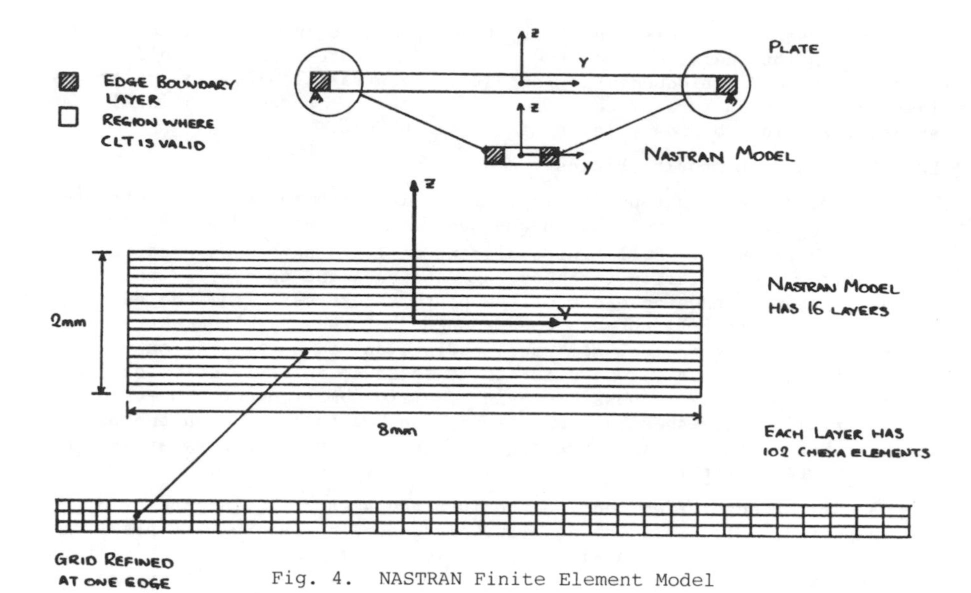

Fig. 4. NASTRAN Finite Element Model

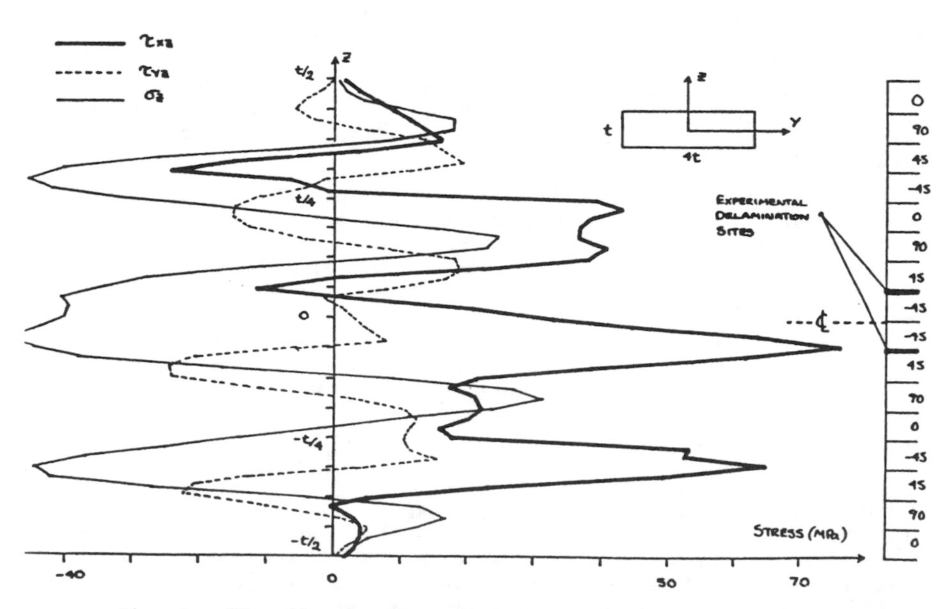

Fig. 5. Edge Stresses from Finite Element Model

(N_{xx} = -523N/mm; M_{xy} = 130N)

predictions and the experimental results (see Fig. 7). It was found that M_{xy} and N_{xx} contributed equally to the interlaminar shears for large values of b/t, but that as b/t was reduced to 20, the postbuckling strength then exceeded the initial buckling load by less than 10% and the role of the twisting moment reduceed to about a half that of the compressive force. A typical picture of the complex boundary layer stress field is shown in Fig. 5 for one lay-up. The shear stress τ_{xz} clearly peaks at one delamination site (the other would occur if the sign of M_{xy} was switched) at a value of about 80 MPa. This compares with an interlaminar shear strength of 85 MPa measured from a short beam test for this layup, and which failed at the same interface. A summary of all failure predictions for all panels tested is shown in Table 1 and expressed as a Reserve Factor [actual strength divided by predicted]. The agreement is very satisfactory considering the complex nature of the stress field and the simplified maximum-shear-stress failure criterion which ignores any coupling between τ_{xz} and σ_{zz} for example. Particularly striking is the comparison between the predictions using the maximum-fibre-strain criterion which are excessively optimistic for most stacking sequences.

Aspect Ratio	b/t	Stacking Sequence	Reserve Factor	
			1	2
3	57	(0/90/±45)2S	1.16	1.58
3	57	(0/±45/90)2S	1.14	1.54
3	57	(45/0/-45/90)S2	0.80	1.32
5	35	(0/90/±45)2S	1.02	1.43
5.5	31	(0/±45/90)2S	0.99	1.35
6.9	25	(0/∓45/90)2S	0.90	1.27
7.8	22	(0/∓45/90)2S	1.03	1.57

1. Using interlaminar shear criterion
2. Using maximum fibre strain criterion

Table 1. Prediction of Panel Strength

Simplified Analytical Predictions

The finite element results seem to have correctly identified the failure interfaces, however the associated computing costs are high, and industry is more likely to use a simpler estimation based on arguments similar to Pipes and Pagano (6,7) for estimating edge effects due to N_{xx}. In the Appendix a simple boundary layer analysis is presented for M_{xy} loading. It is shown that it is necessary to know the profile of $\tau_{xy}(y)$ and a method is given for evaluating this in terms of the two shear moduli. The resulting predictions for τ_{xz} using equation (4) of the Appendix are shown in Fig. 6 and agree well with finite element values. Also presented are the usual simplified predictions for N_{xx} loading using the Pipes and Pagano arguments. However it was necessary to assume a profile again and here we took a bilinear approximation having the same maximum gradient as the finite element results. Other assumptions can easily change the predictions by a factor of 2 either way, and an analysis similar to the torsion approach would be preferable. It should also be borne in mind that there is a limit to the precision sought in estimating gradients at the edge, where the idealised discontinuities in section properties of laminars would theoretically predict singular behaviour. It will probably be better to ignore spurious singularities and use an energy-release argument for predicting propagation of an assumed interlaminar flow (12).

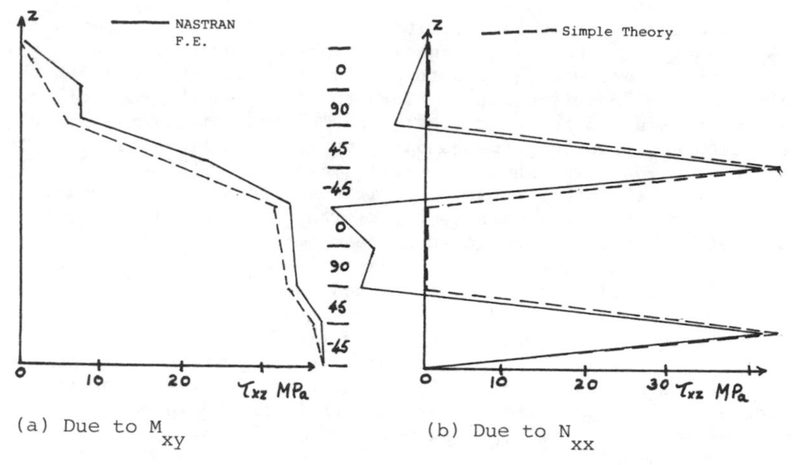

(a) Due to M_{xy} (b) Due to N_{xx}

Fig. 6. Interlaminar Shear Stress at Failure

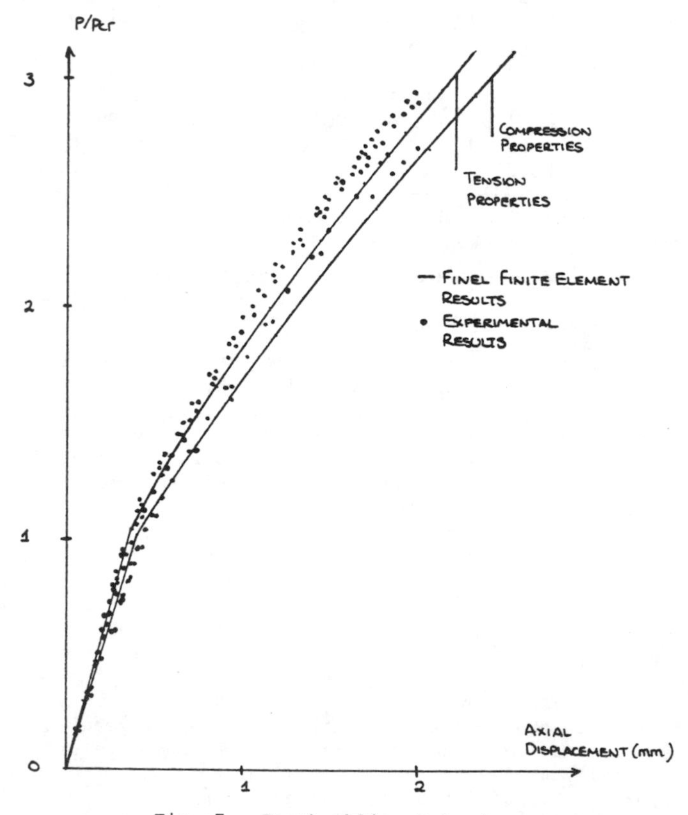

Fig. 7. Postbuckling Behaviour

Finally, however the edge effects are analysed, it is necessary to know the postbuckling values of N_{xx} and M_{xy}. We have tried unsuccessfully to predict these using simplified methods, and eventually resorted to a non-linear finite-element analysis, using the I.C. system FINEL, on the grounds that this technique is now readily available to industry anyway. A 20-noded Mindlin-type element was used with appropriate modification to the shear stiffness terms (11). The results in Fig. 7 show this to be an acceptable prredictor, the short-fall of less than 10% is largely due to taking constant material moduli.

Conclusions

This work has shown - not for the first time - that tests designed to evaluate the maximum strengths of composite plates may give unrepresentative results if edge effects are allowed to precipitate premature failure. The interlaminar shear stresses due to twisting moments decay through a Kirchoff boundary layer which is significantly wider than for isotropic materials, and this fact should be noted in any analysis.

Edge effects are weakening, and in complex composite structures should be eliminated by attention to detailed design. In simple laboratory test specimens this could be possible by local thickening of the panel edges.

Acknowledgements

The authors are indebted to the M.o.D. (P.E. Farnborough) for supporting this work under research agreement 2037/0230/XR/STR, and for the help given in fractographic analysis and computing.

Appendix. Simplified Kirchoff Boundary Layer

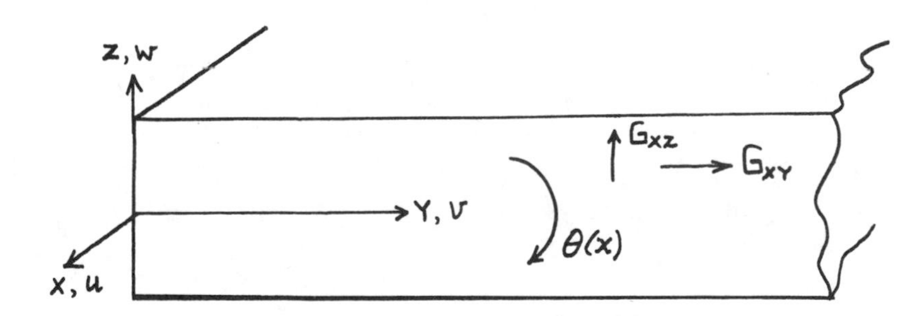

Fig. 8. Torsion of Anisotropic Plate

A simple model to assess the extent of the edge effects under pure torsion can be deduced by assuming firstly that the plate is a homogeneous anisotropic material having two different shear moduli G_{xy} and G_{xz}. This approximates to a laminate having only $\pm 45^o$ angle ply, and clearly $G_{xy} >> G_{xz}$. We simply modify the usual Saint-Venant theory and assume a displacement field, due to uniform twist ($\partial\Theta/\partial x = \Theta'$) as

$$u = zf(y)\Theta'; \qquad v = z\Theta; \qquad w = -y\Theta \qquad (1)$$

The warping function contains f(y) which will assymptote to f(y) = y for large y. The strains from (1) are

$$\varepsilon_{xy} = z\Theta'[1 + f'(y)]; \quad \varepsilon_{xz} = \Theta'[f(y) - y]; \quad \varepsilon_{zy} = 0 \qquad (2)$$

Now consider an arbitrary displacement field generated solely by $\bar{f}(y)$, keeping Θ' constant, then the Principle of Virtual Displacements (10) becomes

$$\int_{-t/2}^{t/2} \int_{0}^{\infty} (\tau_{xy}\, \bar{\varepsilon}_{xy} + \tau_{xz}\, \bar{\varepsilon}_{xz})\, dy\, dz = 0$$

On substituting (2), integrating with respect to z and $\bar{f}'(y)$ by parts, and equating to zero the coefficients of the arbitrary $\bar{f}(y)$, we obtain the differential equation and boundary condition for f(y). The solution is

$$f(y) = \frac{2}{\alpha} e^{-\alpha y} + y \; ; \qquad \text{where } \alpha^2 = 12\, G_{xz}/G_{xy}\, t^2$$

The in-plane shear strain is then

$$\varepsilon_{xy} = 2z\Theta'(1 - e^{-\alpha y}) \tag{3}$$

The length of the boundary layer is estimated by putting $\alpha y = 3.5$, say. $[e^{-3.5} = 0.03]$ In an isotropic plate where $G_{xz} = G_{xy}$, $\alpha^2 = 12/t^2$, and the boundary layer length is of order t. In angle-ply laminates however $G_{xy} = \frac{1}{4}E_{11}$, and G_{xz} is approximately $2G_{12}/5$. For the laminates of this paper G_{xy} and G_{xz} are 39 GPa and 3.4 GPa, so $\alpha = 1.02/t$. Thus we expect a boundary layer wider than 3t.

To deduce the important shear τ_{xz} from the classical laminate shear τ^o_{xy} away from the edge, we write (3) as

$$\tau_{xy} = \tau^o_{xy}\, [1 - e^{-\alpha y}]$$

and use the equilibrium equation

$$\frac{\partial \tau_{xz}}{\partial z} = - \frac{\partial \tau_{xy}}{\partial y} = - \alpha\, \tau^o_{xy} \text{ at } y = 0.$$

For a homogeneous plate τ^o_{xy} varies linearly with z and the above gives a parabolic variation for τ_{xz}. In our composite plate we convert the integral to a summation, and summing from the free surface inwards,

$$\tau_{xz} = - \alpha \sum \tau^o_{xy}\, \Delta t$$

On putting $\alpha = 1.02/t$, and $\Delta t = t/16$,

$$\tau_{xz} = - 0.638 \sum \tau^o_{xy} \tag{4}$$

References

1. Starnes, J.H. and Rouse, M. Postbuckling and failure characteristics of flat rectangular graphite epoxy plates loaded in compression. *Proc. AIAA/ASME/AMS 22nd Structures, Structural Dynamics, and Matls Conf*. 1, April 1981, p423.

2. Ridgard, C. The effects of postbuckling behaviour upon the performance of stability-limited composite structures. B.Ae-MSM-R-GEN-0509. Aug. 1983.

3. Buskell, N., Davies, G.A.O. and Stevens, K.A. Postbuckling failure of composite panels. *3rd Int. Conf. on Composite Structures, Paisley*. Elsevier App. Science. 1985.

4. Stevens, K.A., Buskell, N. and Davies, G.A.O. Failure of buckled composite plates. *Proc. Int. Symp. on Composite Materials and Structures*, Beijing, June 1986, p345-350.

5. Purslow, D. Some fundamental aspects of composite fractography. *R.A.E., TR 81127*, Farnborough, U.K.

6. Pipes, P.B. and Pagano, N.J. Interlaminar stresses in composite laminates under uniform axial extension. *J. Comp. Mats.*, **4**, 1970. p538

7. Pagano, N.J. and Pipes, P.B. Some observations on the interlaminar strength of composite laminates. *Int. Jnl. Mech. Sci.*, **15**, 1972, p679-688.

8. Friedrichs, K.O. and Dressler, R.F. A boundary layer theory for elastic plates. *Comm. in Pure & App. Maths.*, **14**, 1961.

9. Bartholomew, P. and Mercer, A.D. Analysis of an anisotropic beam with arbitrary cross-section. *R.A.E. TR 84058*, Farnborough, UK, June 1984.

10. Davies, G.A.O. *Virtual Work in Structural Analysis*, J.Wiley. 1982.

11. Whitney, J.M. Stress analysis of thick laminated composite and sandwich plates. *Jnl. Comp. Matls.*, **6**, 1972, p426-440.

12. Wang, A.S.D. and Crossman, F.W. Initiation and growth of transverse cracks and edge delamination in composite laminates. *Jnl. Comp. Matls. Supplement.* **14**, 1980, p11-108.

EDGE TRACTIONS AND INTERLAMINAR STRESSES IN MULTILAYERED PLATES

Jean-Claude MOUSSION René SOUCHET

Laboratoire de Mécanique Théorique

40 avenue du Recteur Pineau - 86022 POITIERS - FRANCE

Summary :

A new plate theory is generalized so as to analyse stresses in multilayered elastic plates. This improved plate theory preserves the three-dimensional nature of the elasticity problem. As an example, a bimaterial disk subjected to a uniform pressure on the upper face is studied in detail for clamped boundary conditions.

INTRODUCTION

A new elastic plate theory of three-dimensional nature |1|, is generalized so as to analyse stresses in a bi-material elastic plate. Structural components consisting of two distinct material layers find widespread application. Pioneering work on this type of structure was carried out by Grigolyuk |2|. More recently, Turvey has investigated the yielding of bimetallic disks |3|, and more general results on laminated plates can be found in |4|.

The present approach seecks to examine the effects of the constituent material properties on the distribution of edge tractions and interlaminar stresses. It is well known that the interlaminar stresses play an important role ; by example, delamination between layers is a direct result of these interlaminar stresses. However, actually, because of the three-dimensional nature of the problem, the previous analyses are not altogether satisfactory.

As an example, the simple problem of a bimaterial disk subjected to a uniform pressure on the upper face is studied in detail for clamped boundary conditions. Fairly comprehensive results are presented for the edge tractions, and the interlaminar stresses |5|.

NOTATION

Let $\{z, x_1, x_2\}$ be a cartesian coordinate system. A linear elastic body (S) is bounded by the planes $z = (\varepsilon+1)h$, $\varepsilon = \pm 1$, and generators parallel to the z-axis through a boundary (C) in the (x_1, x_2) - plane. The plate (S) consists of several elastic layers, viz (S_α), $\alpha = 1, \dots N$, ($2h_\alpha$: thickness and a_α : middle z-coordinate of the layer S_α).

We use the following notation, with $\alpha = 1, \dots, N$;

$(u^\circ_\alpha, \underset{\sim}{u}_\alpha)$, $(g^\circ_\alpha, \underset{\sim}{g}_\alpha)$, $(f^\circ_\alpha, \underset{\sim}{f}_\alpha)$: z and (x_1, x_2) -components respectively of the displacements $\underset{\sim}{U}_\alpha$, the body forces $\underset{\sim}{G}_\alpha$, and the edge tractions $\underset{\sim}{F}_\alpha$.

λ_α, μ_α : Lamé coefficients.
E_α, ν_α : Young modulus and Poisson ratio.

$(p^\circ_\varepsilon, \underset{\sim}{p}_\varepsilon)$: z and (x_1, x_2) - components respectively of the tractions on the faces $z = (\varepsilon+1)\,h$. The following classical operators, such as grad, rot, div, Δ, represent derivations in the (x_1, x_2) - coordinates and we use the notation : $\dot{u}^\circ = \partial u^\circ/\partial z$. At each point of the cylindrical boundary (Σ), we introduce the exterior unit normal $\underset{\sim}{n}$, the tangent $\underset{\sim}{t} = \underset{\sim}{z} \wedge \underset{\sim}{n}$ and the components $u_n = \underset{\sim}{u}.\underset{\sim}{n}$, $u_s = \underset{\sim}{u}.\underset{\sim}{t}$.

Finally we recall that

$$(u^\circ/v^\circ) = \int_d u^\circ\, v^\circ\, dx_1\, dx_2 \quad , \quad \ldots,$$

where (d) is the area within the contour (C).

THE BASIC EQUATIONS OF THE THEORY

In an isotropic homogeneous plate, it is natural to consider linear assumptions for the edge tractions and the edge displacements.

In order to write the equations for the composite plate, we suppose that each isotropic homogeneous layer (S_α) satisfies analogous linear assumptions, viz

$$\left|\begin{array}{l} u^\circ_\alpha\,(z) = u^*_\alpha \quad , \quad \underset{\sim}{u}_\alpha(z) = \underset{\sim}{u}^*_\alpha - (z-a_\alpha)\;\;(\underset{\sim}{z} \wedge \underset{\sim}{\alpha}^*_\alpha) \\ f^\circ_\alpha\,(z) = \dfrac{1}{2h_\alpha}\;r^*_\alpha \;,\; \underset{\sim}{f}_\alpha\,(z) = \dfrac{1}{2h_\alpha}\,\underset{\sim}{r}^*_\alpha + (z-a_\alpha)\;\dfrac{3}{2h^3_\alpha}\,\underset{\sim}{m}^*_\alpha\;, \end{array}\right. \tag{1}$$

with $d_{\alpha-1} < z < d_\alpha$. It is clear that u^*_α, r^*_α, ... have the classical meaning used in plate theories.

So we can write the following equations, first with $d_{\alpha-1}<z<d_\alpha$, for $\alpha = 1, \ldots, N$ |1|

$$\left|\begin{array}{l} (\lambda_\alpha + 2\mu_\alpha)\,\dfrac{d^2}{dz^2}\,(u^\circ_\alpha/v^\circ) + (\lambda_\alpha+\mu_\alpha)\,\dfrac{d}{dz}\,(\text{div}\;\underset{\sim}{u}_\alpha/v^\circ) - \mu\,(\text{grad}\;u^\circ_\alpha/\text{grad}\;v^\circ) \\ \qquad = -\,(g^\circ_\alpha/v^\circ) + (\mu_\alpha\;\underset{\sim}{\alpha}^*_\alpha.\underset{\sim}{t} - \dfrac{1}{2h_\alpha}\;r^*_\alpha/v^\circ)_C \end{array}\right.$$

$$\left|\begin{array}{l} \mu_\alpha\,\dfrac{d^2}{dz^2}\,(\underset{\sim}{u}_\alpha/\underset{\sim}{v}) - (\lambda_\alpha+\mu_\alpha)\,\dfrac{d}{dz}\,(u^\circ_\alpha/\text{div}\;\underset{\sim}{v}) - (\lambda_\alpha+2\mu_\alpha)\,(\text{div}\;\underset{\sim}{u}_\alpha/\text{div}\;\underset{\sim}{v}) \\ \quad -\mu_\alpha(\text{rot}\;\underset{\sim}{u}_\alpha/\text{rot}\;\underset{\sim}{v}) = -(\underset{\sim}{g}_\alpha/\underset{\sim}{v}) + (2\mu_\alpha\;\underset{\sim}{z}\wedge\dfrac{d}{ds}\,\underset{\sim}{u}^*_\alpha - \dfrac{1}{2h_\alpha}\,\underset{\sim}{r}^*_\alpha/\underset{\sim}{v})_C \\ \qquad + (z-a_\alpha)\;(2\mu_\alpha\,\dfrac{d}{ds}\;\underset{\sim}{\alpha}^*_\alpha - \dfrac{3}{2h^3_\alpha}\,\underset{\sim}{m}^*_\alpha/\underset{\sim}{v})_C \end{array}\right. \tag{2}$$

and then with $z = (\varepsilon+1)\,h$, $\alpha = 1$ if $\varepsilon = -1$, $\alpha = N$ if $\varepsilon = +1$

$$\left|\begin{array}{l} (\lambda_\alpha + 2\mu_\alpha)\,\dfrac{d}{dz}\,(u^\circ_\alpha/v^\circ) + \lambda_\alpha\;(\text{div}\;\underset{\sim}{u}_\alpha/v^\circ) = \varepsilon(p^\circ_\varepsilon/v^\circ) \\ \mu_\alpha\,\dfrac{d}{dz}\,(\underset{\sim}{u}_\alpha/v^\circ) - \mu_\alpha\;(u^\circ_\alpha/\text{div}\;\underset{\sim}{v}) = \varepsilon\;(\underset{\sim}{p}_\varepsilon/\underset{\sim}{v}) - \mu(u^*_\alpha/v_n)_C \end{array}\right. \tag{3}$$

In the above relations, the trial fields $(v^\circ, \underset{\sim}{v})$ are arbitrary, but they do not depend on the index α.

Now we consider a portion of the plate (S) composed of the αth and $(\alpha+1)$ th laminar. Assuming that the two layers are perfectly bonded along the interface, one can establish the continuity conditions of the displacements and stresses along the interface $z = d_\alpha$, respectively as

$$\left| \begin{array}{l} u^\circ_\alpha = u^\circ_{\alpha+1} \\ \underset{\sim}{u}_\alpha = \underset{\sim}{u}_{\alpha+1} \end{array} \right. \tag{4}$$

and

$$\left| \begin{array}{l} (\lambda_\alpha+2\mu_\alpha)\ \dot{u}^\circ_\alpha + \lambda_\alpha \operatorname{div} \underset{\sim}{u}_\alpha = (\lambda_{\alpha+1} + 2\mu_{\alpha+1})\ \dot{u}_{\alpha+1} + \lambda_{\alpha+1} \operatorname{div} \underset{\sim}{u}_{\alpha+1} \\ \mu_\alpha \dot{\underset{\sim}{u}}_\alpha + \mu_\alpha \operatorname{grad} u^\circ_\alpha = \mu_{\alpha+1} \dot{\underset{\sim}{u}}_{\alpha+1} + \mu_{\alpha+1} \operatorname{grad} u^\circ_{\alpha+1} \end{array} \right. \tag{5}$$

In order to use these relations (4), (5), together with the equations (2), (3), we write down the equivalent weak form of (5)

$$\left| \begin{array}{l} \{(\lambda_\alpha+2\mu_\alpha) \frac{d}{dz} (u^\circ_\alpha/v^\circ)+\lambda_\alpha(\operatorname{div} \underset{\sim}{u}_\alpha/v^\circ)\} - \{(\lambda_{\alpha+1}+2\mu_{\alpha+1}) \frac{d}{dz} (u^\circ_{\alpha+1}/v^\circ) \\ \qquad + \lambda_{\alpha+1} (\operatorname{div} \underset{\sim}{u}_{\alpha+1}/v^\circ) = 0 \\ \{ \mu_\alpha \frac{d}{dz} (\underset{\sim}{u}_\alpha/v) - \mu_\alpha(u^\circ_\alpha/\operatorname{div} \underset{\sim}{v})\}-\{\mu_{\alpha+1} \frac{d}{dz} (\underset{\sim}{u}_{\alpha+1}/\underset{\sim}{v})-\mu_{\alpha+1}(u^\circ_{\alpha+1}/\operatorname{div} \underset{\sim}{v})\} \\ \qquad = - \mu_\alpha(u^*_\alpha/v_n)_c + \mu_{\alpha+1}(u^*_{\alpha+1}/v_n)_c \end{array} \right. \tag{6}$$

It is noteworthy that the equations (2), (3), (4) (6) contain the significant mechanical quantities $(u^*_\alpha, \underset{\sim}{u}^*_\alpha\ \underset{\sim}{\alpha}^*_\alpha)$ and $(r^*_\alpha, \underset{\sim}{r}^*_\alpha, \underset{\sim}{m}^*_\alpha)$, defined along the edge of the αth layer (S_α).

As in the homogeneous case, |1|, the above equations enable us to solve boundary value problems, when the body forces and face tractions are known. If we use trial eigenfunctions connected with the equation $(-\Delta v = \sigma v)$, the z-dependance is first obtained from equations (2), (3), (4), (6). Then, if by example we consider a clamped edge

$$u^*_\alpha = 0\ , \qquad \underset{\sim}{u}^*_\alpha = 0\ , \qquad \underset{\sim}{\alpha}^*_\alpha = 0\ , \qquad \alpha = 1, \ldots N,$$

the unknown quantities r^*_α, $\underset{\sim}{r}^*_\alpha$, $\underset{\sim}{m}^*_\alpha$ are determined from the following conditions

$$u^\circ_\alpha\ (a_\alpha) = 0\ , \qquad \underset{\sim}{u}_\alpha\ (a_\alpha) = 0\ , \qquad \dot{\underset{\sim}{u}}_\alpha\ (a_\alpha) = 0$$

In the following sections, for illustrative purposes, we restrict our attention to the problem of a clamped circular plate which consists of two layers. Thus we shall concentrate our study on the stresses throughout the interior of the plate.

THE BI-MATERIAL CIRCULAR PLATE

In order to show the ability of the improved theory to approach the three-dimensional elasticity problem, we consider the following example |5| :

- The body (S) consists of two homogeneous isotropic layers S_1, S_2.
- The cylindrical surface (Σ) is circular (radius R).
- The edge is clamped

$$\overset{*}{u}_\alpha = 0 , \quad \overset{*}{\underset{\sim}{u}}_\alpha = 0 , \quad \overset{*}{\underset{\sim}{\alpha}}_\alpha = 0 , \quad \alpha = 1, 2.$$

- The loading q_o is assumed uniform, normal to the upper face (for convenience : $q_o = 1$).

The unusual nature of this problem would not seem to be amenable to an analytical solution. Since only numerical processes are available, the test $\{\lambda_1 = \lambda_2, \mu_1 = \mu_2\}$ was used as a control of the numerical process.

Now, for convenience, we represent the displacements $(u^\circ_\alpha, \underset{\sim}{u}_\alpha)$ as

$$u^\circ_\alpha(z,r) = \sum_{i=1}^{\infty} h^\circ_{i\alpha}(z)\; v^\circ_i(r) , \quad (\underset{\sim}{u}_\alpha)_r\;(z,r) = \sum_{i=1}^{\infty} h^1_{i\alpha}\;(z)\;\frac{d}{dr}\, v^\circ_i(r)$$

where

$$v^\circ_i(r) = \frac{1}{\sqrt{\pi}}\;\frac{J_o(\mu_i\; r/R)}{R\;|J_1\;(\mu_i)|}$$

(J_o, J_1) : Bessel Functions ; μ_i : zeros of J_o ; r : radial coordinate)

Since the non-trivial edge conditions are only

$$\overset{*}{u}_\alpha = 0 \quad , \quad (\overset{*}{\underset{\sim}{u}}_\alpha)_r = 0 \quad , \quad (\overset{*}{\underset{\sim}{\alpha}}_\alpha)_s = 0 \quad , \quad \alpha = 1, 2,$$

there results only the following unknown quantities : $(\overset{*}{\underset{\sim}{r}}_\alpha)_n$, $(\overset{*}{\underset{\sim}{m}}_\alpha)_n$, $\alpha = 1,2$. Then, the z-dependance of the displacements, viz the functions $h^\circ_{i\alpha}$, $h^1_{i\alpha}$, involves these quantities, yet to be determined. To obtain further equations for the determination of those four unknown values, use has to be made of the four nontrivial conditions :

$$\left|\begin{array}{l} (\underset{\sim}{u}_\alpha)_r\;(a_\alpha,R) = \displaystyle\sum_{i=1}^{\infty} h^1_{i\alpha}(a_\alpha)\;\frac{\mu_i}{R^2\sqrt{\pi}}\;\frac{J'_o(\mu_i)}{|J_1\;(\mu_i)|} = 0 \\[2em] (\dot{\underset{\sim}{u}}_\alpha)_r\;(a_\alpha,R) = \displaystyle\sum_{i=1}^{\infty} \dot{h}^1_{i\alpha}(a_\alpha)\;\frac{\mu_i}{R^2\sqrt{\pi}}\;\frac{J'_o\;(\mu_i)}{|J_1\;(\mu_i)|} = 0, \end{array}\right.$$

that are approximations on each layer S_α, $\alpha = 1, 2$, for the edge conditions

$$(\overset{*}{\underset{\sim}{u}}_\alpha)_r = 0 , \qquad (\overset{*}{\underset{\sim}{\alpha}}_\alpha)_s = 0 ;$$

we note that the condition $u^\circ_\alpha(z,R) = 0$, and a fortiori $u^\circ_\alpha(a_\alpha,R) = 0$, are fulfilled, because we have $J^\alpha_o(\mu_i) = 0$.

The first step of the present work is intended to compute the four values $(\underset{\sim}{r}^*_\alpha)_n$, $(\underset{\sim}{m}^*_\alpha)_n$ $\alpha = 1, 2$.

We have considered several cases of materials such as Aluminium-Epoxy or Aluminium-Steel, and obtained the results for different values of thicknesses of each layer |5|. For the sake of brievity, we only make the following observations concerning the global edge bending moment m^*_n :

1. The bending moment m^*_n acting on the edge (Σ) is obtained by elementary procedure from the resultant forces $(\underset{\sim}{r}^*_1)_n$, $(\underset{\sim}{r}^*_2)_n$ and moments $(\underset{\sim}{m}^*_1)_n$, $(\underset{\sim}{m}^*_2)_n$ acting on each layer.

2. The value of the bending moment m^*_n does not depend of the interface position. This result was obtained for different values of R/h and different materials. Finally this moment m^*_n increases with R/h.

THE EDGE TRACTIONS

In the section 4, the values $(\underset{\sim}{r}^*_\alpha)_n$, $(\underset{\sim}{m}^*_\alpha)_n$, $\alpha = 1, 2$, were found. Therefore, the displacements and stresses can be determined throughout the plate (S) and along the edge (Σ). This section is devoted to the study of edge tractions $\underset{\sim}{f}_\alpha$, associated with the computed displacement field.

We recall that the edge tractions $(\underset{\sim}{f}_\alpha)_n$ (z) were assumed linear on each layer (S_α)

$$(\underset{\sim}{f}_\alpha)_n (z) = \frac{1}{2\,h_\alpha} (\underset{\sim}{r}^*_\alpha)_n + (z-a_\alpha) \frac{3}{2\,h^3_\alpha} (\underset{\sim}{m}^*_\alpha)_n , \quad \alpha = 1, 2.$$

Thus, for such bi-material plates, the edge tractions are piecewise linear functions of the z-coordinate. Now, the numerical results presented in figure 1 are discussed in the following paragraph.

We have computed $f_n(z)$ for different materials, different thicknesses and different ratios h_1/h_2. All this cases exhibit the same properties :

1. The function $f_n(z)$ is discontinuous for $z = h_1$.
2. The highest value is obtained in the more rigid material.
3. For moderately thick plate, the distribution is of cubic shape.
4. The bending moment m^*_n can be obtained again by integration of the edge tractions through the laminate thickness. So we have a control of the above results.

THE INTERLAMINAR STRESSES

In this section, the interlaminar stresses will be analysed. It is evident that the disk properties are characterised by the thickness of the plate, the interface position and the nature of materials. Consequently, it is not feasible to describe a complete parameter study. So, only one set of curves is shown in the figure 2, for moderately thick plates (R/h = 5).

We make the following observations concerning the data presen-

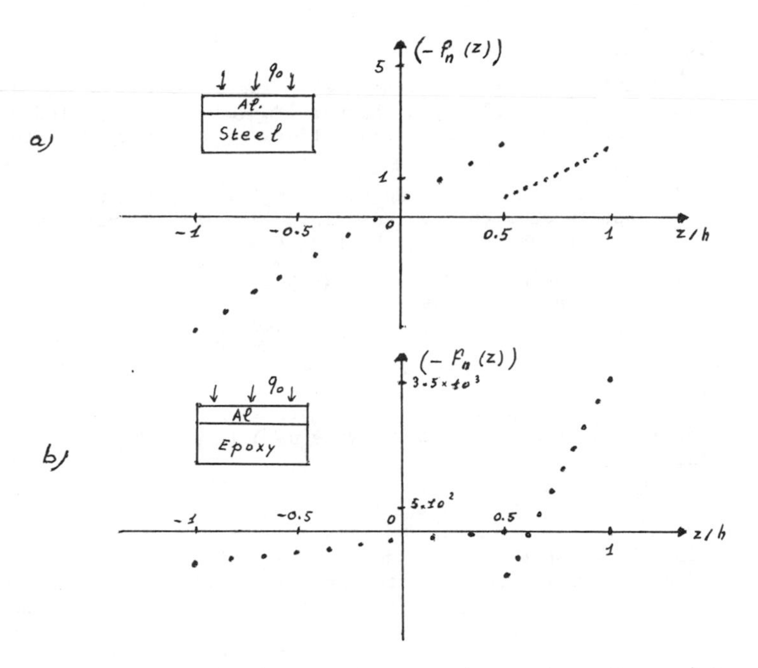

Fig. 1 : Distribution of edge tractions : $h_1 = 3h_2$; a) R = 5h, b) R = 80h.

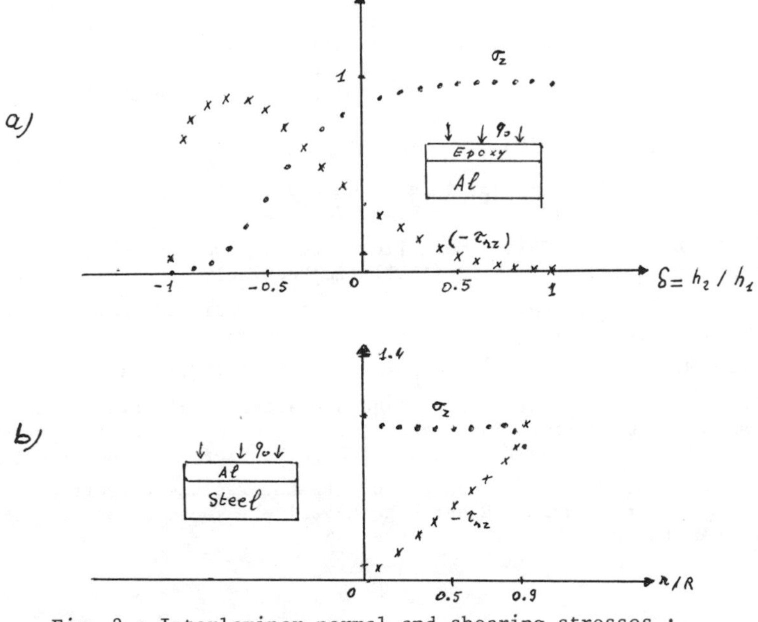

Fig. 2 : Interlaminar normal and shearing stresses ;
a) r = R/2 , b) $h_1 = 3h_2$.

ted in the figure 2.

1. The interlaminar normal stress σ_z increases nonlinearly as the upper layer decreases. We recall that the uniform load q_o is apllied on the upper face. On the other hand, this normal stress remains very near to a constant value along the radius.

2. The interlaminar shearing stress τ_{rz} is maximum when the plate is mainly constituted with the less rigid material. We note that this shearing stress varies linearly along the radius.

3. The interlaminar stresses present an edge effect, that was not analysed in this parameter study.

4. Finally, the interlaminar stresses were computed from both materials (S_1) and (S_2), in order to perform a control on the numerical process. In all observations, these values are the same and are consistent with the continuity conditions of these stresses along the interface. However, we recall that these conditions were used in a weak form by means of the formulae (3-6).

CONCLUSIONS

Using the present theory, an accurate analysis of uniformly loaded clamped bi-material disks has been presented. The effects of the elastic moduli, Poisson ratios and material thicknesses have been examined by conducting a parameter study. The numerous results obtained from the study have been controled by means of the homogeneous plate |5|, and have been presented in non-dimensional graphical form. Several observations have been made on the edge tractions, and the interlaminar normal and shearing stresses.

In conclusion, it seems that it is entirely possible by means of the present theory to obtain more accurate solutions than the ones obtained by classical or various improved plate theories. This study gives a fundamental basis for the development of composite plate and shell theory.

REFERENCES

|1| SOUCHET R. - "Théorie des plaques en Elasticité Linéaire" - Journal de Mécanique - Vol. 17 - n° 1 - p. 53-76 (1978).

|2| GRIGOLYUK E.I. - "Thin bimetallic plates and shells" - Vol. 17 - pp. 69-119 - Inzhenernyi, Sbornid (1953) (in Russian).

|3| TURVEY G.J. - "On the yielding of bimetallic disks", Computers and Structures, Vol. 14, n° 1-2, pp. 1-8, (1981).

|4| JONES R.M. - Mechanics of Composite materials, Mac Graw-Hill (1975).

|5| MOUSSION J.C. - "Analyse des plaques élastiques constituées de deux couches de matériaux. Applications au calcul des contraintes à l'interface " - Thèse 3è cycle - Université de Poitiers - France - 6/12/1983.

|6| EL-HAGGAR N. - "Théorie des plaques transversalement isotropes et des coques cylindriques en Elasticité-Linéaire", Thèse de 3è cycle, Université de Lille I, France, Décembre 1982.

Impact Wave Response and Failure in Composite Laminates

C. T. Sun

School of Aeronautics and Astronautics
Purdue University
West Lafayette, Indiana 47907, U.S.A.

and

T. Wang

LTV Vought Aero Products Division
Dallas, Texas 75265, U.S.A.

Abstract

The dynamic response of composite laminates subjected to the impact of a rigid sphere was investigated. An experimentally obtained contact law was used in conjunction with beam and plate finite elements to calculate contact force as well as dynamic impact response, and then compared with experimental results. Effects of boundary conditions, impact velocity, and thickness of the specimen were studied. The suitability of beam solutions was discussed.

Impact damage modes were reviewed. Experimental results showing the effects of impact on residual strength were presented.

Introduction

Impact produces damage and consequently reduces the strength of composite laminates. Damage can be caused by local contact stress as well as global deformation of the composite plate. Thus, the prediction of contact force and wave motion of the composite plate becomes the key issue in solving impact problems.

Impact involves the contact of two bodies, i.e., the impactor and target composite laminate. Contact force develops as the impactor indents the target. This contact law is usually expressed in terms of the relation between contact force and indentation. The nonlinearity of this relation makes the impact problem nonlinear even if the laminate is assumed to be linearly elastic.

Moreover, there are many impact parameters beside the contact law, e.g., impact velocity, mass of the impactor, contact area, rigidity of the impactor, and stiffness of the laminate. All these additional parameters may alter the characteristics of the impact behavior. The common industrial practice of using a single parameter, i.e., the kinetic energy of the impactor ($1/2mv^2$), to gage the severity of the impact is of questionable validity.

In this paper, an approach developed by Sun et al [1-4], for impact analysis is reviewed and additional results are obtained.

Contact Law

The Hertzian contact law $F=k\alpha^{3/2}$, where F is the contact force, α the indentation, and k the contact rigidity, cannot be directly used for contact between an elastic sphere and a half-space. The reasons are:

(1) Most laminated composites cannot be adequately represented by a half-space.

(2) The anisotropic and nonhomogeneous properties of laminates have not been taken into account.

(3) The inelastic behavior needs to be modeled.

Analytical solutions for a contact problem of this type are very difficult. Moreover, it would require sophisticated experiments to characterize the inelastic constitutive equations for laminated composites. Thus, a direct measurement of the coefficients in the simple contact power law seems most efficient.

An experimental procedure for determining the contact law for an elastic sphere in contact with flat laminates has been proposed by Yang and Sun [2]. The procedure involves loading, unloading, and then reloading a laminate specimen. The experimental data is then used to fit contact laws given by

Loading

$$F = k\alpha^{n} \tag{1}$$

Unloading

$$F = F_m[(\alpha-\alpha_o)/(\alpha_m-\alpha_o)]^q \tag{2}$$

Reloading

$$F = k_1(\alpha-\alpha_o)^p \tag{3}$$

The values of n, q, and p that best fit the experimental results were found in [2] to be

$$n = 1.5, \qquad q = 2.5, \qquad p = 1.5 \tag{4}$$

for different composite material systems.

With q set equal to 2.5, the only coefficient in the unloading law given by Eq. (2) is α_o. This coefficient can be regarded as the permanent indentation in a loading-unloading cycle. For a given F_m and α_m, we define an unloading rigidity

$$s = \frac{F_m}{(\alpha_m-\alpha_o)^{5/2}} \tag{5}$$

so that the unloading law becomes

$$F = s(\alpha-\alpha_o)^{5/2} \tag{6}$$

Yang and Sun [2] discovered that the relation

$$k/s = \alpha_{cr} \tag{7}$$

seemed to hold for the experimental data. Using Eqs. (7), (1), and (5), the following was found.

$$\alpha_o/\alpha_m = 1 - (\alpha_{cr}/\alpha_m)^{2/5} \tag{8}$$

$$\alpha_o = 0 \qquad \text{if } \alpha_m \leq \alpha_{cr}$$

This equation can be used to calculate α_o as a function of α_m. For the graphite/epoxy composite used in this study we have

$$\alpha_{cr} = 8.03\text{x}10^{-2}\ \text{mm}\ (3.16\text{x}10^{-3}\ \text{in.}) \tag{9}$$

Using the above values to calculate α_o in Eq. (2), unloading curves were adequately described, as shown in Fig. 1.

In the reloading process, indentation began with α_o. Since $p = 1.5$, only k_1 needed to be determined in the unloading law, Eq. (3). Yang and Sun [2] observed from the experimental results that, when the loading level is not too high, the reloading curve always returns to where unloading begins. If such a condition is imposed on the reloading law, then

$$k_1 = F_m/(\alpha_m - \alpha_o)^{3/2} \tag{10}$$

Thus the reloading rigidity k_1 can be determined if the unloading condition (F_m, α_m, and α_o) is specified. In other words, there is no need to perform reloading experiments to find the reloading rigidity k_1.

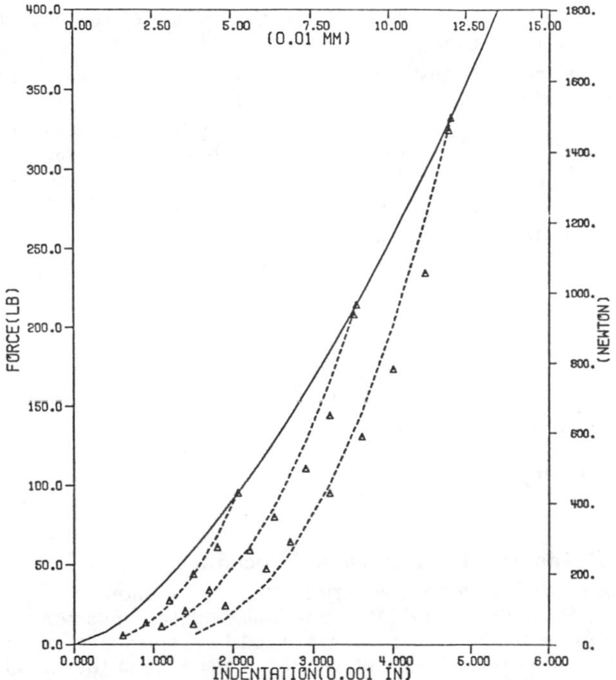

Fig. 1 Unloading curves for graphite/epoxy with 50.8 mm(2") span, 25.4 mm(1") width, and 12.7 mm(0.5") indentor using $\alpha_{cr} = 0.0803$ mm(0.00316"). Ref. [2].

Finite Element Models

Two types of finite elements were used in the present study. A beam finite element with six degrees of freedom was developed by Sun and Huang for dynamic response of elastic isotropic beams subjected to impulsive loading [5]. For the present analysis, the bending rigidity of the laminate was used.

The plate finite element used in this study was a quadrilateral isoparametric element with nine nodes, based on the Mindlin plate theory. The details of this laminated plate theory were given by Whitney and Pagano [6].

For both elements, the stiffness and mass matrixes were derived in the usual manner. The assembled equations of motion were

$$[M]\{\ddot{\delta}\} + [K]\{\delta\} = \{P\} \tag{11}$$

where $\{P\}$ indicates the external load vector. These equations of motion of the plate must be coupled with the impactor through the contact law.

Let the impact point be denoted by (x_o, y_o) and the transverse displacement of the impactor measured from the instant of contact be denoted by w_s. The indentation was then given by

$$\alpha = w_s - w(x_o, y_o) \tag{12}$$

The equation of motion governing the impactor during the loading process was

$$m_s\ddot{w}_s = -k[w_s - w(x_o,y_o)]^{3/2} \tag{13}$$

where m_s is the mass of the impactor. Thus, the motion of the impactor was nonlinearly coupled with that of the plate.

Equations (11) and (13) must be solved simultaneously. The time variable was integrated numerically by the finite difference method.

When laminates have high aspect ratios, one may be inclined to treat them with the beam theory. The beam solution is accurate only for narrow laminates that possess a greater bending rigidity in the width direction than that in the longitudinal direction. For example, the beam theory can model $[90/45/90/-45/90]_{2s}$ laminates better than $[0/45/0/-45/0]_{2s}$ laminates.

The elastic constants for the graphite/epoxy composite considered here were given as

$$\begin{aligned} E_1 &= 120 \text{ GPA} \\ E_2 &= 7.9 \text{ GPA} \\ G_{12} &= 5.5 \text{ GPA} = G_{23} = G_{13} \\ \nu_{12} &= 0.30 = \nu_{23} = \nu_{13} \end{aligned} \tag{14}$$

Impact Wave Response - Experimental Procedure

The schematic diagram for the experimental setup is shown in Fig. 2. A pendulum with a steel ball of 12.7 mm diameter was used as the impactor for low velocity impact (below 5 m/sec.), and an air gun was used to shoot a steel ball for high velocity impacts. Adjusting the pressure of the compressed air in the chamber of the air gun produced a projectile velocity ranging from 20 m/sec to 100 m/sec. All specimens were impacted at the center by the ball.

For high velocity impact, two light emitting diodes (LED's) and two photo detectors were used to find the velocity of the projectile. When the projectile interrupted the first light beam, a pulse was generated to start the time counter. As the projectile cut the second light beam, another pulse was generated to stop the time counter. The velocity of the projectile was obtained by dividing the distance between the two LED's by the time registered on the counter. For low velocity impacts, the velocity of the ball was estimated by the height that it drops.

Two boundary conditions were used in the impact experiments, clamped-clamped and free-free conditions. For the clamped-clamped condition, the specimen was tightly gripped to a massive stand. In the case of free-free end condition, the

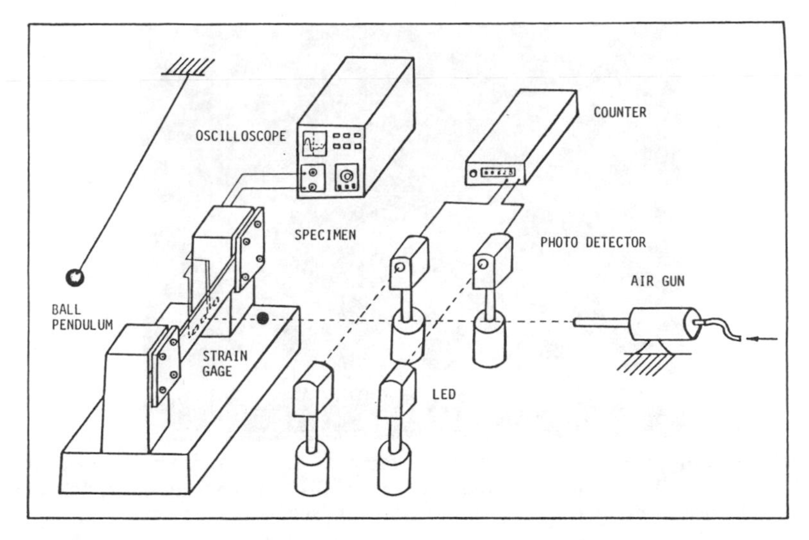

Fig. 2 Schematic diagram of experimental apparatus.

specimen was suspended on two thin strings. Strain gages were mounted on the specimen at various locations. One gage was placed exactly behind the impact point to trigger the oscilloscope, which then recorded the strain signals from other gages. The other strain histories need to be corrected by adding the time lag. The strain gages used were Type EA-13-062 AQ 350 marketed by Micro Measurement Company. Eastman 910 adhesive was used as the bonding glue. Signals from the gages were amplified by a 3A9 Tektronix amplifier and displayed on the screen of the oscilloscope.

After having been impacted by the steel ball, the specimens were tested on an MTS machine to find the residual tensile strength due to the impact damage. Impact damage is discussed in a later section.

Impact Wave Response - Beam-like Specimens

Beam-like $[90/45/90/-45/90]_{2s}$ laminate specimens were used in the impact experiment. These specimens were about 26 mm wide, and the length ranged from 178 mm to 381 mm. The thickness was about 2.7 mm. The beam finite element was used in the analysis.

Daniel et al [7] concluded from their transverse impact experiments that the in-plane membrane deformation was negligible. To verify this in the present study, a series of tests were conducted with two strain gages mounted on opposite faces of the composite beam. The longitudinal strain histories at these locations were recorded. Figure 3 shows a typical strain history of a specimen impacted at the center by a 6.35 mm diameter steel ball traveling at 31.46 m/sec. The strains on opposite sides of the beam had the same magnitude but opposite signs. This result indicates that the deformation resulting from the impact is dominated by bending and extensional deformation can be neglected.

In Figs. 4-5 the longitudinal surface strains at different locations on a clamped-clamped $[90/45/90/-45/90]_{2s}$ laminate subjected to impact of a 12.7 mm diameter steel ball are presented. The impact velocity was 3.16 m/sec. The finite element solutions agree very well with the experimental data during the initial period after

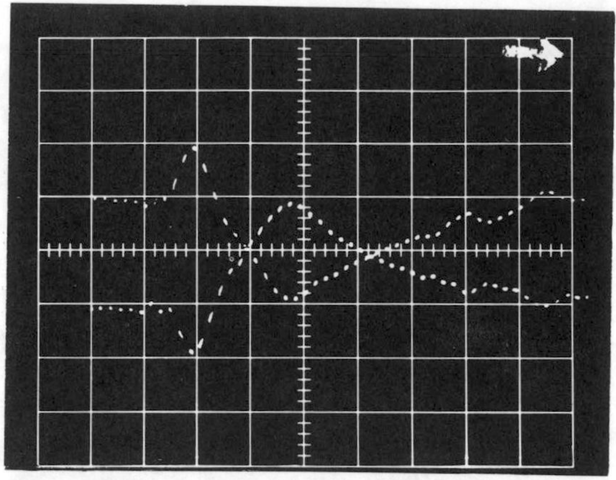

Fig. 3 Strain-gage signals on opposite faces at 38.1 mm from the impact point on the $[90/45/90/-45/90]_{2s}$ beam (2.62x27.74x228.6 mm) impacted with a 6.35 mm diameter steel ball at 31.46 m/sec.

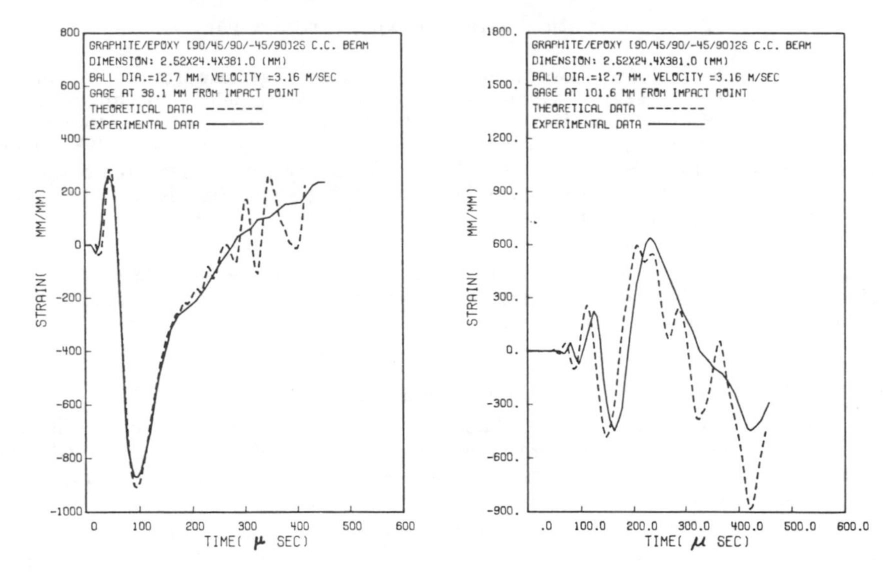

Figs. 4 & 5 Experimental and theoretical strain responses at 38.1 mm (Fig. 4) and 101.6 mm (Fig. 5).

the wave arrives. The agreement is especially good at points closer to the impact point. After the initial wave train passes the gage location, discrepancies between the finite element solutions and experimental results are noted.

Initially, such discrepancies were thought to have originated from numerical instability in the finite element program. However, a study of the numerical stability indicated that the finite elements had already converged. This led to the reexamination of the boundary conditions. It was suggested that the clamped end condition as modeled by the finite element model was not actually realized in the experiment. The finite element solution predicts that the wave will be totally reflected, and exhibits a strong oscillatory behavior due to the wave reflections. In reality, part of the wave penetrated into the grips. To verify this point, a free-free laminated beam (2.82mm x 27.9mm x 177.8mm) was used for an additional experiment. The dynamic strain history at 38.1mm from the impact point is shown in Fig. 6. Excellent agreement between the finite element solution and the experimental data up to 400 μsec is noted.

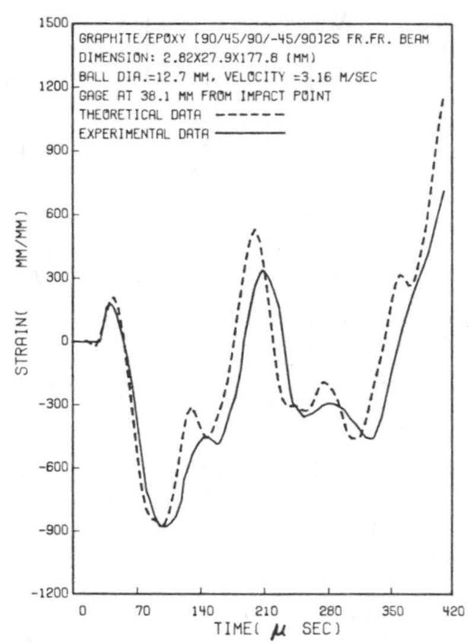

Fig. 6 Experimental and theoretical strain responses for a free-free $[90/45/90/-45/90]_{2s}$ beam at 38.1 mm from the impact point.

More experiments with a free-free beam were then conducted. This beam was substantially longer (381mm) than the previous beam (177.8mm). The experimentally obtained strain histories at two locations are shown in Figs. 7-8. It is evident that the experimental results also show a pronounced oscillatory behavior as predicted by the finite element solution, although of a lesser magnitude. This smaller strain magnitude at the later time could be due to material damping that was not taken into account in the finite element solution. It is also noted that the wave predicted by the finite element solution travels at a slightly higher velocity than the measured value. This could be due to the fact that the displacement-formulated finite element tends to be stiffer than the actual structure.

High impact velocities cause higher modes of vibration (waves of shorter wavelengths) to occur. This makes the plate effect more pronounced and the plate model more suitable for analysis. Figure 9 shows the strain response in the free-free laminate impacted with a 6.35mm steel ball at the velocity 31.13 m/sec. The plate finite element solution agrees better with the experimental data than the beam finite element solution.

Impact Damage Modes

The modes of failure in a laminate produced by impact of a hard object are quite complex. In general, there are three major modes of failure, matrix cracking in

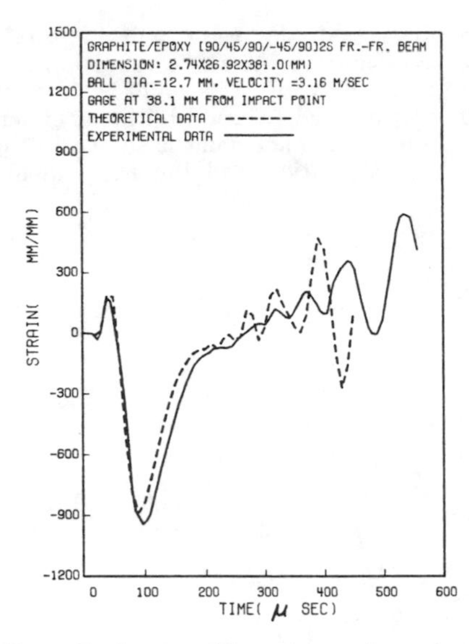

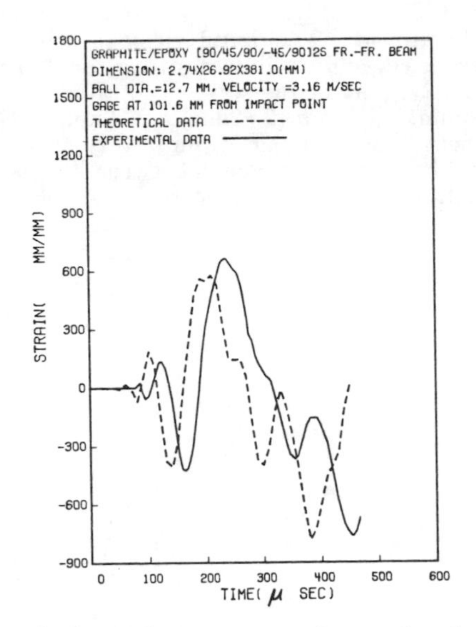

Figs. 7 & 8 Experimental and theoretical strain responses for a free-free $[90/45/90/-45/90]_{2s}$ beam at 38.1 mm (Fig. 7) and 101.6 mm (Fig. 8) from the impact point.

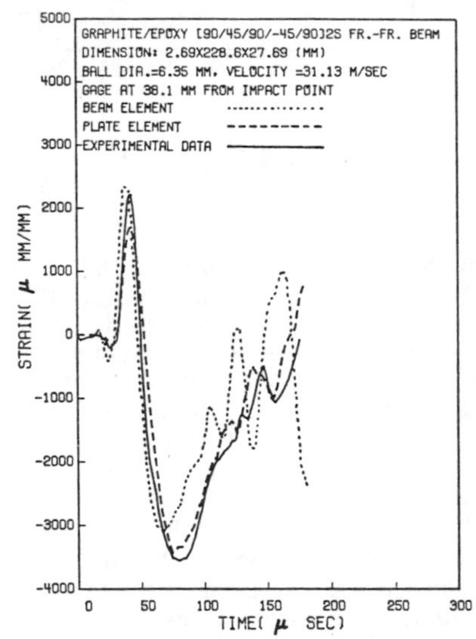

Fig. 9 Beam and plate finite element strain solutions for a free-free $[90/45/90/-45/90]_{2s}$ laminate at 38.1 mm from the impact point.

the lamina, delamination, and fiber breakage. For low impact velocities, matrix cracking and delamination are more pronounced. It was observed by Sun and Rechak[8] that delamination is induced by matrix cracks in the lamina and never appears by itself. For a $[0_5/90_5/0_5]$ laminate, the typical damage pattern due to impact is sketched in Fig. 10.

There are two types of matrix cracks, transverse shear and bending. The former occurs near the impact site due to high transverse shear stresses, and the latter is due to bending stresses. The high magnitude of the transverse shear stress is directly related to the magnitude of the contact force and the size of the contact area. The bending stress is closely related to the flexural motion of the laminate. More discussions on matrix cracks in laminates can be found in [9].

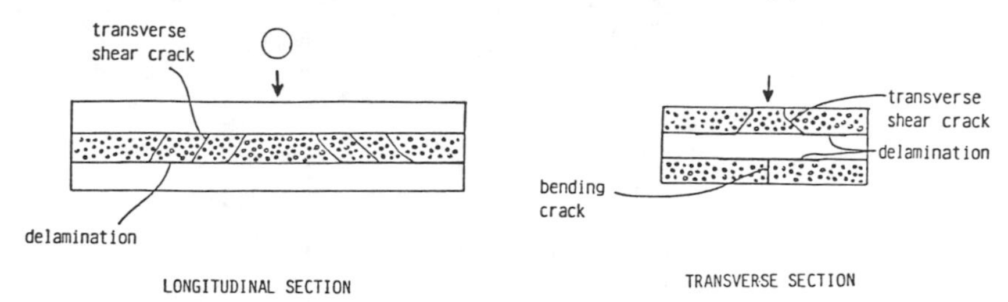

Fig. 10 Impact damage modes in a $[0_5/90_5/0_5]$ graphite/epoxy laminate. Ref. [8].

Energy is absorbed by the laminate from the impactor during the contact period. This energy is absorbed in two forms, vibrational energy and energy dissipated in the material failure process causing permanent indentation in the laminate. If the impactor is relatively small, impact in general will not cause severe damage due to the wave motion. Instead, damage is expected to be localized and to result from the direct contact between the impactor and the laminate. It is then reasonable to use the energy dissipated at the impact zone as a variable to quantify the destructive energy.

Assuming that there are no other failure mechanisms that will consume energy, the total energy (W_T) transferred from the projectile to the target lamina is the sum of the vibration energy (W_v) and the energy dissipated at the contact zone (W_p).

$$W_T = W_v + W_P \tag{15}$$

Explicitly, these energies can be expresses in terms of the contact force F(t), the laminate displacement at the contact point w(t), the displacement of the projectile $w_s(t)$, and the indentation $\alpha(t)$ as

$$W_T = W_v + W_P = \int_0^T F(t)\dot{w}_s(t)\, dt \tag{16}$$

$$W_v = \int_0^T F(t)\dot{w}(t)\, dt \tag{17}$$

$$W_P = \int_0^T F(t)\dot{\alpha}(t)\, dt \tag{18}$$

where T is the contact duration. The total "plastic energy" W_P is the area enclosed by the loading and unloading curves.

The plate finite element model was used in the analysis. Figure 11 presents the total energy W_T and the plastic energy W_P versus impact velocity for the $[90/45/90/\text{-}45/90]_{2s}$ clamped-clamped composite beam. The impactor was a 6.35mm diameter steel ball. From Fig. 11, it is seen that the plastic energy increases almost linearly with respect to impact velocity, while the total energy imparted to the laminate from the impact increases almost exponentially. In other words, a much greater share of the energy will turn into vibration energy at higher impact velocities.

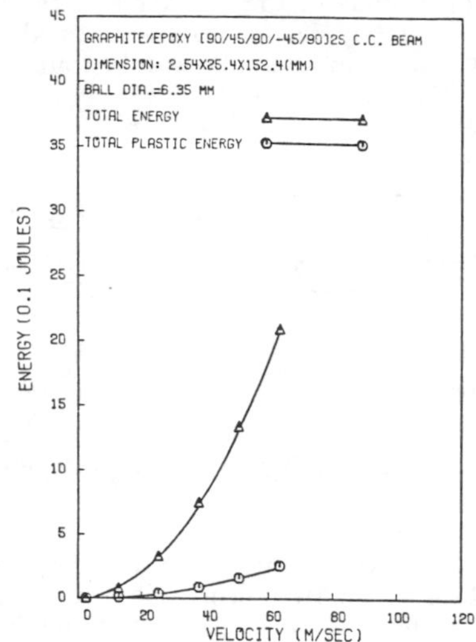

Fig. 11 The total energy and the total plastic energy imparted from a 6.35 mm steel ball to a $[90/45/90/-45/90]_{2s}$ graphite/epoxy laminate.

Effects of Specimen Geometry

To study the effect of laminate thickness, laminates of repeating groups of layers were considered. The laminate layups were $[90/45/90/-45/90]_{ns}$, where n is the number of the repeating groups of layers. For example, when n=1, the laminate has 10 layers and when n=2, the laminate has 20 layers.

For an impact velocity of 25.4 m/sec, the energies were plotted versus thickness in Fig. 12. It is noted that as the thickness of the laminate increases, the total amount of energy transferred from the projectile to the target decreases. However, the plastic energy apparently increases as the thickness increases. Thus, for thick laminates, damage induced by flexural waves could be less severe as compared with that caused by local contact.

Further results for the effects of span and width of the laminate on the energies W_T and W_P have been obtained. In general, a laminate with a short span absorbs less total energy from the impactor but a greater percentage of this energy is dissipated in the indentation process. Also, both values of W_T and W_p become independent of the span beyond a certain point. This means, that for large laminates, impact damage will not depend on the size of the laminate.

Reduction in Strength

Static strength is reduced after a laminate composite is subjected to impact. How to correlate this reduction in strength to pertinent parameters is an important task.

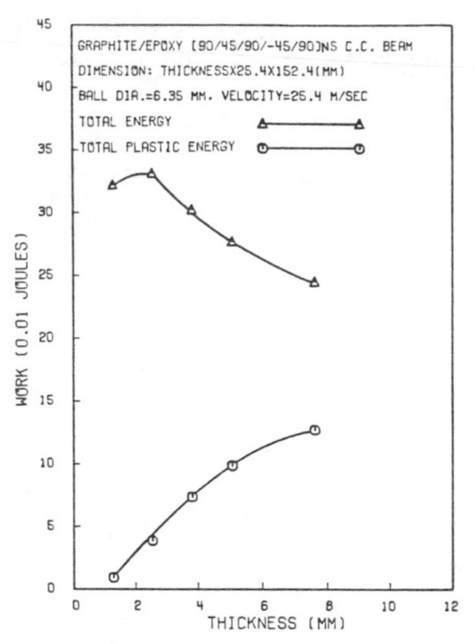

Fig. 12 Thickness effect on the ratio of the total energy and plastic energy.

There are many variables that can affect the amount of impact damage. In the past, many researchers considered the impact velocity as the key variable. However, the relation between the degradation in material properties and impact velocity is meaningful only if the mass of the projectile and the structural properties of the target remain unchanged. Other researchers[10-12] used the initial kinetic energy of the projectile as the variable. Again, conclusions beyond the specific test conditions can not be drawn from these results.

To the target laminate, the contact force is the only interaction with the impactor. Different impact velocities produce different contact forces. Of course, contact force is also affected by the flexibility of the structure and the local contact behavior. It is conceivable that contact force, which produces local indentation and wave motion, causes all the damage in the laminate. Hence, it is reasonable to assume that the damage in the laminate, and thus the reduction in strength, must be a function of the contact force.

When the impactor is relatively rigid and small, the wave motion induced in the laminate by the impact is not large enough to cause much damage, and the damage is usually confined to a small region surrounding the impact point where energy is dissipated in the contact process. It is assumed that the amount of plastic energy dissipated in the contact zone is a good parameter to measure the degree of damage received by the laminate.

The specimens tested were $[90/45/90/\text{-}45/90]_{2s}$ laminates with the dimension 2.67x27.7x152.4 mm. Seven specimens were tested to determine the virgin static strength, which yielded an average strength of 2.15×10^8 Pascal.

Eighteen specimens were impacted with a 6.35 mm diameter steel ball at the mid-span of the laminate at various velocities. The impacted specimens were then tested on an MTS servo-hydraulic material testing machine to determine their residual strengths.

Six specimens were impacted with a large steel ball of 12.7 mm diameter. Five specimens exhibited little reduction in strength. When a slightly higher impact velocity was used, laminate failure occurred due to excessive bending.

In Figs. 13 and 14, the residual strengths corresponding to both impactors are plotted versus the total energy W_T and the plastic energy W_P, respectively. These

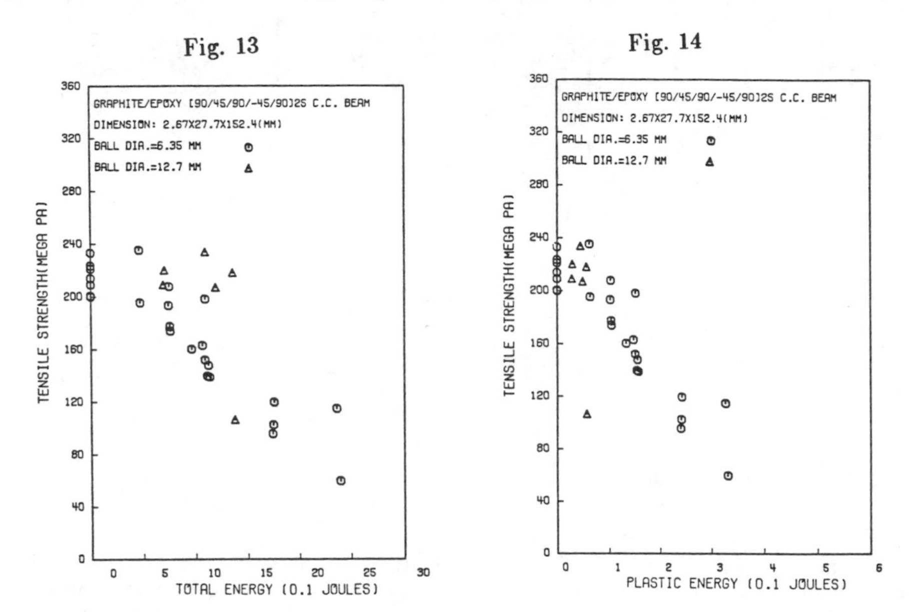

Fig. 13 Residual tensile strength vs total energy for 6.35 and 12.7 mm impactors.

Fig. 14 Residual tensile strength vs plastic energy for 6.35 and 12.7 mm impactors.

two sets of data do not show any common trend when W_T is used. However, both sets of data show a consistency in that there is no strength reduction when the plastic energy W_P incurred in the impact is less than 0.1 Joule.

Conclusions

The statically determined contact laws for steel balls indenting laminated composites have been shown to be reasonably adequate in the dynamic impact analysis. The dynamic response of the laminate can be accurately predicted with a plate finite element program that incorporates the static contact laws.

For beam-like laminated composites, beam theory can be used only if the laminates are very narrow or the transverse bending rigidity is much greater than the longitudinal one. In general, lower impact velocities would reduce the error incurred in the use of beam theories to model beam-like laminates. It has been found that the contact force is more sensitive to the beam modeling error than is the dynamic strain in the laminate.

The energy dissipated at the impact zone (W_P) appears to be a suitable parameter to measure impact damage and can be used to predict degradation of strength of laminated composites after impact. A critical value of the plastic energy seems to exist below which there would be no reduction of strength.

References

1. C.T. Sun, "An Analytical Method for Evaluation of Impact Damage Energy of Laminated Composites", ASTM STP 617, 1977, pp. 427-440.

2. S.H. Yang and C.T. Sun, "Indentation Law for Composite Laminates", ASTM STP 787, 1982, pp. 425-449.

3. T.M. Tan and C.T. Sun, "Use of Statical Indentation Laws in the Impact Analysis of Laminated Composites", *Journal of Applied Mechanics,* Vol. 107, 1985, pp 6-12.

4. C.T. Sun and J.K. Chen, "On the Impact of Initially Stressed Composite Laminates", *Journal of Composite Materials,* Vol. 19, No. 6, 1985, pp. 490-504.

5. C.T. Sun and S.N. Huang, "Transverse Impact Problems by Higher Order Beam Finite Element", *Journal of Computer and Structures,* Vol. 5, 1975, pp. 297-303.

6. J.M. Whitney and N.J. Pagano, "Shear Deformation in Heterogeneous Anisotropic Plates", *Journal Applied Mechanics,* Vol. 37, 1970, pp. 1031-1036.

7. I.M. Daniel, T. Liber, and R.H. La Bedz, "Wave Propagation in Transversely Impacted Composite Laminates", *Experimental Mechanics,* 1979, pp. 9-16.

8. C.T. Sun and S. Rechak, "Effect of Adhesive Layers on Impact Damage in Composite Laminates", to appear in ASTM STP.

9. S.P. Joshi and C.T. Sun, "Impact Induced Fracture in a Laminated Composite", *Journal of Composite Materials,* Vol. 19, 1985, pp. 51-66.

10. G. Dorsey, G.R. Sidy, and J. Hutchings, "Impact Properties of Carbon Fiber/Kevlar 49 Fiber Hybrid Composites" *Composites,* 1978, pp. 25-32.

11. G.E. Husman, J.M. Whitney, and J.C. Halpin, "Analytical Treatments and Studies of Material Response to Impact Residual Strength Characterization of Laminated Composites Subjected to Impact Loading", *Foreign Object Damage to Composites,* ASTM STP 568, American Society for Testing and Materials, 1975, pp. 92-113.

12. J.C. Carlisle, R.L. Crane, W.J. Jaiques, and L.T. Montulli, "Impact Damage Effects on Boron-Aluminum Composites", *Composite Reliability,* ASTM STP 580, American Society for Testing and Materials, 1975, pp. 458-470.

EFFECT OF VISCOELASTIC CHARACTERISTICS OF POLYMERS FOR A COMPOSITE BEHAVIOR UNDER DYNAMIC AND IMPACT LOADING

C. BURTIN Chercheur CERMAC INSA Lyon
P. HAMELIN Directeur CERMAC

Centre d'Etudes et de Recherches sur les matériaux composites
INSA Lyon 20 avenue Einstein 69621 Villeurbanne Cédex France

ABSTRACT

The viscoelastic properties of epoxy and vinyl-ester resins are measured as function of temperatures and frequencies. Dynamic mechanical behavior and impact loading behavior of composites plates using these matrixs are studied. We establish correlations between these two methods and we try to justify the influence of viscoelastic properties of the polymers for the evaluation of plates stiffness.

INTRODUCTION

The impact behavior of composite materials and particularly that of laminated plates (fiber reinforced plastic) is required to be studied for several reasons. A lot of data is available and shows that the dynamic factor and the resistance rise are particularly appreciable in the case of materials having an organic matrix reinforced with either glass or carbon fiber (1) (2).

The principal factors which influence impact results are loading conditions (3) (4) and testing methods. We distinguish traditional testing methods issued of qualification procedures of metalic materials such as instrumented charpy-izod impact test (5), drop weight impact test (6) green pendulum (7), compressed air canon (8) and the experimental set ups which simulate behavior laws such as hopkinson bar (9).

The properties and material characteristics of composite materials are particularly sensitive to the matrix nature (10), to the mineral phase presence (11)to the nature of interface or the interphase (12).

The studies carried out by G. MAROM and E. DRUKKER, A. WEINBERG, J. BANBAJ (2) show the important role of the fibers nature and their lamination.

The studies of GRIFFITHS L.J., MARTIN D.J. (13), Z.G. LIU, C.Y. CHIEM (14) show the different behavior laws obtained as a fonction of strain rate.

It can be also noted that the criteria at the rupture mechanism (15) as well as the crack mode can be found in the case of impact loading.

The works carried out by BERRY J., HULL D (16), P.H. THORNTON (17), E. BURTIN, P. HAMELIN (18) show the difficulty to analyse the strain behavior of structural elements under impact loading.

The aim of this study is to show the effect of matrix and its viscoelastic properties with respect to the impact behavior or the dynamic behavior of structural elements such as plates.

I STUDIED MATERIALS

Several types of resin defined by the following table have been studied

References	Nature	References	Nature
B8 ELF	Epoxyde	470-36	Vinyl-ester
B8 ELF + flexible element	Epoxyde	XD80	Vinyl-ester + flexible element

The plates were made from cloth (Verester BD500-39) or unidirec tional fibers (Verester).

The plates contain 8 and 10 layers respectively with a thickness of about 3 mm.

The polymerisation control is realized by differential thermal analysis and by micro-calorimetry after an heating period of three hours at 140°C.

II VISCOELASTIC PROPERTIES MEASUREMENT AND DYNAMIC BEHAVIOR OF COMPOSITE PLATES

The experimental study is carried out in collaboration with C. CHESNEAU and C. TREMY (19) from METRAVIB Society Ecully (France).

2.1. The experimental set-up is shown in figure (1). The materials are loaded in tension-compression by points of frequencies (7 Hz to 500 Hz) at a constant temperature (10°C - 100°C).

We exploit particularly the measurements at glass plate level in order to obtain the characteristics which will be used in the study of impact behavior.

This concerne a frequency domain of 10^3 to 10^4 Hz at a temperature zone ranging between 20 and 40°C.

2.2. Principe of measurement :

The dynamic stiffness can be expressed as follows :

K',E',tg s = f (T,t)

VISCOELASTIC CHARACTERISTICS

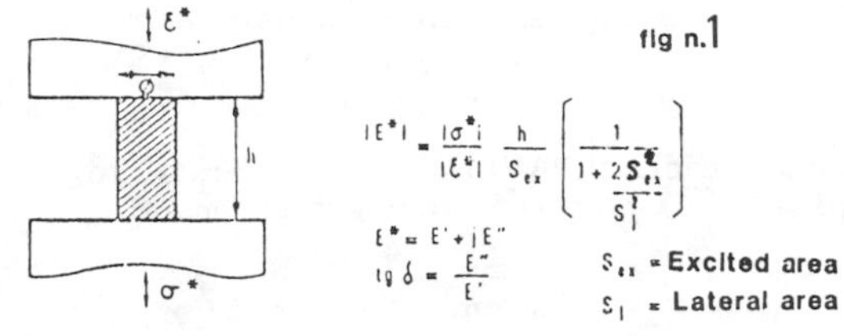

MEASUREMENT TRACTION - COMPRESSION

$$K^* = \frac{F^*}{U^*} \quad \begin{cases} F^* = F_0 e^{i(wt+\delta)} \\ U = U_0 e^{i(wt)} \end{cases}$$

$$K^* = \frac{F_0}{U_0} e^{i\delta}$$

$$\begin{cases} K' = \frac{F_0}{U_0} \cos\delta \\ K'' = \frac{F_0}{U_0} \sin\delta \end{cases}$$

2.3. Results and discussion :

2.3.1. Equivalence principle time-temperature - temperature-frequency

According to the equivalence principle time-temperature (21) (20) obtained from isotherms (20 to 100°C) between 7,8 Hz and 500 Hz. we can obtain a translation factor a_T, defined for each temperature, which allows the construction of a master curve.

We can express the variation of E',E", and tgδ as a fonction of frequency at a fixed temperature 20°C

2.3.2. Determination of apparent energy activity is given according to the following equation :

$$E_A = R \frac{d(\log a_T)}{d(I/T)}$$

$$\begin{cases} B8 \longrightarrow 120 \text{ KJ/mole} \\ 470\text{-}36 \longrightarrow 130 \text{ " "} \\ XD80 \longrightarrow 140 \text{ " "} \end{cases}$$

The tangent to the curve log of a_T is determined as a function of 1/T at each point. At glass plate level E_A is independant of T.

Fig. (2) (3) (4) show the curves obtained for the three resins and the following table gives an estimation of the energy activation.

2.4. Analysis of dynamic behavior of viscoelastic plastes :

2.4.1. Hypothèsis

a) The behavior law in the case of bidirectional orthotropic corps is definited by :

$$\begin{Bmatrix} \sigma_x \\ \sigma_y \\ \tau_{xy} \end{Bmatrix} = \begin{bmatrix} D_{11} & D_{12} & D_{16} \\ D_{21} & D_{22} & D_{26} \\ D_{61} & D_{62} & D_{66} \end{bmatrix} \begin{Bmatrix} \varepsilon_x \\ \varepsilon_y \\ \gamma_{xy} \end{Bmatrix} \quad \{1.1\} ; \quad [D] = \begin{bmatrix} \frac{E_1}{1-\nu_{12}\nu_{21}} & \frac{\nu_{21}E_1}{1-\nu_{12}\nu_{21}} & 0 \\ \frac{\nu_{12}E_2}{1-\nu_{12}\nu_{21}} & \frac{E_2}{1-\nu_{12}\nu_{21}} & 0 \\ 0 & 0 & G \end{bmatrix} \quad \{1.2\}$$

- in an other system D is determinated in functun of T (rotational matrix)

$$[D]_{(x,y)} = [T^{-1}]\,[D]_{(1,2)}\,[T^{-1}]^t \qquad \{1.3\}$$

- in the case of an elastic behavior of composite material the different terms of matrix's stiffness can be expressed from characteristics of resin and fibers using the expresson proposed by PUCK (22), TSAI (23), GRESZCUK (24)

- in the case of viscoelastic behavior, M.PAVEN (25), Z.HASHIN (26), R.G.WHITE and E.M.V.ABDIN (27) proposed other expressions of these constants.

$$\{1.4\} \quad \begin{aligned} E_1^* &= V_f E_f + V_m E_m^* & \qquad E_2^* &= E_f E_m^* / (V_m E_f + V_f E_m^*) \\ \nu_{12}^* &= V_f \nu_f + V_m \nu_m^* & \qquad G^* &= G_f G_m^* / (V_m G_f + V_f G_m^*) \end{aligned}$$

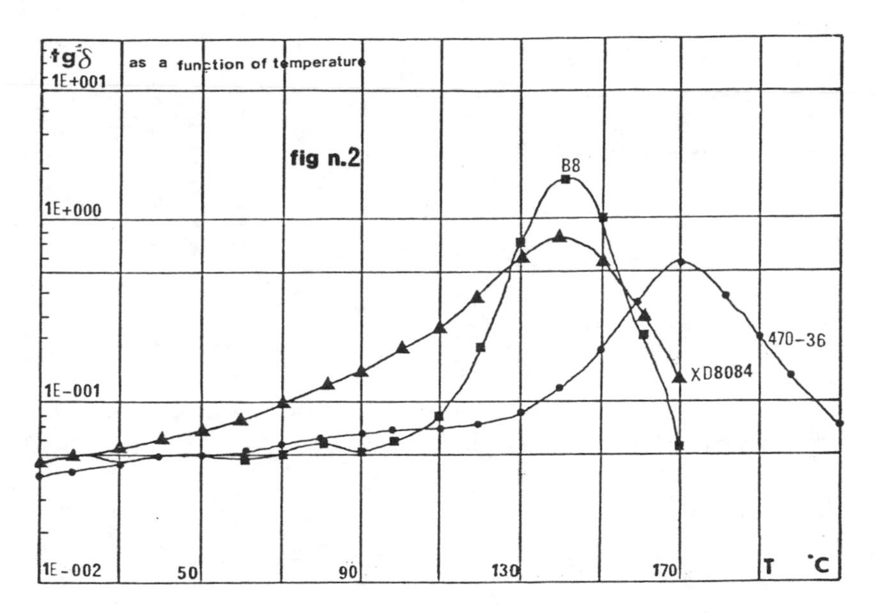

b) Hypothesis on thin plates behavior = considering that the displacement and the strain are small and that the hypothesis of love kirchoff are respected. The deplacement field expressions are the following :

$$\{1.6\} \quad u_i = u_i^0 - x_3 w_{,i} \quad i \in (1,2) \qquad w = u_3$$

$$\varepsilon_i = \varepsilon_i^0 + x_3 K_i \; ; \; K_i = -w_{,ii} \; ; \; \varepsilon_i^0 = u_{i,i}^0 \qquad \gamma_{12} = -2x_3 w_{,12}$$

. The deplacements are : $u_i = u_i^0 - x_3 w_{,i}$

. The equilibrium equation of an elastic plate :

$$\begin{Bmatrix} N \\ M \end{Bmatrix} = \begin{bmatrix} A & B \\ B & C \end{bmatrix} \begin{Bmatrix} \varepsilon^0 \\ \chi \end{Bmatrix} \qquad \{1.7\}$$

where $A = \sum_{K=1}^{n} [D]_{\alpha K} \cdot (z_{K+1} - z_K)$

where $B = \sum_{K=1}^{n} \frac{1}{2} [D]_{\alpha K} (z_{K+1}^2 - z_K^2)$ $\qquad \{1.8\}$

$C = \sum_{K=1}^{n} \frac{1}{3} [D]_{\alpha K} (z_{K+1}^3 - z_K^3)$

in the case of a viscoelastic behavior :

$$\begin{Bmatrix} N(t) \\ M(t) \end{Bmatrix} = \begin{bmatrix} \hat{A} & \hat{B} \\ \hat{B} & \hat{C} \end{bmatrix} \begin{Bmatrix} \varepsilon^0(t) \\ \chi(t) \end{Bmatrix} \qquad \{1.9\}$$

$\{1.10\}$ where $\hat{A} = \sum_{K=1}^{n} [\hat{D}]_{\alpha K} (z_{K+1} - z_K) \quad \hat{B} = \sum_{K=1}^{n} \frac{1}{2} [\hat{D}]_{\alpha K} (z_{K+1}^2 - z_K^2) \quad \hat{C} = \sum_{K=1}^{n} \frac{1}{3} [\hat{D}]_{\alpha K} (z_{K+1}^3 - z_K^3)$

with the operators : $\hat{D}[f(t)] = f(t) D(0) + \int_{-\infty}^{t} f(t-s)\, dD(s) \qquad \{1.11\}$

. under a harmonic loading $e^{i\omega t}$, we have the equilibrium equation.

$$\{1.12\} \qquad \begin{Bmatrix} N^* \\ M^* \end{Bmatrix} = \begin{bmatrix} A^* & B^* \\ B^* & C^* \end{bmatrix} \begin{Bmatrix} \varepsilon_0^* \\ \chi \end{Bmatrix}$$

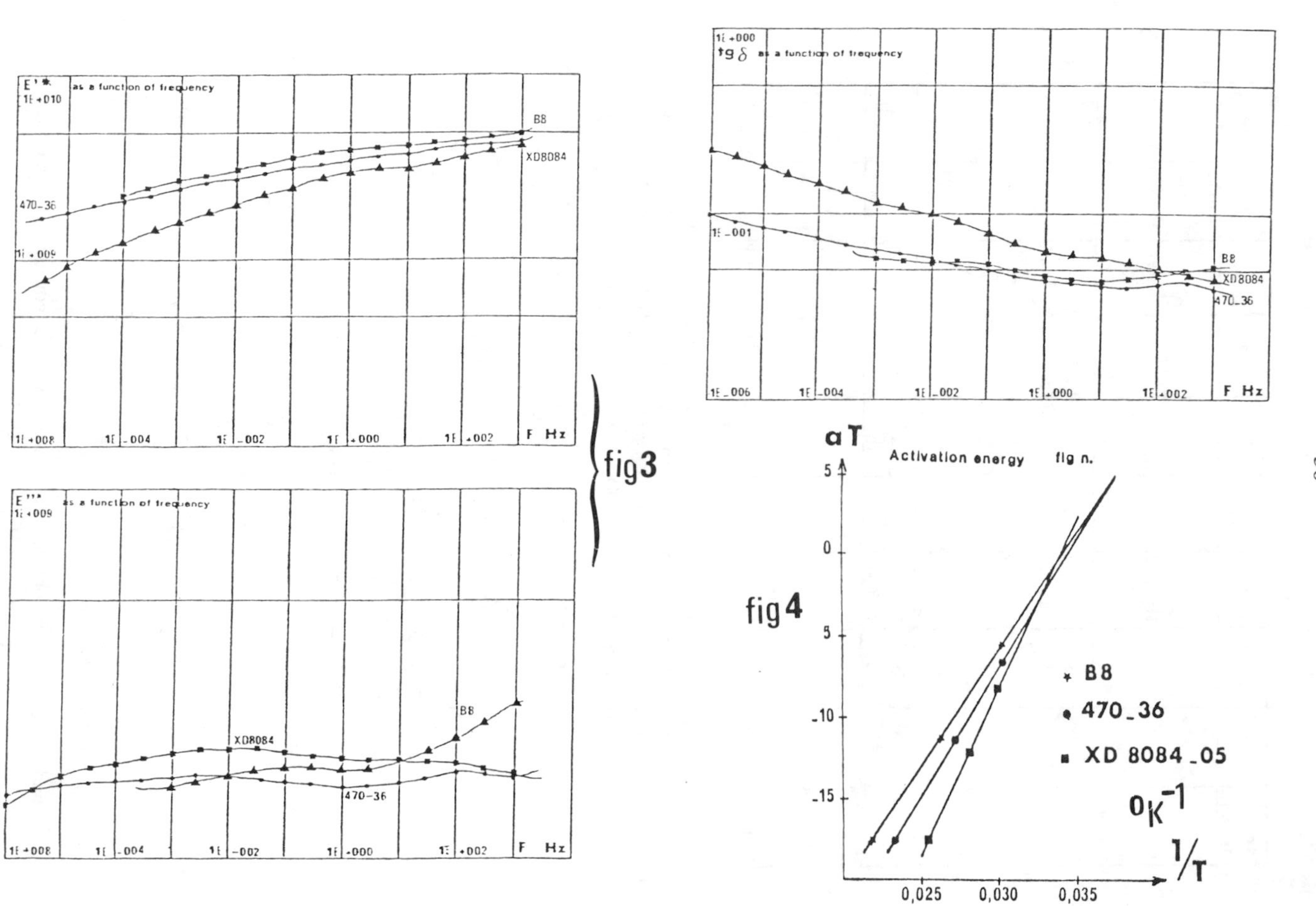

E'* as a function of frequency
1E+010
1E+009
1E+008
1E-004
1E-002
1E+000
1E+002
F Hz
B8
XD8084
470-36
E''* as a function of frequency
1E+009
tg δ as a function of frequency
1E+000
1E-001
1E-006
fig3
aT
Activation energy fig n.
5
0
-10
-15
fig4
* B8
• 470_36
■ XD 8084_05
0K-1
1/T
0,025
0,030
0,035

in functon of correspondant principle proposed by HASHIN (27), A*,B* and C* are determined by the formulas (1,2), (1,4) and (1,8).

2.4.2. Lagrange method :

- generalized force corresponding to coordinates: Qi

{1,13} $Q_i = \partial V / \partial q_i$ where V is the work carried out during displacement δq_i

- cinetic energy : $T = \frac{1}{2} \rho \int (\dot{x}^2 + \dot{y}^2 + \dot{z}^2)\, d\Omega$ {1,14}

$$x = u_1(t, q_i)\ ,\ y = u_2(t, q_i)\ ;\ z = u_3(t, q_i)$$

- potential energy :

$$V = \frac{1}{2} \int_\Omega \sigma_{ij} . \varepsilon_{ij}\, \delta\Omega \quad \{1,15\}$$

- lagrange equation :

$$\frac{d}{dt}\left(\frac{\partial T}{\partial \dot{q}_i}\right) - \frac{\partial T}{\partial q_i} + \frac{\partial V}{\partial q_i} = Q_i \quad \{1.16\}$$

. in the case of an orthotropic homogenious plate fixed at its ends (section a x b x h), the deflection expression is :

$$\{1.17\} \quad w = \sum_{m=1}^{\infty} \sum_{n=1}^{\infty} q_{mn}(t)\left(1 - \cos \frac{2m\pi}{a} x\right)\left(1 - \cos \frac{2n\pi}{b} y\right)$$

. the potential energy is :

$$\{1.18\}\ V = \frac{h^3}{12} . 2ab\pi^4 \left[\frac{D_{11}}{4} \sum\sum m^4 q_{mn}^2 + \frac{D_{22}}{4} \sum\sum n^4 q_{mn}^2 + \frac{2D_{12} + 4D_{66}}{a^2 b^2} \sum\sum n^2 m^2 q_{mn}^2\right]$$

. the cinetic energy is : $T = \frac{5\rho hab}{8g} \sum\sum \dot{q}_{mn}^2$ {1.19}

. we obtained lagrange equation for each generalized coordinate :

$$\{1.20\} \quad \ddot{q} + \alpha_{mn}^2 q = \beta_{mn} Q_{mn}$$

$$\{1,21\} \quad \alpha_{mn}^2 = \frac{h^3}{12} . \frac{48\pi^4 g}{15\rho h} . \left[D_{11} \frac{m^4}{a^4} + D_{22} \frac{n^4}{b^4} + (2D_{12} + 4D_{66}) \frac{m^2 n^2}{a^2 b^2}\right]$$

$$\{1,22\} \quad \beta_{mn} = \frac{4g}{5\rho hab}$$

. the resonal frequency is $\frac{\alpha_{mn}}{2\pi}$ and the period of free vibration is

$$\tau_{mn} = \frac{2\pi}{\alpha_{mn}}$$

. under the action of a concentrated load Qmm = Po coswt, the deflection is expressed by : {1,23}

$$w = \sum\sum \left[A_{mn} \cos\alpha_{mn} t + B_{mn} \sin\alpha_{mn} t + \frac{P_o \beta . \cos\alpha_{mn} t}{\alpha_{mn}^2 - \omega^2}\right] \times \left(1 - \cos \frac{2m\pi}{a} x\right)\left(1 - \cos \frac{2n\pi}{b} y\right)$$

. in the case of vibration of a viscoelastic plate after a long delay the free vibration are neglected and only rests the vibration forced by the load

$$w_f = P_o \beta \cos\omega t \sum\sum \frac{1}{\alpha_{mn}^2 - \omega^2}\left(1 - \cos \frac{2m\pi}{a} x\right) \times \left(1 - \cos \frac{2n\pi}{b} y\right) \quad \{1,24\}$$

. the displacement vector will be late with an angle δ at the center of the plate.

$$\frac{W_0 \cos \omega t}{P_0 \cos(\omega t+\delta)} = \beta \sum\sum \frac{1}{\alpha_{mn}^2-\omega^2} \qquad \{1,25\}$$

. Separating the real part and the imaginary part we can calculate :

$$\frac{W_0}{P_0} = \sqrt{A'^2 + A''^2} \qquad \{1,26\}$$
$$tg\delta = \frac{A''}{A'}$$

2.4.3. Application : We study firstly the variation of the three first resonal frequencies in function of the fiber orientation θ . Fig (5).

. we find that the first resonal frequency is maximum at θ = 45° and the second resonal frequency is max at θ = 0° and minimum at θ = 90°. On the contrary, the third resonal frequency is minimum at θ = 0° and maximum at θ = 90°.

. at a second time, we study the amplitude and the phase shift variations of composites plates as a function of the frequency for the differents resins (B8, 470-36, XD80).

. the results are presented by the fig. (6) (7) and we found that the viscoelastic properties influence the dynamic characteristics of plates.

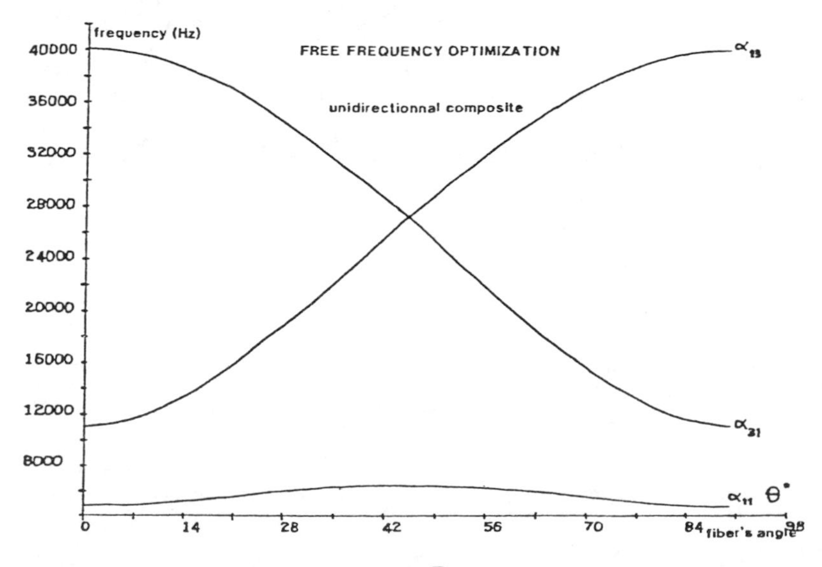

fig n. 5

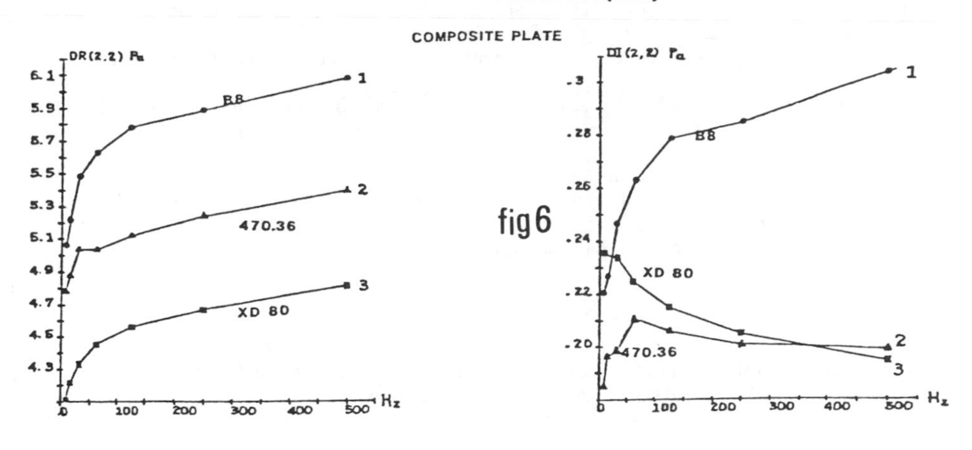

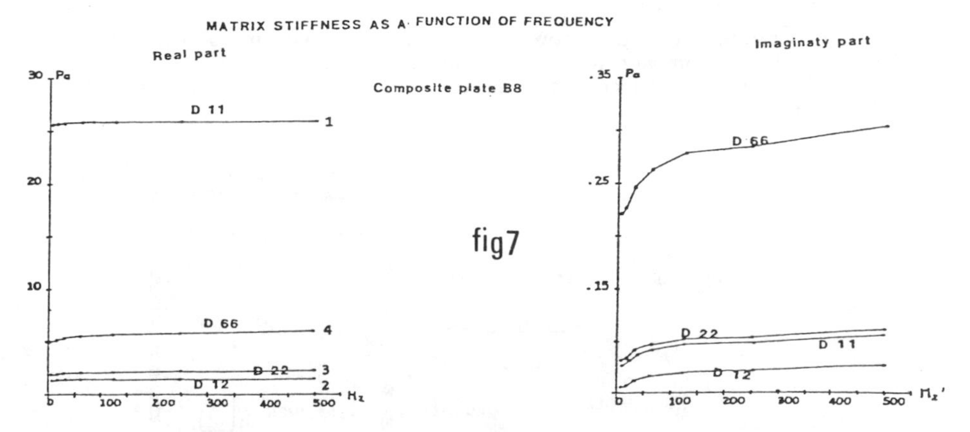

III IMPACT BEHAVIOR MEASUREMENT OF COMPOSITE PLATES

The study is carried out on the impact behavior analysis of composite plates fixed at their ends and loaded at their centre.

The experimental study has been realized at DT/DAT/PL - PSA Etudes et Recherches Peugeot La Garenne Colombes 92250 France.

3.1. Experimental principle :

Description of rheometrics RIT 8000 Fig. (8)

The rheometric apparatus is composed for four parts :

a) Hydraulic unit :

Through a bondage systeme with opened circuit the hydraulic unit prints a constant velocity at the impact head during impact duration. This velocity can be chosen between 0,013 m/s and 13 m/s.

b) Impact system :

In the present study, the impact head used has a hemispheric form of 17 mm diameter, equiped at its base with an effort captor.

The forces measurement can be done in a domain of 500 to 2500.{N}
On the other hand a photoelectric cell placed on one of striker guiding arm permits to adjust the length of strike.

The striker velocity can be controled at a length varied from 75 to 280 mm.

The fixation system of samples lets an available circular surface of 52 mm diameter at the centre.

c) Environnement room unit

The room has a volume of 0,5 m3. It allows to control and to carry out the tests at temperatures between 76°C and at + 176°C.

The relative humidity can be varied from 20 to 95 %.

d) Calcul unit

The stress-strain curve is registed in memory and can be seen instantly on a cathodic screen.

The dynamic stiffness and the energy absorbed are calculated automaticly from the curve points designed by the operator.

Materials studied

Stress-strain curves are shown in figure. The maximum force Fmax registered during the impact test is directly accessible on the curve. The max energy Emax absorbed from the impact begining until Fmax is reached is calculated.

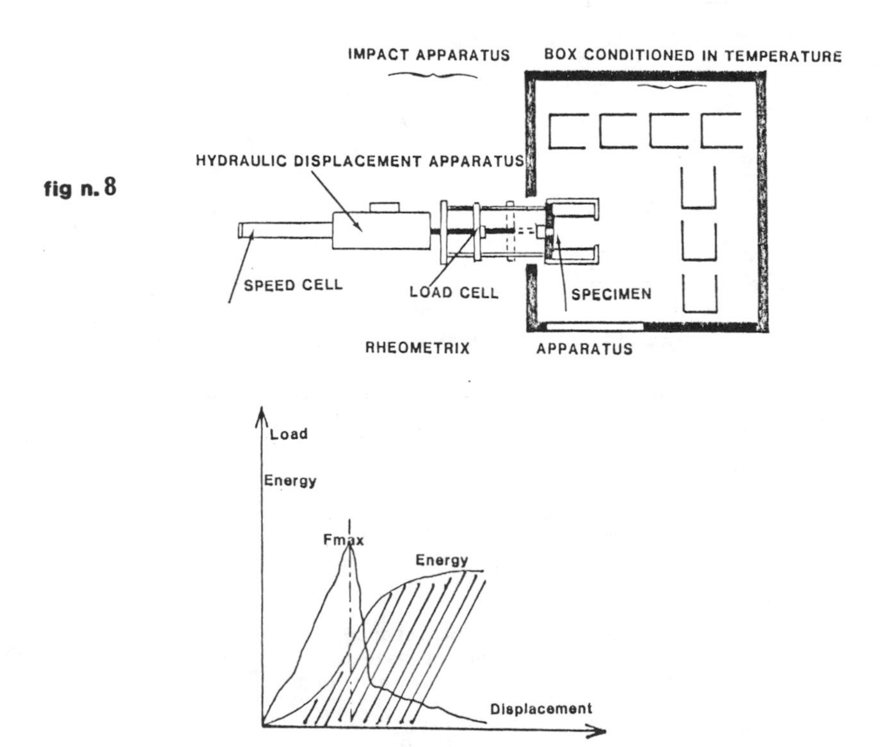

fig n. 8

fig n. 9

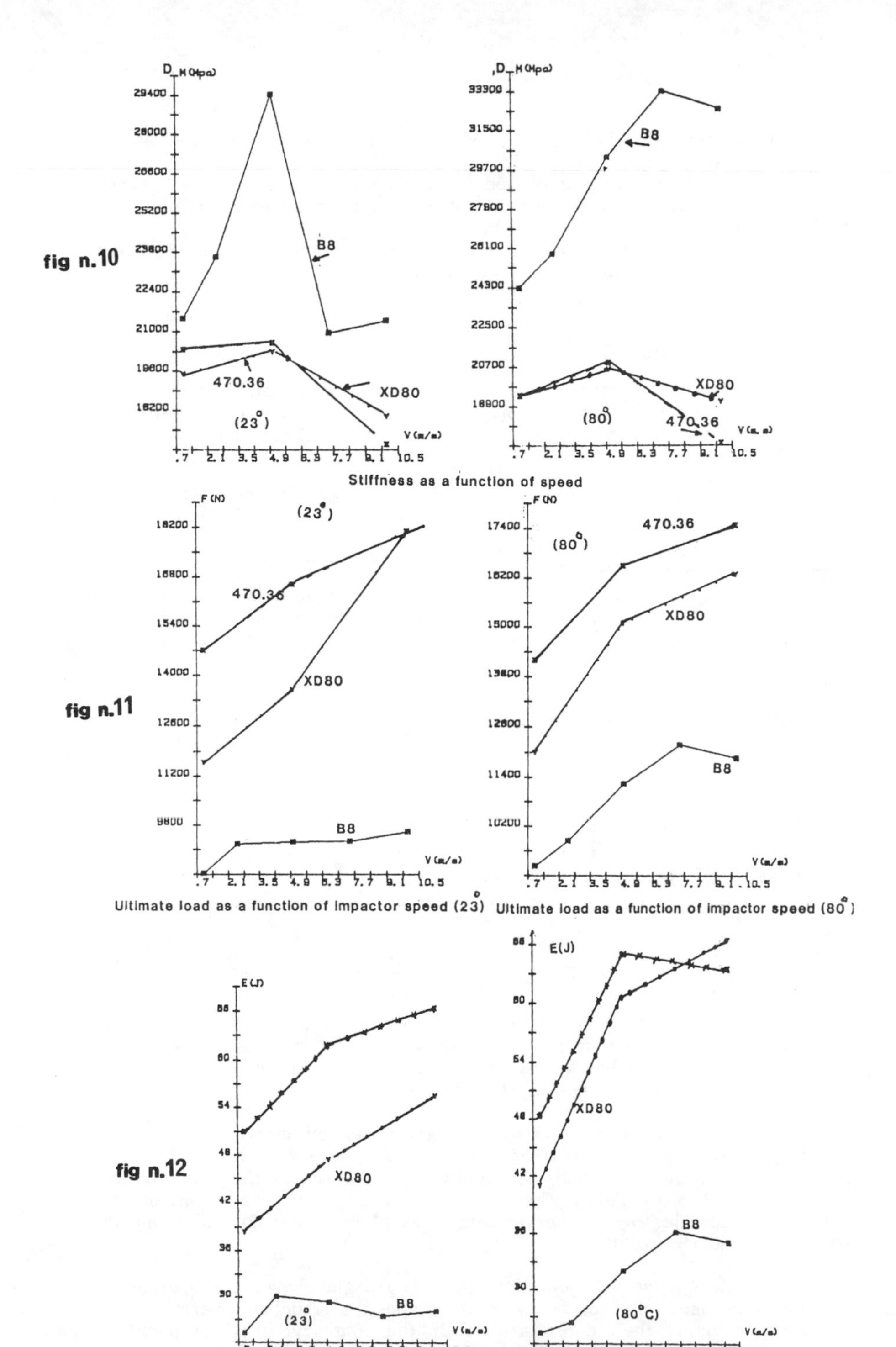

Stiffness as a function of speed

Ultimate load as a function of impactor speed (23°)

Ultimate load as a function of impactor speed (80°)

Energy as a function of impactor speed

The dynamic moduli E of studied materials is evaluated from stiffness calculation at the origin.

A particular study is developed for damaged surface at impact time.

The parameters Fmax - Emax and E have been studied for all retained materials.

The tests are carried out either at a room temperature of 23°C or at 80°C. The velocity varied from 1 to 10 m/s.

SYNTHESIS AND CONCLUSIONS

The comparaison between calculation results which give values of first frequency of unidirectionnal composite plates as function of fibers orientation and impact loading results proposed by MALLICK and BROUTMAN shows that their critical values are localized at an angle of 45°.

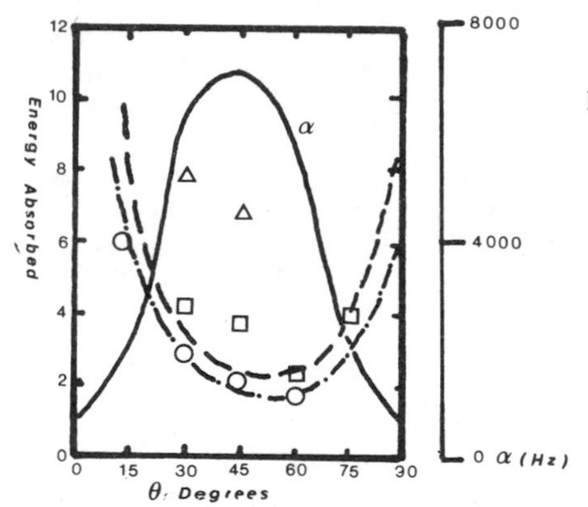

we remark that orthotropic composite plates present the same vibrations law for an orientation of fibers $-\theta, +\theta$

The direct comparison between shift factors or E" values for polymers as function of either temperatures or frequencies, and stiffness of impacted plates shows that the higher values are obtained for resins which present higher values of $tg\delta$ as a function of temperature.

In the case of plates vibrations, as a function of matrix's viscoelastic characteristics and as a function of frequency we remark that the first resonal frequency increases and the amplitude decreases as the ultimate impact load of the composites plates. The phase shift factors vary as a funcion of viscoelastic properties of polymers similarly as plates stiffness under impact loading as a function of the speed.

In conclusion, we consider that viscoelastic and dynamic properties of polymers influence impact behavior of composites structures permit the interpretation of their deformations. On the contrary, it is not possible to analyse and to modelize the fracture mechanisms of materials or structures.

BIBLIOGRAPHY

(1) CHIEM C.Y. and LIU Z.G. "Etude du comportement des composites à fibres de verre/epoxy en cisaillement à grandes vitesses de déformation" J de Phy. Colloque C5 supplément au N°8 Tome 46 (août85)

(2) MAROM G., DRUKKER E., WEINBERG A., BANBAJI J., Impact behavior of carbon/kevlar hybriol composites Composites revue Volume 17 N°2 April 86.

(3) HUGUES BP., WATSON AJ. "Compressive strength and ultimate strain of concrete under impact loading" Concrete research Vol.30 N°105 1978

(4) WATSTEIN D. "Effect of straining rate on the compressive strength and elastic properties of concrete" Journal of ACI April 1953

(5) MERLE G., YONG SOK O, PILLOT C., SAUTEREAU H., 1985 "Instrumented and temperature controlled charpy impact tester", Polymer testing Vol.5 pp.37-45.

(6) HERITIER J., THERY S., BALLADON P. 1985 "Comportement au choc de plaques stratifiées" Annales des composites-AMAC Vol.4-1985.

(7) GREEN A. "The impact testing of concrete" Mechanical properties of Non metallic brittle materials Editeur WALTON, W.H., London 1958

(8) GARY G. "Modelisation et étude expérimentale du flambement dynamique" Thèse de Doctorat d'Etat en Sciences Physiques PARIS 4 Octobre 1980

(9) DUFFY J., CAMPBELL J.D., HAWLEY R.H. On the use of a torsional spli-Hopkinson bar to study rate effects in 1100-0 Aluminium. J.Appl. Mech., 38, pp83-91, 1971

(10) LIN YG., SAUTEREAU H., PASCAULT JP., 1986 BDMA catalysed DDA-epoxy resin system : temperature and composition effects on curing mechanism". Journal of applied polymer science 1986

(11) LIN YG., SAUTEREAU H., PASCAULT JP., 1986 Résistance à l'impact de composites modèles epoxydes - billes de verre - introduction d'une interphase élastomère - Annales des composites AMAC Déc.1986

(12) LIN YG. 1986 "Formation et caracterisation d'un réseau époxyde. Effet renforçant d'une interphase élastomère entre une charge et ce réseau". Thèse Doctorat d'état. Lyon 1986 Université Lyon I.

(13) GRIFFITHS L.J., MARTIN D.J. A study of the dynamic behavior of a carbon fiber composite using the split-hopkinson pressure bar. J.Phys. D. Appl. Phys Vol 7 1974

(14) Z.G. LIU, C.Y. CHIEM Analyse du comportement et de la rupture microstructural des composites carbone/epoxy soumis aux grandes vitesses de déformation Receuil "le comportement au choc des matériaux et des structures composites" Annales des composites AMAC-12 décembre 1985

(15) DOREY G., SIDEY G.R., HUTCHINGS J., 1978 Impact properties of carbon fibre/kevlar 4,9 Fibre hybrid composites, composites 9 N°1 (jannary 1978) pp25-32.

(16) HULL D. "Energy absoption of composite materials under crash conditions" 83 A 40216 NASA

(17) THORNTON P.H. "The interplay of geometric and materials variables in energy absorption" I. Engineering materials and technology P114,1977

(18) BURTIN C. HAMELIN P. 1985 "Etude du comportement au choc de tubes composites" Annales des composites 1985/4 AMAC PARIS

(19) TREMY C. CHESNEAU C 1985 "Influence des propriétés viscoelastiques sur le comportement au choc des matériaux composites" Annales des composites 1985/4 AMAC PARIS

(20) FERRY JD. 1970 "Viscoelastic properties of polymers" John WILLEY and Sin Inc New York

(21) HAMELIN P. 1979 "Contribution à l'étude du comportement rhéologique de liants viscoélastiques en vue de l'analyse du fluage des matériaux composites" Thèse Etat Université LYON I 1979

(22) PUCK "Zur Beany ruchung and verformung GFK Mehrschichtenverbund Bauclementen Kunstoffe 1967 Allemagne

(23) TSAI S.W. HANN H.J. "Introduction to composites materials" technomie Pub. Co Westport 1980

(24) GRESZCUK L.B. "Theorical and experimental studies in properties and behavior of filamentary composites"Douglass Report N°3550 1966

(25) PAVEN H., DUBRESCU U. "Viscoelastic models in the rheology of hybrid Polymeric composites of (Phase in phase) in phase type" Rheology Plenum Press-New York

(26) HASHIN Z. "Complex moduli of viscoelastic composites" Int . J. Solid. Structures, G, 1970

(27) WHITE R.G., ABEDIN EMY. "Dynamic properties of aligned of short carbon fiber-reinforced plastic in flexure and torsion" J.Composite Vol.16 N°4 October 1985.

HIGH STRAIN-RATE BEHAVIOUR OF CARBON FIBER COMPOSITES

CHIEM C.Y. and LIU Z.G.
Ecole Nationale Supérieure de Mécanique
Laboratoire des Sciences des Matériaux de la Mécanique
Groupe ENSM-IMPACT, GIS Composites, "Sous-Groupe CHOC"
1 rue de la Noë, 44072 NANTES Cédex, France

ABSTRACT. The torsional split Hopkinson bar technique is used to produce shear strain rates ranging from 10^3 to $10^4 s^{-1}$ in dynamic and from 10^{-4} to $10^{-3} s^{-1}$ in quasi-static loading. The geometry of specimens versus the other results is analysed. The effects of strain rate on yield stress and shear strength in the woven carbon/epoxy composite, compared with other investigations, is determined and the fracture surfaces are examined by Scanning Electron Microscopy.

Key Words. high strain-rate, carbon/epoxy, impact, Hopkinson bar, torsional shear, dynamic behaviour.

I. INTRODUCTION

Development in recent years have clearly indicated that the class of composite materials based on fibers reinforcement have considerable potential for use in engineering components and structures. In order to achieve this potential a reorientation of design approach may be necessary in which due consideration must be given to the distinctive characteristics that these materials possess on load-bearing applications.

In considering loading effects it is required to distinguish clearly between 'material' response, the response of the material unaffected by the geometry of the test-piece or by the method of load application, and the 'structural' response, where the mechanical behavior is determined as well by the geometry of the test-piece as by its fundamental material properties. Of course, it is the final structural response of the designed component which is important but in the design process the material properties will be needed. At high-rate loading, the 'material' response becomes more important because the influence of loading is more local in the structure.

The split Hopkinson-bar system has been used to measure the stress-strain properties of composite materials at very high strain rates in compression /1/, tension /2-4/ and torsional shear /5-7/. Relatively, the torsional type of deformation has certain advantages over other test geometries for studing material behavior in the plastic region. This is easy to describe with precision, and most nearly duplicates the basic process for plastic deformation such as pure shear. The absence of geometric effects, such as necking and barrelling, permits deformation to much larger strains than are possible in tension or compression.

The work presents why the geometry of specimens is choosen, how to experimentate, which results are obtained and how far these results are significant compared to other results.

II. EXPERIMENTAL DETAILS

2.1. Geometric Analysis of Specimens

There are several types of shear loadings : direct (double) shear in a rivet, shear in a beam and shear produced by torsional loading. In dynamic loading, the compressive split Hopkinson bar apparatus are used for testing composite materials in interlaminar shear and transverse shear with rectangular specimens (fig. 1) /8/, in the punching test with disc specimen (fig. 2a) /9/ and with double-notch shear specimen (fig. 2b) /9/. With the torsional split Hopkinson-bar, Parry and Harding /5/ adopted the tubular specimen (fig. 3) and, Chiem and Liu /10-11/ the cuboïd-shape specimens (fig. 4).

In a solid cylindrical bar in torsion, stress-strain relation is different between the elastic range and plastic range (fig. 5) /12/, within the elastic range, the stress varies linearly from zero at axis of twist to a maximum at the extreme fiber and is linearly proportional to the angle of twist (fig. 5b). When the torsional loading is above the proportional limit, if straight-line variation of strain is assumed, the actual stress variation is similar to that shown by the solid line in fig. 5c. Thus the effect of yielding of the surface fibers during their early stage of plastic action is masked by the resistance of the remainder of the section. This difficulty is overcome by the use of properly designed tubular specimens, or cuboïd specimens which can give more accurate measurement of the shear strength since all fibers are at about the same stress. However, if a thin sheet is subjected to shear or a thin tube to torsion, before the shear strength of material is reached failure may occur by buckling due to the compressive stresses that act at 45° to the planes of maximum shear /5/. Thus, in tubular specimens for torsion tests, the relative thickness of the wall is greater than some critical value if shear failure is to be assured. Unfortumately, the capacity of the torque produced by torsional split Hopkinson-bar requires this thickness of the wall as small as possible to reach a high rate of strain. And also, the small thickness of the wall of composite material specimen is not representative because of the anisotropy characteristics of the composite and the manufacturing of tubular specimens with composite materials is more expensive than that used for cuboïd specimens. This is the reason why we adopt more frequently the cuboïd specimens than the tubular specimens in shear testing of composite materials by dynamic loading, in particular when torsional split Hopkinson bar is used.

2.2. Test Procedure

The torsional split Hopkinson-bar apparatus used in the present experiments is essentially the same as that described by Chiem and Liu /13/. Since a complete description of the apparatus and experimental procedure has been detailled previously, only a brief account is given here. As referred to figure 6, the sudden opening of the clamp releases the torque stored in the input bar between the rotating head and the clamp. The torsional pulse propagates from the clamping point and interacts with the specimens which are sandwiched between the input and the output bars. The wave interaction at the specimens results in a portion of input pulse being transmitted through the specimens with the remainder being reflected back into the input bar. The strain gages acquire the incident, reflected and transmitted pulses (see the Lagrange diagram in fig. 6) which are recorded with a digital oscilloscope. From the transmitted wave, of magnitude T_T, we can calculate the average shear stress in the specimens

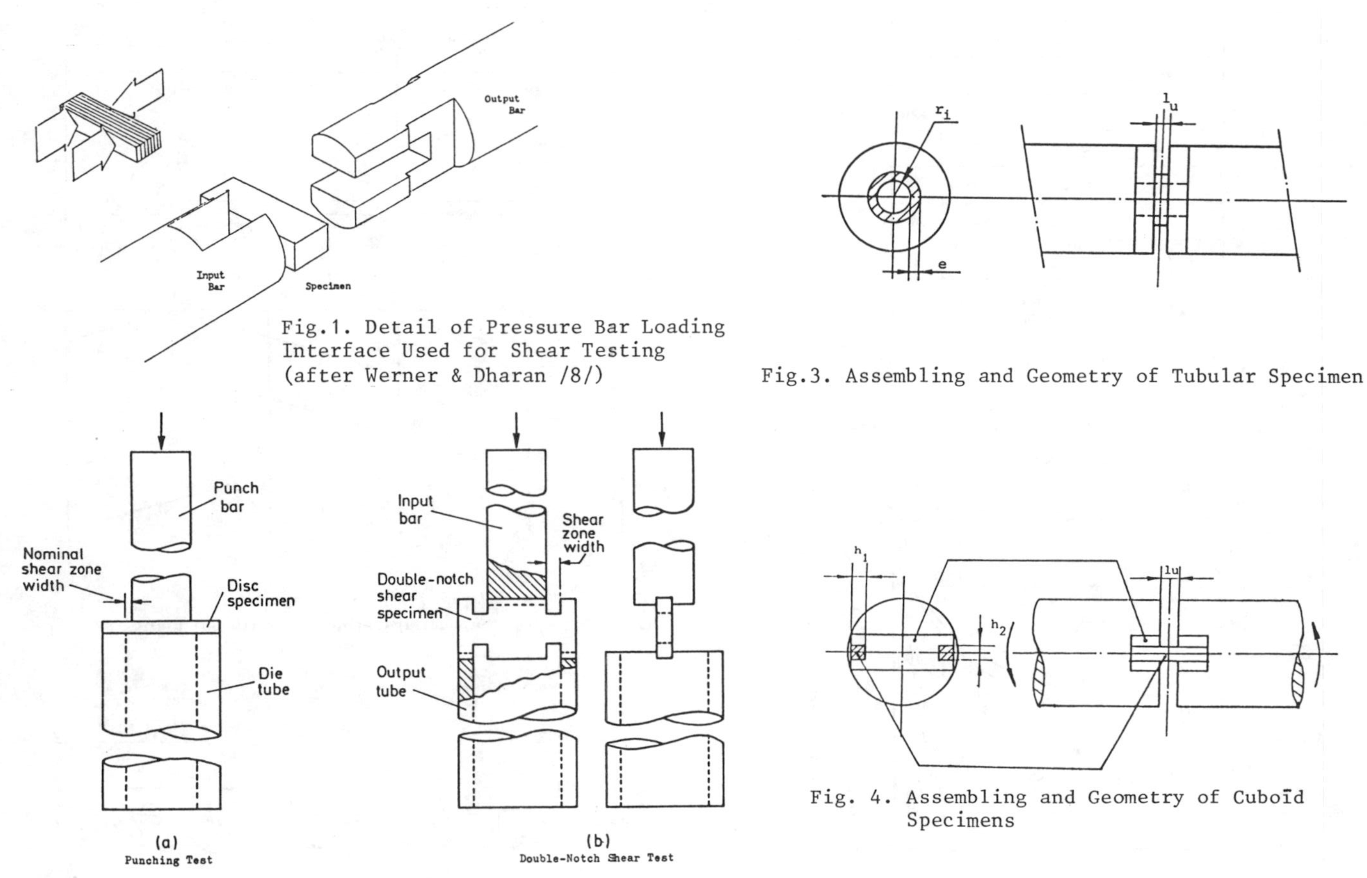

Fig.1. Detail of Pressure Bar Loading Interface Used for Shear Testing (after Werner & Dharan /8/)

Fig. 2. Shear-Loading Versions of SHPB (after Harding /9/)

Fig.3. Assembling and Geometry of Tubular Specimen

Fig. 4. Assembling and Geometry of Cuboïd Specimens

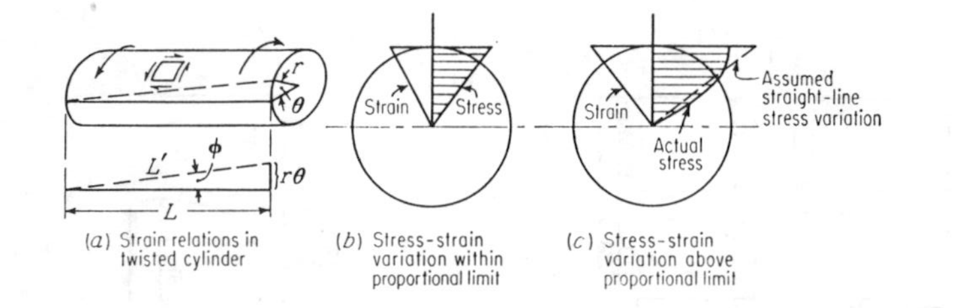

***Fig.* 5 Strain and stress relationships in torsion.** /12/

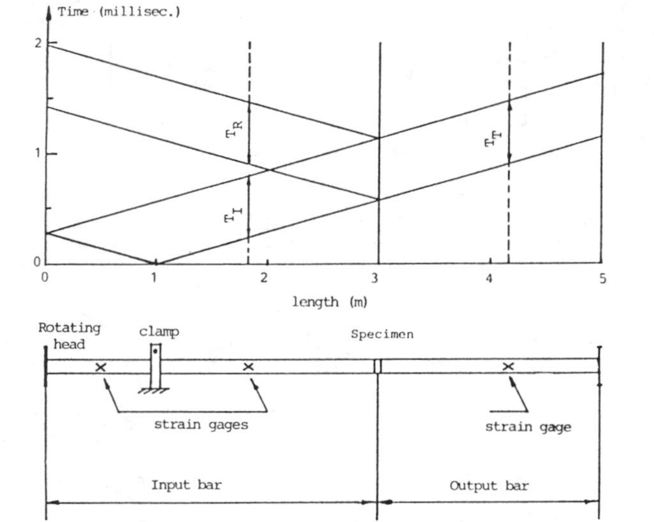

Fig.6 - Schematic of the torsional Hopkinson bar and Lagrange diagram.

Fig. 7 - Courbes contrainte-déformation en cisaillement.
(Shear stress-strain curves.)

Tableau 1 : valeurs relevées de la figure 7
(values taken from figure 7)

N°	$\dot{\gamma}\,(s^{-1})$	τ_{max} (MPa)	τ_Y (MPa)
1	2623	127,6	116,5
2	3631	139,1	138,0
3	5166	163,2	151,6
4	5410	172,8	153,6
Q1	$3,37.10^{-3}$	10[illegible]7	62,0
Q2	$5,90.10^{-4}$	110,3	62,4

$$\tau = T_T/(nh_1h_2V_m) \tag{1}$$

where n is the number of specimens, V_m the mean radius of specimens (c.f. /14/) and, h_1 and h_2 being the length and the width of specimens, respectively.

The relative velocity between the ends of the specimens is then given by

$$\dot{\theta}(t) = 2\ T_R/J\rho c \tag{2}$$

where T_R is the reflected wave magnitude, J being the polar moment of inertia of the bar, ρ the density and c the shear wave velocity. The mean shear strain-rate in the specimens may be calculated from

$$\dot{\gamma} = \dot{\theta}(t)\ .\ \frac{V_m}{h_3} \tag{3}$$

where h_3 is the useful length of the specimens. The shear strain is then obtained by integration of relation (3)

$$\gamma = \frac{V_m}{h_3} \int_o^t \dot{\theta}(t)\ dt \tag{4}$$

Finally, the shear stress-strain relation is carried out from (1) and (4)

$$\tau = f(\gamma) \tag{5}$$

2.3. Material Tested

The material tested is a woven carbon/epoxy composite. It is reinforced 70 % in weight by 10 layers. The specimens were cut by a truncating machine. Araldit AW-106 and Hardener HV 953 U are used for cementing. The technique detail was given in reference /13/.

III. EXPERIMENTAL RESULTS

3.1. Shear Stress-Strain Curves

The shear stress-strain curves of the tested material are presented in fig. 7. The strain rate varies from $10^{-4}s^{-1}$ to $10^{-3}s^{-1}$ for quasi-static loading and from $10^{3}s^{-1}$ to $10^{4}s^{-1}$ for dynamic loading. In all tests some inelastic deformation of the specimens was observed beyond the region of approximately linear stress-strain response. At the quasi-static loading rate audible cracking of the specimens was noticed at around the limit of the elastic region. The subsequent deformation of the specimens is elasto-plastic up to a maximum strain, about 75 %. We characterize the results by determining, for each test, the yield stress and the maximum stress supported by the specimens, (or yield point) τ_y and shear strength τ_m (see Table 1) and, the strain to fracture defined as the strain corresponding to the maximum stress. Because it is not possible to describe the mechanical behavior of the composite in terms of an average stress-strain curve at a given strain rate. The shear strain to fracture is about 70 % for the quasi-static tests and, about 20 % and no important plastic deformation for dynamic tests. However, the maximum shear stress is much higher in dynamic loading than in quasi-static loading. This difference between the quasi-static and dynamic tests will be discussed in the following section.

3.2. Microscopic Observations

The fracture appearance of the specimens is observed under the Scanning Electron Microscope. The figures 8a and b represent the fracture surface of the specimens subjected to quasi-static rate and dynamic rate loadings respectively. In the former case, the failure in the specimens was characterized by the irregular cracks in the matrix both longitudinal and transverse to the fibers. This is the reason why the audible cracking is sound. There are no debris of broken fibers and pieces of matrix on the fracture surface of specimen. The crescent split debonding between the fibers and the matrix implies that the failure of the specimens is on ductile fracture. In the later case, the debris of the shattered specimens are found. There is no sign of fibers-matrix debonding. This implies that before the fracture of the specimens, the bond between the fibers and matrix permits both components to support together the loading. Generally, the mechanical behavior of the resin is rate dependent. For example, its shear strength increases with the rise of strain-rate. This explain that the composite is rate-sensitive to dynamic loading. Finally, in both cases, no fiber is pulled out of matrix.

IV. DISCUSSIONS

4.1. Effect of Strain Rate on the Shear Strength

The variations of shear strength and yield stress with strain rate are shown respectively in fig. 9a and b. For the woven carbon/epoxy composite, a significant increase in both shear strength and yield stress are observed as strain rate was increased from $10^3 s^{-1}$ to $10^4 s^{-1}$. It fits an empirical relationship of the form

$$\tau = A + B \log \dot{\gamma} \tag{6}$$

where τ_o is a material constant, B being the strain-rate sensitivity factor. The values for A, B and R, the correlation coefficient, that were calculated with the least square method, are given in table 2.

Table 2. Material constants and their correlation coefficient

Shear Stress	A	B	R
τ_y	- 275.4	115.2	0.9907
τ_m	- 350.5	139.0	0.9765

To identify the physical and mechanical significance of the material constants A and B more investigations are required. This is one of our future subjects /15/.

In the quasi-static loading, both the yield stress and shear strength are independent of strain rate over the range of $10^{-4}s^{-1}$ to $10^{-3}s^{-1}$. Because of the absence of data at the intermediate strain rates, we connect this range with dot line. However, we can predict that there is no marked increase in both the yield stress and shear strength between the quasi-static and the intermediate rate tests and, that there is a transition of the failure mechanism of the composite between the intermediate and dynamic (or impact) rate tests.

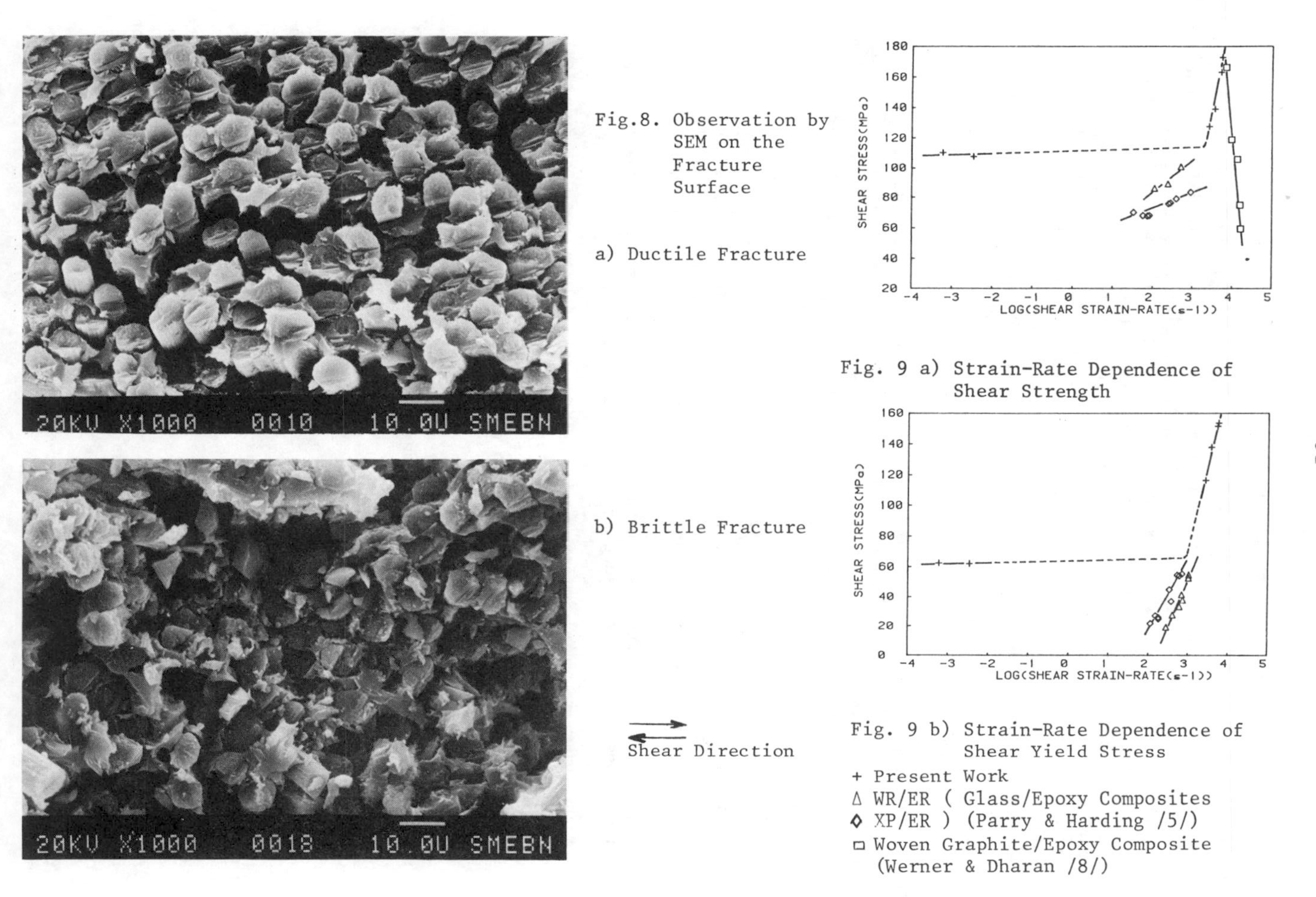

Fig.8. Observation by SEM on the Fracture Surface

a) Ductile Fracture

b) Brittle Fracture

Shear Direction

Fig. 9 a) Strain-Rate Dependence of Shear Strength

Fig. 9 b) Strain-Rate Dependence of Shear Yield Stress

+ Present Work
Δ WR/ER (Glass/Epoxy Composites
◇ XP/ER) (Parry & Harding /5/)
□ Woven Graphite/Epoxy Composite (Werner & Dharan /8/)

4.2. Comparison with other results

The results had been obtained by Parry and Harding /5/ with two epoxy-based composites : woven-roving glass mat reinforcement (WR/ER) and non woven cross-ply glass reinforcement (XP/ER). The glass/resin volume fraction is 40 % for WR/ER and 41 % for XP/ER composites. Both composites show very similar behavior to the carbon/epoxy composite that we have tested with a significant increase in τ_y and τ_m as the strain-rate is raised (see fig. 9 and fig. 3 for the geometry of the specimens). However, the results of Werner and Dharan /8/ on woven graphite/epoxy composite in transverse shear deformation (see fig. 1) show that the increasing strain rates resulted in a decrease in the shear stress (see fig. 9a). They concluded that the decrease in shear stress with strain rate was probably due to an increasing density of interlaminar shear failures between the plies as the strain rate is increased.

It is very difficult to compare quantitatively the various results because of the difference of the material properties, the testing conditions and the experimental methods. However, it is useful to bring out qualitatively the effects of strain rate on the composite yield point and strength.

V. CONCLUSIONS

1. The use of the torsional split Hopkinson-bar is a reliable technique for determining the high strain rate response of composite material in shear. The shear strain rates ranging from 10^3 to $10^4 s^{-1}$ in dynamic and from 10^{-4} to $10^{-3} s^{-1}$ in quasi-static loading were achieved in this investigation.

2. The use of the cuboïd specimens for this technique shows advantages over other types of specimens such as easy manufacturing.

3. The effect of strain rate on the shear stress is significant. The increase in shear stress with strain rate is probably due to the rate sensitivity of matrix material. The microscope observations on the fracture surfaces show the difference of failure of the specimens between the quasi-static and dynamic loading.

4. The comparison with other results for composite material in dynamic shear behavior give a qualitative indication in the engineering applications.

Acknowledgements

The authors are endebted to the French Ministry of Research and Technology for sponsorship of these investigations through "G.I.S. Composite/Sous-Groupe CHOC". They are also grateful to Professor S. Offret, Chairman of Laboratoire des Sciences des Matériaux de la Mécanique, for her continuous interests and suggestions.

REFERENCES

/1/ Griffiths, L.J. & Martin, D.J., "A study of dynamic behaviour of a carbon fibre composite using the split Hopkinson pressure bar", J. Phys. D. : Appl. Phys., vol. 7, 1974, pp. 2329-2341.
/2/ Harding, J. & Welsh, L.M., "A tensile testing technique for fibre-reinforced composites at impact rates of strain", J. of Materials Science, 18 (1983), pp. 1810-1826.

/3/ Ross, C.A., Cook, W.H. & Wilson, L.L., "Dynamic tensile tests of composite materials using a split Hopkinson pressure bar", Exp. Mech. (nov. 1984), pp. 30-33.
/4/ Harding, J. & Welsh, L.M., "Impact testing of fibre-reinforced composite materials", 4th Internl Conf. on composite materials, Tokyo, oct. 1982.
/5/ Parry, T. & Harding, J., "The failure of glass-reinforced composites under dynamic torsional loading", Proc. of the Colloque Internl. du C.N.R.S. n° 319, June 1981.
/6/ Chiem, C.Y. & Liu, Z.G., "Behaviour of the woven reinforced glass fiber-epoxy composite materials subjected to dynamic shear-loading", Proc. of the Internl. Symp. on Intense Dynamic Loading and its Effects, Peking, June 1986, pp. 584-591.
/7/ Liu, Z.G. & Chiem, C.Y., "Analysis of the behaviour of the unidirectional (U.D.) Carbon/epoxy composites subjected to different shear strain-rates", (in French), JNC5, Paris, sept. 1986, pp. 741-755.
/8/ Werner, S.M. & Dharan, C.K.H., " Behavior of woven graphite-epoxy at very high strain rates", 29th National SAMPE Symposium, April 3-5, 1984.
/9/ Harding, J., "High rate straining and Mechanical properties of materials", OUEL Report n° 1502/83, Oxford Univ.
/10/ Chiem, C.Y. & Liu, Z.G., "Adaptation of the Kolsky torsion bars to the study of the dynamic behavior of composite materials (in French), Annales des Composites (Déc. 1985), pp. 189-206.
/11/ Liu, Z.G. & Chiem, C.Y., "Analysis of the behaviour and the fracture microstructural aspects of carbon/resin composites subjected to high strain rates", (in French), Déc. 1985), pp. 207-221.
/12/ Davis, H.E., Troxell, G.E. & Wiskocil, C.T., "The testing and inspection of Engineering Materials", Mc Graw-Hill Book Company, 1964.
/13/ Chiem, C.Y. & Liu, Z.G., "Le comportement et la modélisation des matériaux composites carbone/époxyde soumis aux grandes vitesses de déformation", Rapport final n° CL/86.03, Mars 1986.
/14/ Chiem, C.Y., Thèse de Docteur-ès-Sciences (Juin 1980), ENSM de Nantes, France.
/15/ Liu, Z.G., Thèse de nouveau doctorat (Ph.D.), to appear, (1986), ENSM de Nantes, France.

The Influence of Processing Conditions on Mechanical and Fracture Properties of GRP Plates

C.A.C.C. Rebelo*, A.J.M. Ferreira+, A.T. Marques+, P.M.S.T. de Castro+

ABSTRACT

A study of the mechanical properties and fracture toughness of two commonly used resins is presented. The resins are CRYSTIC 272 (a polyester isophtalic resin) and 600 (a bisphenolic resin). The dependence of maximum strength, hardness, Young's modulus and elongation at rupture with cure conditions is described. The dependence of fracture toughness with the specimen thickness is presented, for both resins. A fractographic study of the fracture surfaces is presented, using the scanning electron microscope. Some initial results concerning the use of those resins in GFRP are presented.

INTRODUCTION

The characterization of the mechanical behaviour of composites requires the knowledge of the behaviour of the resin matrix and of the fibres. The present authors are conducting a systematic investigation of the behaviour of glass fibre reinforced plastics (GFRP), a material commonly used in many engineering applications. The use of this material in structures subjected to ever more critical loading and service conditions requires a quantitative knowledge of its performance, which must be obtained by careful relevant mechanical testing.
The present paper gives a detailed description of a series of tensile tests conducted on two resins, CRYSTIC 272 and 600, together with hardness measurements, for different cure conditions. The effect of cure time was studied. The effect of applying, or not applying, a post cure was also studied.
The extensive compilation of experimental data now presented includes a few preliminary results that were presented in an earlier report [1].
Regarding fracture toughness, the measurements were made on single edge notched tensile specimens and the dependence of K_C with cure time and thickness was studied. Some earlier results, for the 272 resin, were also already presented in an earlier report [2].
A fractographic examination of the fracture surfaces of the tensile test and fracture toughness specimens was carried out using a scanning electron microscope.

EXPERIMENTAL WORK: RESINS

Material and specimens for the resins characterization

The materials tested are the CRYSTIC 272 and 600 resins, supplied by Quimigal (Portugal). The 272 resins is a polyester isophtalic resin. The cure system used was a MEKP catalyst (Butanox M50) and a cobalt accelerator (TP 395 VZ). The catalyst and the accelerator were used in a weight fraction of 2% and 0.1% respectively. The CRYSTIC 600 resin is a bisphenolic resin, and the cure system used was 2% of MEKP catalyst and 1% of accelerator.
The tensile tests were carried out using ISO R 527 type I specimens, cut from 300x400mm^2 plates of 4mm nominal thickness, casted in special purpose

* Faculdade de Ciências e Tecnologia, SAEM, Universidade de Coimbra, Portugal

+ Departamento de Engenharia Mecânica, Universidade do Porto, Portugal

built rigs. Twelve specimens were taken from each plate. The specimens were tested at a displacement rate of 2mm/min, in an Hounsfield tensometer, with an instrumented 5kN load cell, and an inductive type RDP D5-100 LVDT transducer was used. The testing temperature was 21 ± 1°C. Load versus displacement signals were fed, after adequate conditionning, to a Bryans XY recorder. Measurements of Barcol hardness were taken on all the samples tested, typically 50 measurements for each plate.
The fracture toughness testing was conducted on single edge notched tensile specimens, cut from plates identical to those already mentioned. The specimens are of 120x30mm^2 nominal dimensions and 4mm nominal thickness pin-loaded by 10mm diameter pins, and it was possible to obtain 10 specimens from each plate. From these 10 specimens, 5 were post cured and 5 were not post cured. The a/W ratio was in the range 0.45 to 0.55.

Results

The maximum stress (obtained from the engineering stress/strain curve) is presented in Figures 1 and 2 for the 272 and 600 resins, respectively. These show the measurements as function of the time for cure, for post cured material. A trend of increasing maximum stress for the non post cured materials is clearly displayed. It is interesting to notice the higher scatter of the results for the post cured material. In both cases (272 and 600 resins), the maximum stress increases almost linearly when plotted on a log time basis, for the non post cured material, up to the level of the post cured resin after around 2 to 3 months of cure time. Post cure consisted of 3 hours at 80°C.
The measurements of Young's modulus, maximum stress, elongation at rupture and Barcol hardness are colected together in Table 1. Young's modulus, maximum stress and elongation at rupture values presented are typically averages of 10 or more specimens.
Regarding the fracture toughness data, the trends for the 272 resin were already presented in[2]. The actual data for each of the five specimens tested, for each condition, is presented in Figures 3 and 4, as a function of cure time for non post cured and post cured material, respectively. The same type of data, concerning the 600 resin, is presented in Figures 5 and 6, where the same trends detected for the 272 resin are again displayed.

Discussion

The higher scatter of the maximum stress data recorded for the post cured materials, is probably related to the corresponding higher tensile strength and lower ductility of the material in that condition. This implies, of course, an higher sensitivity to the presence of any scratches or voids on the material's surface, which may always be present, notwithstanding the great care that was taken in their preparation.
The trends of increasing Young's modulus and hardness, and decreasing elongation with the increase in maximum stress, must be related to the decrease of fracture toughness measurements, and this is well confirmed by the fracture toughness data now presented.
The scatter lines presented in Figures 1 and 2 display the arithmetic mean value, the maximum and minimum. It should be mentioned that the Chauvenet criterion was used to reject or include data,[3].
Regarding thickness effect studies, single edge notched tensile specimens of the same planar overall nominal dimensions were tested, with a range of thicknesses from 4 to 16mm. The resulting K_C data for 272 and 600 resins is presented in Figures 7 and 8 respectively . For each thickness

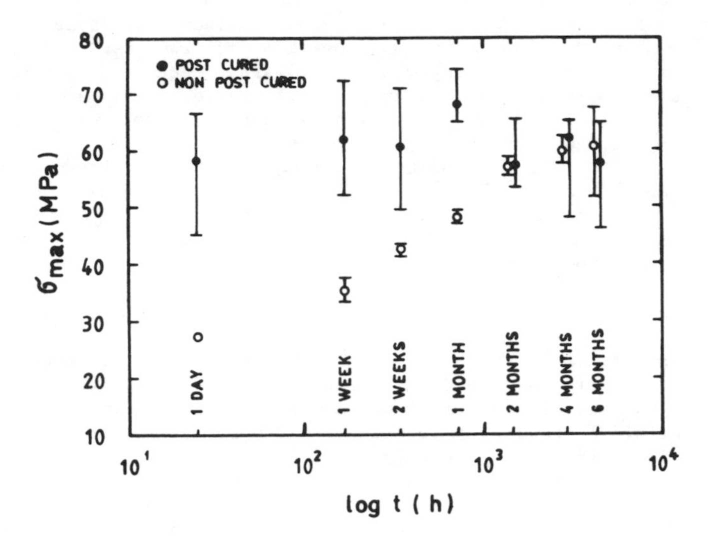

Fig. 1 - maximum stress as a function of time for cure, 272 resin

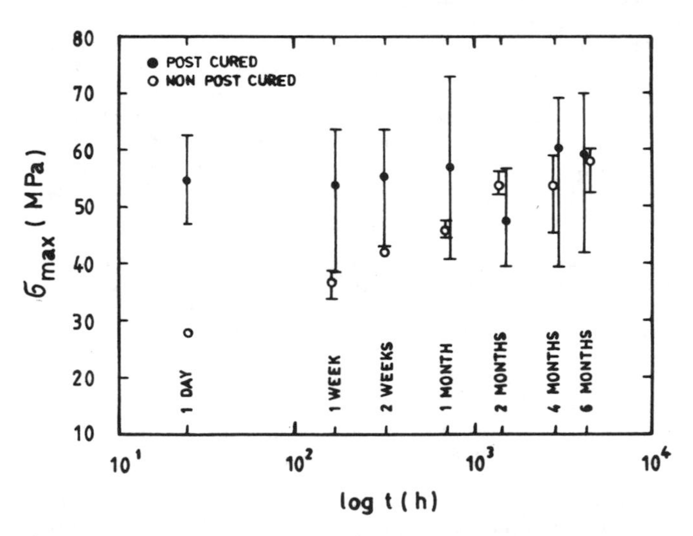

Fig. 2 - maximum stress as a function of time for cure, 600 resin

TABLE 1 - TENSILE AND HARDNESS DATA

RESIN	CURE CONDITIONS	YOUNG'S MODULUS (MPa)	MAXIMUM STRESS (MPa)	ELONGATION AT RUPTURE (%)	BARCOL HARDNESS
CRYSTIC 272	1 DAY	1720	27.5	12.45	12
	1 WEEK	1970	35.4	12.94	18
	2 WEEKS	2110	42.5	8.31	24
	1 MONTH	2410	47.9	8.45	28
	2 MONTHS	2460	57.0	5.01	32
	4 MONTHS	2830	59.3	2.73	37
	6 MONTHS	2840	60.7	2.55	37
	1 DAY + P.C.	3100	58.0	2.14	41
	1 WEEK + P.C.	3170	61.8	2.20	39
	2 WEEK + P.C.	3140	60.7	2.30	41
	1 MONTH + P.C.	3300	68.1	2.59	39
	2 MONTHS + P.C.	3210	57.2	2.02	41
	4 MONTHS + P.C.	3230	62.1	2.19	40
	6 MONTHS + P.C.	3250	57.6	2.04	41
CRYSTIC 600	1 DAY	1230	27.8	7.41	6
	1 WEEK	2160	36.6	6.13	16
	2 WEEKS	2270	42.0	4.72	22
	1 MONTH	2560	45.9	4.58	24
	2 MONTHS	2400	53.6	4.15	28
	4 MONTHS	2680	53.6	2.52	33
	6 MONTHS	2960	57.9	3.23	33
	1 DAY + P.C.	3090	54.8	2.07	38
	1 WEEK + P.C.	3140	53.5	2.04	38
	2 WEEKS + P.C.	3090	55.2	2.15	40
	1 MONTH + P.C.	3020	56.9	2.50	39
	2 MONTHS + P.C.	3060	47.6	1.74	37
	4 MONTHS + P.C.	3130	60.0	2.32	40
	6 MONTHS + P.C.	2860	59.1	2.73	40

P.C. - Post-cure of 3 hours at 80°C

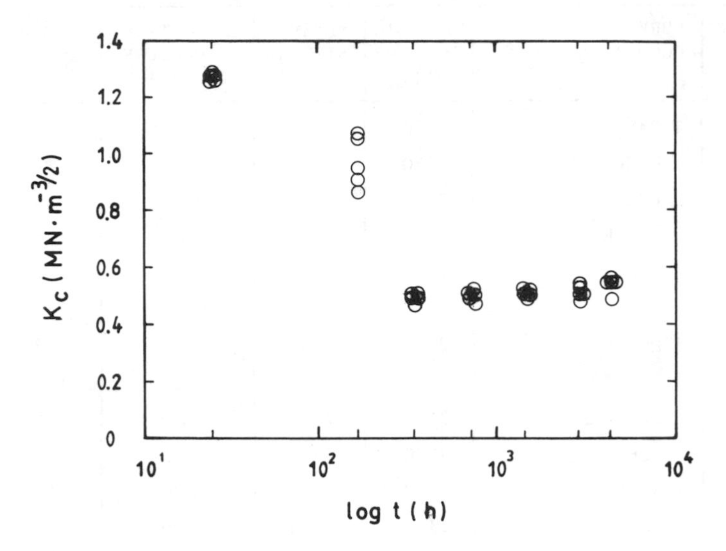

Fig. 3 – fracture toughness as a function of cure time, non post cured 272 resin

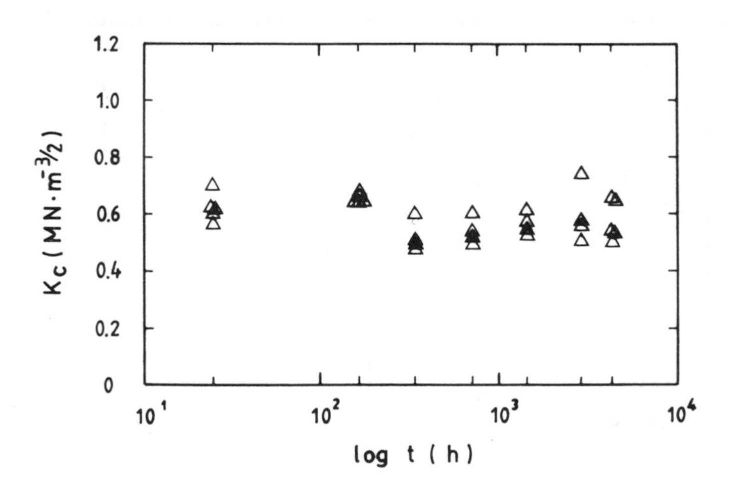

Fig. 4 – fracture toughness as a function of cure time, post cured 272 resin

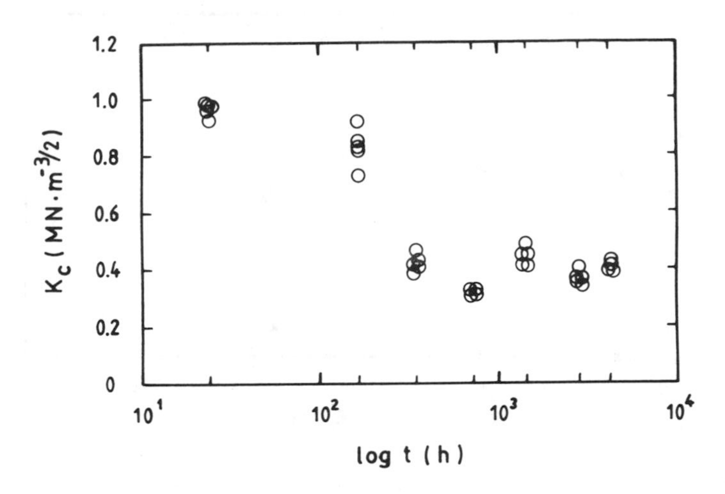

Fig. 5 – fracture toughness as a function of cure time, non post cured 600 resin

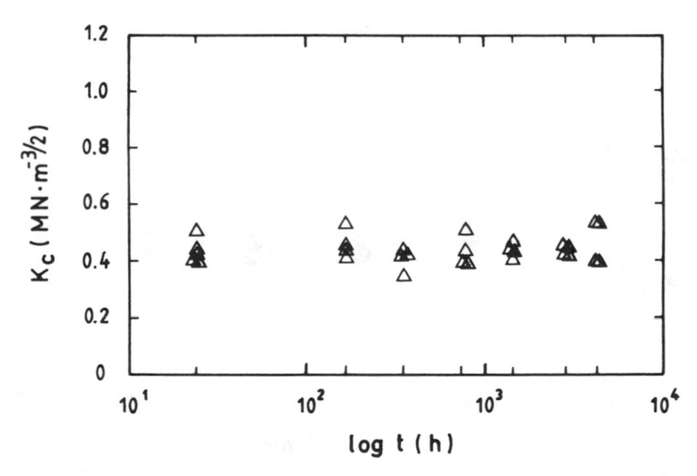

Fig. 6 – fracture toughness as a function of cure time, post cured 600 resin

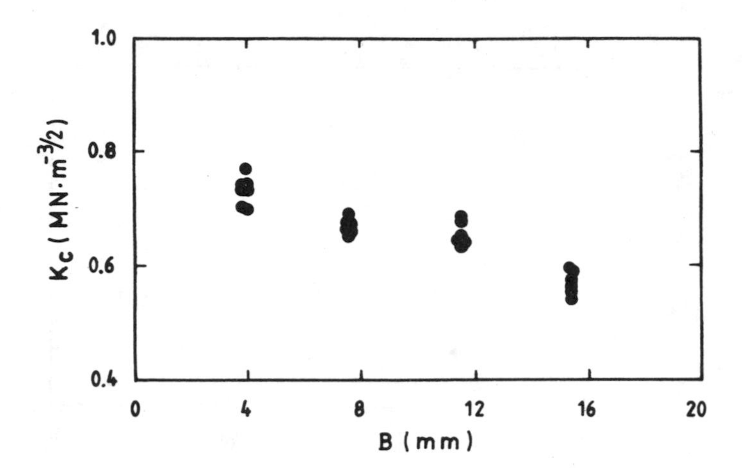

Fig. 7 - fracture toughness as a function of thickness, 272 resin

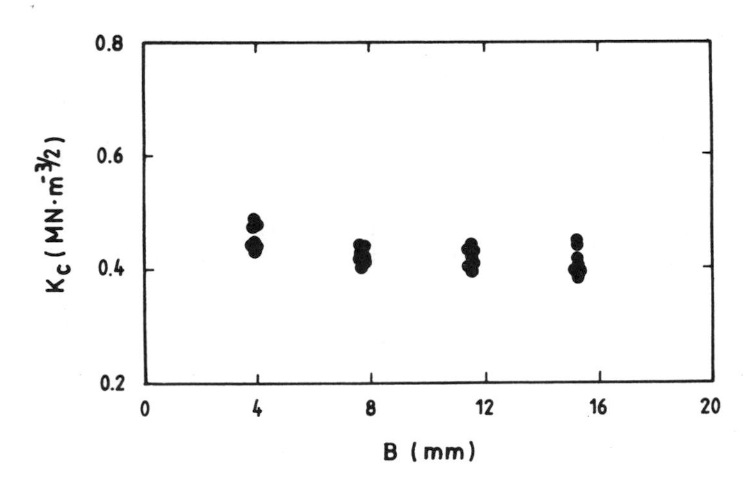

Fig. 8 - fracture toughness as a function of thickness, 600 resin

7 specimens were tested, and all the resulting data is included on the figures.
The well known ASTM E399 rule for ensuring plane strain conditions at the crack tip are verified in both cases. The criterion is,[4]:

$$B \geqslant 2.5 \left(\frac{K_{Ic}}{\sigma_y}\right)^2$$

where K_{Ic} is plane strain fracture toughness and σ_y is yield stress. In our case, instead of yield stress, the maximum engineering stress recorded was used. Taking K_{Ic} as the fracture toughness recorded for 4mm, the minimum B values for the 272 resin is therefore 0.36mm whereas for the 600 resin it is 0.17mm. It is obvious that in both cases the ASTM condition is verified. However, the fracture toughness data for the 600 resin is approximately independent of thickness (although slightly decreasing), whereas for the 272 resin a marked decrease in fracture toughness is noticed; this decrease is around 30% when the thickness increases from 4 to 16mm. This result is somewhat unexpected. Since both sets of specimens are well above the minimum ASTM thickness, an independence should be expected in both cases, whereas a dependence is found in both, although marked in the 272 resin, and just noticeable in the other resin.
The present results suggest the need for further study of a fracture toughness testing standard for polymers since it appears that the ASTM E399, developed for metals, is not directly applicable without further discussion to these materials.

EXPERIMENTAL WORK: COMPOSITES

Material and specimens for the composites' characterization (GFRP)

Two types of glass fibre reinforced (GFRP) laminates were tested. The matrix was the CRYSTIC 272 resin already mentioned and the cure system was also, in weight fraction, of 2% and 0.1% for the MEKP catalyst and the cobalt accelerator respectively. A post-cure of 3h at 80° C was used. Two types of reinforcements, woven-roving and chopped strand mat, were used. A nominal weight fraction in fibre content of 40% and 45% was obtained for the chopped strand mat/woven-roving laminates respectively.
The tensile tests were carried out using the specimen suggested by the Guidelines for the Generation and Use of Data in Military Handbook 17A - Part I,[5], machined from 300x400mm² plates.
The plates were made by the hand-lay up process, although special care has been taken in order to keep the larger surfaces flat, parallel and smoth by using a 'Melinex' release film and applying a dead weight of 50N after impregnation. Four specimens were used for each type of laminate. The specimens were tested at a displacement rate of 2mm/min, in an Hounsfield tensometer, with an instrumented 20kN load cell and an inductive type RDP D5-100 LVDT transducer was used. The testing temperature was 21 ± 1°C. Load versus displacements signals were also fed , after adequate conditioning, to a Bryans XY recorder.
The fracture toughness testing was conducted on single edge notched tensile specimens, cut from plates identical to those already mentioned. The specimens are 180x45mm² nominal dimensions and 10mm nominal thickness for the chopped strand mat laminate specimens or 8mm nominal thickness for the woven-roving laminate specimens, pin loaded by 15mm diameter pins. Typically 5 specimens were tested for each type of laminate. The a/W ratio was in the range of 0.45 to 0.55.

Results

The tensile tests were conducted inorder to establish the basic mechanical behaviour data of the laminates used, to observe the stress/strain curve behaviour, and to check the applicability of "The Guidelines and Use of Data in Military Handbook 17A,Part I", ref.[5]. The results discussed are average values of four tensile specimens of the chopped strand mat laminate, and three specimens of the woven-roving laminate. The specimens had an overall length of 290mm, a gauge length of 50mm, a width of 12.7mm (within the gauge length). From the straight-sided centre gauge section to the ends a tapered section of 1° was used, giving a width of 19mm, with nominal thickness of 2.5mm or 2.25mm for the chopped strand mat or woven-roving laminates respectively.
For the chopped strand mat specimens an average tensile strength of $75.7 MN/m^2$, an average strain at rupture of 1.39% and an average modulus of 5320 MN/m^2 were obtained. The tensile strength and strain at rupture values obtained were rather low when compared with values suggested by the materials' suppliers. The observed behaviour was probably due to a specimen size effect. Therefore, further tests were made with other types of specimens. By using straight sided specimens with a 21mm nominal width and nominal thicknesses of 2.5mm or 4.00mm, an increase in tensile strength of 10MPa or 70MPa, respectively,was found. A third specimen was tested having the same geometry specified in the military standard, but with a 4mm nominal thickness, and gave an increase in tensile strength of 14MPa. For the woven-roving specimens average values of 270 MPa, 2.15 and 12600 MPa were obtained for the tensile strength, strain at rupture and modulus, respectively. The same geometric effect described above has been found. By increasing the thickness to 4.0mm, and using the same type of specimen (military standard), an increase of 89MPa was obtained for the tensile strength. The use of a straight sided specimen of 4.0mm thickness and 21mm width gave an increase of 93MPa for the tensile strength.
Although a more detailed study of these geometric effects must be carried out, it can already be said that the specimen geometry suggested by the military standard is not applicable to test the chopped strand mat and woven-roving laminates described in this paper. It seems that an increase in thickness - the thickness effect is probably related to the number of plies in the laminates - and width is needed (values of 4.0mm and over 20mm are suggested for the thickness and the width respectively).
The toughness tests were carried out with the modified single edge notched tensile specimen. Five specimen were used to characterize the chopped strand mat laminates and four specimens were used for the woven-roving laminates. For both laminates it was not possible to define critical stress intensity factors, as the load versus displacement curves were not linear up to the rupture. Instead, maximum load toughness values were recorded. For the chopped strand mat laminates (10mm nominal thickness) the average value of $16.2 MN.m^{-3/2}$ was obtained for the maximum load toughness. For the woven-roving laminates (8mm nominal thickness) the average value of $32.1 MNm^{-3/2}$ was obtained.

Discussion

As far as the toughness tests are concerned, some general comments can already be made. Firstly, the single edge notched tensile specimen has been modified, as compared with the specimen used for resin specimens, in order to obtain better representativity of the composite behaviour.

In order to obtain ~ 20mm uncracked ligament, keeping the a/W ratio between 0.45 and 0.55, a width of 45mm has been used, and the other dimensions were altered accordingly. Secondly, although 8mm and 10mm thick specimen were used, the ASTM E399 thickness rule was not satisfied.
Thirdly, no natural crack has been introduced in the SENT specimens, as it was supposed - and was confirmed- that this would not make any difference in the toughness measurements with GFRP specimens. Fourth, a significant increase in toughness has been obtained, for both laminates, when compared with the resin toughness, but the more significant increase was for the woven-roving laminates because of higher weight fraction of glass and type and orientation of reinforcement.

CONCLUSIONS

i) For both resins, the maximum stress increases almost linearly when plotted on a log time basis, for the non post-cured material, up to the level of the post-cured resin, after around 2 to 3 months of cure time.

ii) Fracture toughness of Crystic 272 and 600 resins decreases with cure time when no post cure is applied, whereas it is practically constant whenever the post cure is used, regardless of cure time.

iii) Fracture toughness of Crystic 272 and 600 resins tends to the value of the post cured resin after a period of two weeks, when post cure is not applied.

iv) A modification of the single edge notched tensile specimen used for the resins was necessary in order to obtain adequate representativity of the composite.

v) A significant increase in toughness is obtained when reinforcements are used, and this effect is more marked with the woven-roving type than with the chopped strand mat type.

REFERENCES

(1) Rebelo, C.A.C.C.; Marques, A.T.; de Castro, P.M.S.T., "The influence of cure conditions on the fracture of non-reinforced thermosetting resins", EUROMECH 204 Colloquium, Poland, 1985 (Proceedings to be published by Elsevier Applied Science)

(2) Rebelo, C.A.C.C.; Marques, A.T.; de Castro, P.M.S.T., "Fracture characterization of composites in mixed mode loading", 6th European Conference on Fracture, ECF6, Amsterdam, June 1986 (to be published by EMAS)

(3) Holman, J.P., "Experimental Methods for Engineers", McGraw Hill, 4th edition, 1984

(4) ASTM E399-83, in: ASTM 1985 Annual Book of ASTM Standards, vol. 03.01, 1985

(5) Maciejczyk, J.A., Slobodzinski, A.E., "Guidelines for the Generation and Use of Data in Military Handbook 17A, Part I", US Army Armament Research and Development Comand, January 1977.

ACKNOWLEDGMENT

The collaboration in the experimental work, of Mr. F.M.F. Oliveira, of the Department of Mechanical Engineering, University of Porto, is gratefully acknowledged

PROBABILISTIC ANALYSIS OF LAMINATE STRENGTH WITH OR WITHOUT DELAMINATION

Zhenlong Gu and Baicheng Wen*

Harbin Institute of Technology

Harbin,China

Abstract

Probability densitiesof ply strength were obtained either by direct testing or by quoting from available literatures.Based on a general probabilistic failure criterion,the failure sequence of plies and final damage state of a laminate were predicted.In the failure criterion,the stresses and strength parameters,α and β,should be in-situ values.With this in mind,stress concentration in a ply due to a matrix crack in its adjacent ply and constraint effect from adjacent plies were taken into consideration. The theoretical predictions were compared with experimental results and influence of free edge delamination on the laminate strength reduction was discussed.

1.Introduction

The fiber reinforced composite laminate is a structure with plies as its components.If there is not any damage in a laminate, it can be homogenized to be an elastic anisotropic plate.This is the foundation upon which the classical laminate theory has been laid.However,matrix crack,delamination and other types of damage inevitably occur even at the stress level far below the laminate strength and they will significantly complicate the stress analysis.The matrix cracks in adjacent ply will cause stress concentration in the ply in question,while the delamination near the matrix crack will cause stress relaxation,so the damage evolution will cause stress redistribution in the plies of a laminate.Furthermore,the damage is random in nature,so it is formidable to think of performing stress analysis in order to relate the damage state with the stress state.

On the other hand,the ply strength in a laminate is different from that of separted ply[1].Plies of different fiber directions have different stiffnesses and plies of different stiffnesses act as constraints to each other.The constraint effect increases the strength of in-situ ply.

Theoretically,the in-situ stresses and in-situ strengths of plies in a laminate are differnet and they should fail one after another.But actually it is not,since when one of the plies fails, its load will be shifted to its adjacent plies.So probably the adjacent plies may fail by chain reaction.

If the failure sequence of plies in a laminate has clean-cut definition,it can be only one of the follows:

°plies fail one by one with increasing load and the last surviving ply dominates the strength of laminate.

------

*Graduate student

°After a part of plies lost their bearing capability the remaining plies and constraints will fail by chain reaction. The laminate strength is dictated by the remaining plies and constraints.

°All the plies and constraints of a laminate reach their respective strengths almost at the same load and fail by chain reaction.

Whichever failure sequence a laminate has,its strength is determined by its final damage state.To find the failure sequence of plies,classical laminate theory was used to find the relation between in-plane stresses and external traction.The interlaminar shear stress τ_{xz} near a matrix crack was related to external traction by shear lag method.

The free edge delamination will not be studied in this paper, since a laminate subjected to edge delamination has no clean-cut definition of failure sequence of plies.Edge delaminations cause matrix cracks and matrix cracks will accelerate delamination.As the load increases,it extends first in longitudinal direction then in transverse direction.In most of the cases,it will not penetrate the width of the laminate,so it makes the failure sequence determination difficult.

Since laminate is a structure,the failure mechanism of different structure is generally different.However,for the laminate with clean-cut failure sequence of plies,many of them have similar failure process,i.e.,matrix cracks appear first in a ply with largest fiber angle with respect to the loading direction. Crack density increases until final saturation.Local through-width delamination will occur at matrix crack tip and it causes stress relaxation and loss of bearing capability of cracked ply. The matrix cracks appear in the plies with less fiber angles later than in the plies with larger fiber angles.However,the matrix cracks reduce the ply stiffness.So the more weakened ply will shift its load to the less weakened plies and makes their crack densities increase.Subsequently,the damage development of more weakened plies will be retarded and in the less weakened it quickened.So the ply failure sequence determined with elastic stress analysis will hopefully hold valid even for the damaged laminates.

The failure criterion of a laminate with clean-cut failure sequence of plies can be written in a very general form:

$$Q=Q_1 \cdot Q_2 \cdot \cdots \cdots Q_N=Q_c \tag{1}$$

where Q......failure probability of a laminate;
Q_k (k=1,2,...N)......failure probability of individual ply and constraint or interface;
Q_c......critical failure probability,in this paper,Q_c=0.99;
Subscript N......total number of plies and constraints.

In other words,failure probability of a laminate can be described with a in-parallel logic diagram(Fig.1).Three types of final damage state resulted from three types of failure sequence of plies can be illustrated with Fig.1a 1b and 1c,where Q_k=1 indicates the corresponding ply or interface has been failed before the total rupture of a laminate.

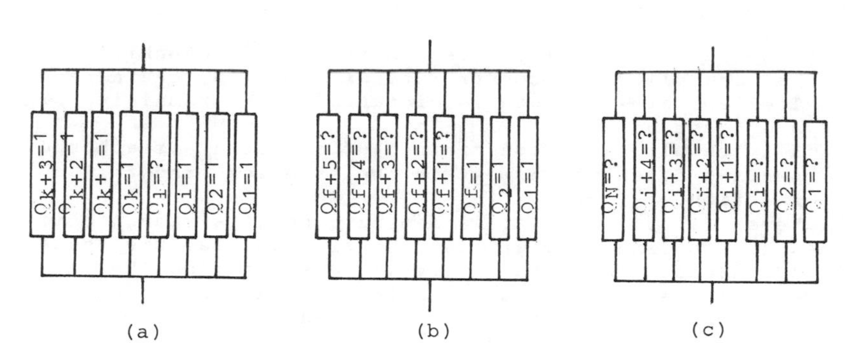

Fig.1 Logic diagram of failure probability of a laminate

2. Experimental data

In order to determine the probability distribution function Q_k in terms of stresses and relative parameters, separated plies with different fiber angles were tested. Fundamental experiments were conducted to obtained the strength probability densities of 0° and 90° plies. T300/648 panels were made of 8 layers of prepreg tape and fabricated according to manufacturer's recommended cure cycle. Here code 648 denotes a kind of epoxy resin commonly used in China. Fig.2 shows the specimen configuration used in this paper. It was cut from panel with impregnated diamond blade. The specimens were tensioned on a Shimadzu testing machine with loading rate of 0.5 mm/min. The damage of specimen was monitored using penetrant enhanced radiography after every load step. Two of the matrix crack patterns are shown in Fig.3 and Fig.4. Failure probability function turns out to be Weibull distribution:

$$Q=1-\exp(-\sigma/\alpha)^{\beta} \qquad (2)$$

The statistical data shown in Table 1 were calculated from experimental results of twenty specimens. The data lacking for

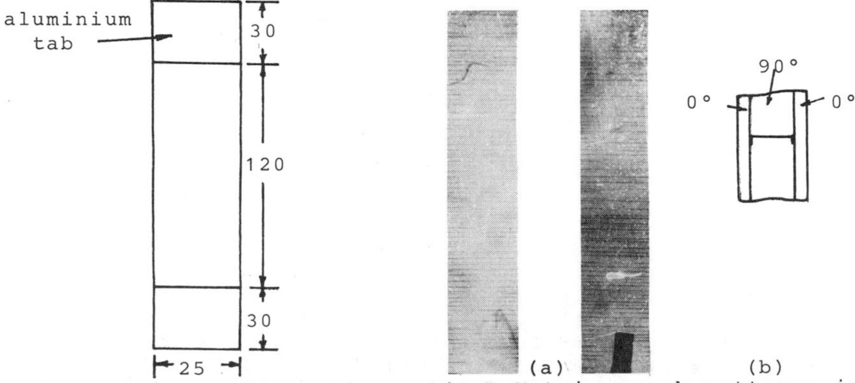

Fig.2 Specimen Configuration

Fig.3 Matrix crack patterns in $[0/90_3]_s$ laminate and local delamination

Table 1. Statistical Data of Unidirectional T300/648 Laminate

data / ply angle	α (kg/cm²)	β	mean strength (kg/cm²)	standard deviation (kg/cm²)	D.C. (%)	95%confidence interval of mean (kg/cm²)
0°	11565.7	16.978	10935.7	849.5	7.8	10745-11125
90°	544.4	16.572	520.7	43.04	8.2	501.8-539.6

Table 2.Statistical Data Used in Strength Prediction

data / cons-traints	α (kg/cm²)			β		
	[±25/90$_n$]s	B.Jones	[0/±45/90]s	[±25/90n]s	Jones	[0/±45/90]s
90°	457.4*	371.9	429.0*	7.54#	7.54	7.54#
25°	686.1** or 457.4**			7.54#		
45°			457.4**			7.54#
0°		11305.3	17600.0*		10.11	10.11#
25/-25	686.1△			8.82△△		
25/90	457.4△			8.82△△		
0/90		371.9△			8.82△△	
0/45		558.0△	1150.0*		8.82△△	8.82△△
45/-45		558.0△	1150.0*		8.82△△	8.82△△
45/90		558.0△	1150.0*		8.82△△	8.82△△

------

*The location parameters α of 25°,45° and 90° plies of [±25/90]s and [0/±45/90]s typedlaminates are assumed to be equal to their respective mean strength.The error thus incurred amounts about -3%,it means the strength prediction is slightly lower than the true one.

**The off-axis ply may be failed either of shear or of transverse tension depending on which one of the ratios τ_{xy}/S and σ_y/Y is greater.And accordingly their respective strengths are taken.

△Interlaminar shear strength S_{xz} will decrease as the angle between two adjacent plies increases.The interlaminar strengths of 0/45,45/90 and 25/-25 are assumed to be equal to in-plane shear strength and those of -25/90 and 0/90 are equal to the transverse tensile strengths.

#The lacking shape parameters β in case of 0° and 90° plies are substituted by known ones of that case.

△△The Shape parameters of interlaminar strengths are taken to be average of those of 0° and 90° plies.

for further analysis were replenished from [2,3,4,7](Table 2)

3.Strength prediction

It should be stressed that the failure probability distribution Q_k in Eq.1 is that of in-situ ply,not that of separated ply shown in Eq.2.To obtain the Q_k of in-situ ply or constraint is impossible.Assume the failure probability density is also of Weibull distribution with the same shape parameter β.Their difference lies in the enhancement of applied stress due to concentration and mean strength or characteristic parameter α due to constraint effect of adjacent plies,i.e.,

$$Q_k=1-\exp(-(k.\sigma)/(m.\alpha))^{\beta} \quad (3)$$

The in-situ ply is strengthened compared with the separated ply in two senses.Firstly,the appearance of matrix crack of an off-axis ply will be delayed to higher stress level due to the constraint effect of adjacent plies.Secondly,a single matrix crack may cause complete failure for a separated ply,while the in-situ ply will not lose its bearing capability until the crack density reaches its saturation and the induced local delamination appears. Even at that time,it still can transmit the shear stress and stress in its fiber direction.Fig.5 shows the relation between crack density and the applied stress,where σ_m is the mean strength of that separated off-axis ply,σ_i is the mean stress at which the first matrix crack appears and σ_s is the mean stress at which the crack density becomes saturated. Fig.5 shows two cases of saturated crack density.Fig.5a shows the local delamination occurs earlier because of its weak interfacial bonding.Fig.5b shows that the cracked ply can continue to transmit the stress higher than σ_s because the interlaminar shear stress has not reached its ultimate strength when matrix crack density becomes saturated.For simplification,assume σ_s is the stress at which the in-situ ply ceases to be effective.Now the strengthening effect from adjacent plies can be characterized by the constraint factor $m=\sigma_s/\sigma_m$.Table 3 shows the values of m in different types of laminates[2,5,6].Usually,the matrix cracks in plies of least fiber angle appear only at the last stage of loading,so for them no m value is needed.It can be seen from Table 3 that constraint effect depends on the relative stiffness of adjacent plies,e.g., value of m will decrease as the thickness of 90° ply increases, and constraint effect of 0° ply on a 90° ply will be greater than that on a 45° ply.The delamination will reduce the constraint to the adjacent ply.In addition,within the delamination region,e.g.,in pre-made delamination region,a single matrix crack may be sufficient to fail that ply.In such case,the value of m should be reduced.However,it is not necessarily to imply that the through-width delamination is bound to reduce the laminate strength.If the premade delamination does not change its final damage state,the laminate strength will not change as in the case of [0/45/90/-45//] ,here // denotes the location of pre-made delamination.

The stress concentration in a ply is very complicated problem because of the randomness of matrix crack distribution in adjacent ply.Here assume that the stress concentration results only from the presence of a single matrix crack in adjacent ply with less fiber angle.Stress concentration factor $k_{p\to q}$ was calculated with shear lag method.Here code $p\to q$ denotes the stress concentration in ply q due to the presence of a matrix crack in ply p. The calculation results of $k_{p\to q}$ are shown in Table 4.

Now the general form of probabilistic strength criterion(Eq. 1) can be specified with following equations for different cases of failure sequence of plies corresponding to Fig.1a,b and c, respectively:

$$Q=(1-\exp(-\sigma_1/\alpha_1)^{\beta_1})^2=0.99 \qquad (4)$$

subscript l is pertaining to the last surviving ply;

$$\begin{aligned} Q&=Q_1.Q_2\ldots Q_{f+1}\ldots\ldots Q_{f+r} \\ &=.(1-\exp(-(k_{f+1}.\sigma_{f+1})/(m_{f+1}.\alpha_{f+1}))^{\beta_{f+1}}). \\ &\quad .(1-\exp(-(k_{f+2}.\sigma_{f+2})/(m_{f+2}.\alpha_{f+2}))^{\beta_{f+2}})\ldots\ldots \\ &\quad .(1-\exp(-(k_{f+r}.\sigma_{f+r})/(m_{f+r}.\alpha_{f+r}))^{\beta_{f+r}})=0.99 \qquad (5) \end{aligned}$$

subscript f......number of completely failed plies,
r......number of surving plies,
for the completely failed plies, $C_{22}=0$;

$$\begin{aligned} Q&=Q_1.Q_2\ldots\ldots Q_N \\ &=(1-\exp((-k_1.\sigma_1)/(m_1.\alpha_1))^{\beta_1}).(1-\exp((-k_2.\sigma_2)/(m_2.\alpha_2))^{\beta_2})\ldots(1-\exp((-k_N.\sigma_N)/(m_N.\alpha_N))^{\beta_N})=0.99 \qquad (6) \end{aligned}$$

In order to compare with the experimental results which is mean values of experimental data,the Q_c should be changed to be the failure probability corresponding to the mean strength.Table 5 shows the value of $\bar{Q}_k$ for separated ply and $\bar{Q}_c$ for the laminate. For the first approximation, $\bar{Q}_k$ of separated ply may be taken as 0.4 and $\bar{Q}_c$ of the laminate as 0.45.The strength prediction was implemented with a computation program developed according to the block diagram shown in Fig.6.The results are shown in Table 6,where the experimental results are also presented for comparison.The strength N_x is in kg/cm.

The computation can be illustrated with follwing examples (N_x is in kg/cm):

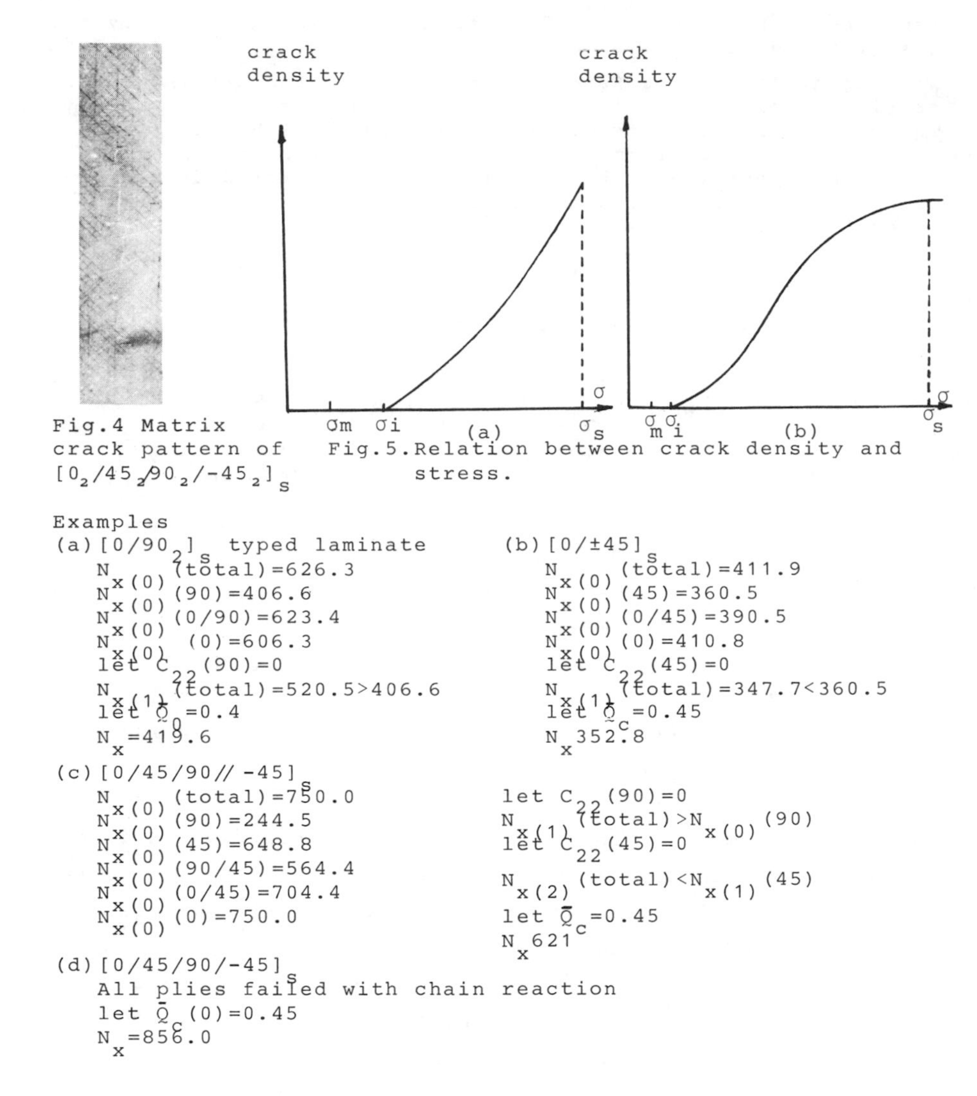

Fig.4 Matrix crack pattern of $[0_2/45_2/90_2/-45_2]_s$

Fig.5. Relation between crack density and stress.

Examples

(a) $[0/90_2]_s$ typed laminate

$N_{x(0)}(total)=626.3$
$N_{x(0)}(90)=406.6$
$N_{x(0)}(0/90)=623.4$
$N_{x(0)}(0)=606.3$
let $C_{22}(90)=0$
$N_{x(1)}(total)=520.5>406.6$
let $\bar{Q}_c=0.4$
$N_x=419.6$

(b) $[0/\pm 45]_s$

$N_{x(0)}(total)=411.9$
$N_{x(0)}(45)=360.5$
$N_{x(0)}(0/45)=390.5$
$N_{x(0)}(0)=410.8$
let $C_{22}(45)=0$
$N_{x(1)}(total)=347.7<360.5$
let $\bar{Q}_c=0.45$
N_x 352.8

(c) $[0/45/90//-45]_s$

$N_{x(0)}(total)=750.0$
$N_{x(0)}(90)=244.5$
$N_{x(0)}(45)=648.8$
$N_{x(0)}(90/45)=564.4$
$N_{x(0)}(0/45)=704.4$
$N_{x(0)}(0)=750.0$

let $C_{22}(90)=0$
$N_{x(1)}(total)>N_{x(0)}(90)$
let $C_{22}(45)=0$
$N_{x(2)}(total)<N_{x(1)}(45)$
let $\bar{Q}_c=0.45$
N_x 621

(d) $[0/45/90/-45]_s$

All plies failed with chain reaction
let $\bar{Q}_c(0)=0.45$
$N_x=856.0$

4. Discussion and Conclusions

(1). Laminates of different constructions have different failure sequence and final damage state. The strength of laminate depends on the final damage state and can be predicted from it.

(2). Stress concentration due to matrix crack in adjacent ply and constraints between plies are two very important factors. The strength of $[90_2/0]_s$ is higher than that of $[0/-90_2]_s$ because the latter has higher stress concentration. The $[0/45/90/-45]_s$ typed laminate ruptures with all its

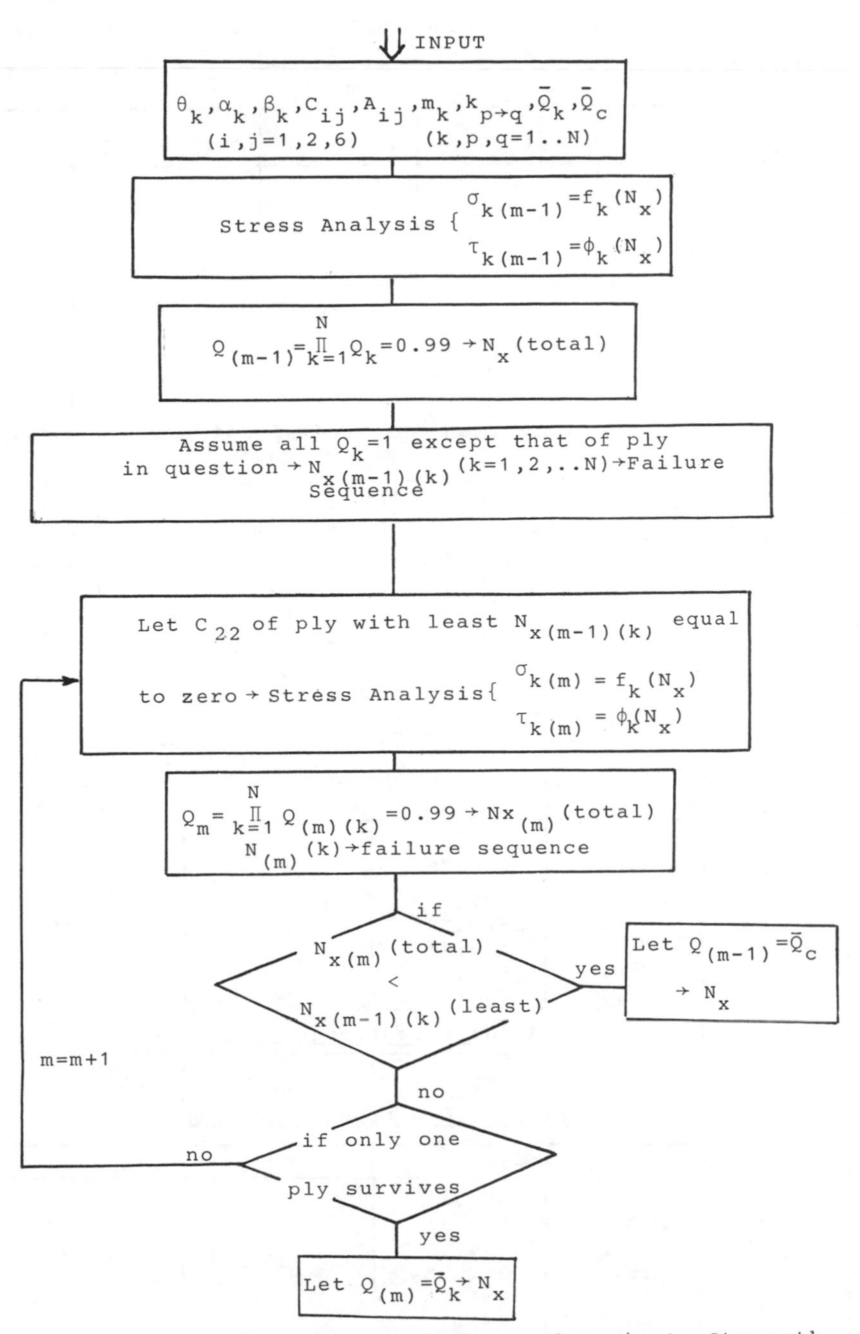

Fig.6 .Block Diagram for prediction of Laminate Strength
(m—number of repeating cycles)

Table 3 Constraint Coefficients m in Different Laminates

ply angles	m					
	n	[±25/90$_n$]s	[0/90$_n$]s	[0/±45/90]s	[0/45/90/-45]s	[0/±45]s
90°	½ or 1	2.0	2.0			
	2	1.9	1.9 (n=1.5)			
	3	1.6				
	4	1.4	1.9	1.9 (n=1)	4.0 (n=1)	
	6	1.2				
	8	1.2				
45°				1.5	2.0	1.5

Table 4. Stress Concentration Factors

Stress Concentration in Ply q induced by a matrix crack in ply p	$k_{p \to q}$
90°→45°	1.114
90° → 0°	1.034
90°→25°	1.046
45° → 0°	1.117

Table 5. Failure Probability Corresponding to Mean Strength

Ply or Laminate	$\vec{Q}_k$ or $\bar{Q}_c$
0°	0.32048
90°	0.38013
[0/ 90/±45]s	0.43363
[0/±45/90]s	0.45049

Table 6 Strength of Laminates

[0/90$_n$]s typed laminates

m	n	[0$_m$/90$_n$]s prediction	[0$_m$/90$_n$]s experimetal results	[0/±45]s prediction	[0/±45]s experimental results
1	½	420.0	396.0*		
		419.8	384.7*		
1	1	259.2	273.5**		
1	1.5	419.7	399.7*	352.8	358.7**
1	2	419.6	429.4*		
2	1	839.9	829.3*		
2	2	839.7	847.7*		
		π/4 Quasi-isotropic Laminates			
		[0/45/90/-45]s		[0/45/90//-45]s	
		856.6	813.0	621.4	643.0
		Laminates with Edge Delamination			
		[±25/90$_n$]s		[0/±45/90]s	
	1	504.1	376.4*		
	2	563.2	383.7*		
	3	566.8	394.1*	436.3	353.2**
	4	538.1	341.3*		
	6	511.3	301.9*		
	8	498.4	273.3*		

*cited from [7]; ** cited from [4]

plies failed with chain reaction,while the laminate $[0/45/90/\!/-45]_s$ ruptures with 90° plies failed first and with strength remarkably reduced.Extension of pre-made delamination in this case did not appear.

(3).The stress analysis is tremendously complicated due to randomness of damage and discreteness of damaged laminates.Simplified engineering method of stress analysis is needed.This method should be able to take the effects of stress concentration,constraints and stress redistribution into account.

(4).For the laminates with edge delamination,from Table 6 it can be seen that although the trends of strength change with thickness change of 90° plies comform to that of experimental results,the deviation between them ranges -23.5% to -82.4%,while the deviation of prediction for the laminates without edge delamination is only about ±5%.Edge delamination will drastically reduce the laminate strength.The possible explanation is that edge delamination renders the possibility to form a continuous crack to separate the laminate into two parts before the matrix crack density reaches its saturation.

5.Acknowledgements

This work is supported by Science Fund of Chinese Academy of Science.The testings of $[0/45/90/-45]_s$ and $[0/45/90/\!/-45]_s$ typed laminates were completed in Composite Laboratory of Purdue University with supports from Professor C.T.Sun whose comments led to some final alteration.

6.References

1.Yamada,S.E. and Sun,C.T.
Analysis of Laminate Strength and its Distribution
J.Composite Materials,Vol.12,(July 1978),p275

2.A.S.D.Wang,N.N.Kishore and C.A.Li
Cracks Development in Graphite-Epoxy Cross-Ply Laminates under Uniaxial Tension
Composite Science and Technology 24(1985) 1-31

3.A.S.D.Wang
Fracture Mechanics of Sublaminate Cracks
AD A130782,1982

4.Liu Xili and Wang Bingquan
Fundamentals of Composite Mechanics
(in Chinese),1983

5.A.S.D.Wang,M.Slomiana and R.B.Bucinell
Delamination Crack Growth in Composite Laminate
ASTM STP 876 W.S.Johnson Ed.American Society for Testing and Materials,Philadelphia 1985,pp135-167

6.R.Y.Kim and R.M.Aoki
Transverse Cracking and Delamination in Composite Materials
Fibre Science and Technology 18(1983) 203-216

7.Brian Jones
Probabilistic Design and Reliability
Composite Materials Vol.8,Structural Design and Analysis Part II,Ed.C.C.Chamis,Academic Press,1975 pp34-73

REDUCED STRESS CONCENTRATION FACTORS IN ORTHOTROPIC MATERIALS

IVO ALFIREVIĆ

Fakultet strojarstva i brodogradnje,
Univerzitet u Zagrebu,Salajeva 5,
41000 Zagreb, Yugoslavia

ABSTRACT

Stress distribution about elliptic holes in orthotropic strips loaded in tension was determined by means of the birefringent coatings. Using the general tensor failure criterion for orthotropic materials tension and compression strength as a function of orientation were determined. Fracture of an orthotropic structure occurs at the point where the local stress to strength ratio reaches its maximum rather than at the place of maximum stress. In order to predict the load at which the fracture starts as well as the position of initial crack, the reduced stress concentration factors were defined and determined.

INTRODUCTION The main objective of this work is to introduce the idea of reduced stress concentration factors for flat orthotropic bars with different geometrical discontinuities (central holes, side notches, fillets etc.) in plane stress. The bars can be loaded in tension, compression or flexure as shown in Fig. 1.

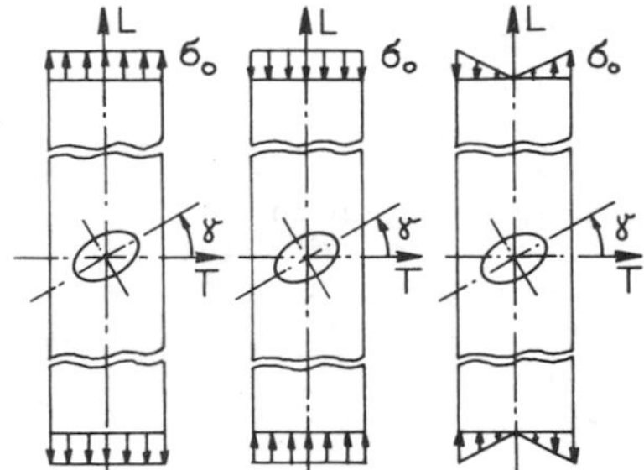

Fig. 1.

In such bars maximum stress occurs at the free boundary of the geometrical discontinuity. The fracture or failure of the isotropic bar will start at a point of maximum stress. However, if the bar is orthotropic, failure will start at a point where the ratio of normal stress to local strength is a maximum rather than at a point of maximum stress. Thus, the stress concentration factors should be defined in a different way in order to be of use to designers. It is necessary to tabulate these concentration factors in engineering handbooks.

ELASTIC PROPERTIES AND FAILURE CRITERION The stress-strain relationship for linear-elastic orthotropis material is given by

$$\sigma_{ij} = C_{ijkm}\,\varepsilon_{km}$$
$$\varepsilon_{ij} = S_{ijkm}\,\sigma_{km} \qquad (1)$$

Where C_{ijkm} and S_{ijkm} are elastic stifnesses and elastic compliances respectively. They constitute tensors of fourth order. For the plane stress problem the above equations written in engineering notation reduce to

$$\varepsilon_x = \frac{1}{E_x}\sigma_x - \frac{\nu_{xy}}{E_x}\sigma_y - \alpha_x \tau_{xy}$$

$$\varepsilon_y = -\frac{\nu_{xy}}{E_y}\sigma_x + \frac{1}{E_y}\sigma_y - \alpha_y \tau_{xy} \tag{2}$$

$$\gamma_{xy} = -\alpha_x \sigma_x - \alpha_y \sigma_y + \frac{1}{G_{xy}}\tau_{xy}$$

If the axes of orthotropy T, L coincide with the coordinate axes x, y, eqs (2) become

$$\varepsilon_x = \frac{1}{E_T}\sigma_x - \frac{\nu_{TL}}{E_T}\sigma_y$$

$$\varepsilon_y = -\frac{\nu_{TL}}{E_L}\sigma_x + \frac{1}{E_L}\sigma_y \tag{3}$$

$$\gamma_{xy} = \frac{1}{G_{TL}}\tau_{xy}$$

The modulus of elasticity E_x for x axis that makes an angle φ with T axis is determined by

$$\frac{1}{E_x} = \frac{\cos^4\varphi}{E_T} + \frac{\sin^4\varphi}{E_L} + \left[\frac{1}{G_{TL}} - \frac{2\,\nu_{LT}}{E_T}\right]\cos^2\varphi\,\sin^2\varphi \tag{4}$$

The general tensor form of the failure criterion for orthotropic material is given by

$$(\Pi_{ij}\,\sigma_{ij})^{\alpha} + (\Pi_{ijkm}\,\sigma_{ij}\,\sigma_{km})^{\beta} + + (\Pi_{ijkmnp}\,\sigma_{ij}\,\sigma_{km}\,\sigma_{np})^{\gamma} + \ldots \leq 1 \tag{5}$$

where α, β, γ, ... are scalar components and Π_{ij}, Π_{ijkm}, Π_{ijkmnp} are strength tensores of second, fourth, sixth ... order respectively. For practical purposes, the failure criterion is used in the form

$$\Pi_{ij}\,\sigma_{ij} + \sqrt{\Pi_{ijkm}\,\sigma_{ij}\,\sigma_{km}} \leq 1 \tag{6}$$

The tensor Π_{ij} is symmetric and all its diagonal components are equal to zero. If a material exhibits the same tensile and compressive strengths, than all components of Π_{ij} vanish. The tensor Π_{ijkm} exhibits the same properties of symmetry as tensor C_{ijkm}, thus it has only nine independent components. In the case of an orthotropic plate in plane stress, the number of independent components of the tensor Π_{ijkm} reduces to four. The components of strengthtensors have no physical meaning and they are not directly measured in a laboratory. Instead of that, engineering strength constants are defined and measured.

The following engineering strength constants will be used throughout this paper:

σ_L^t	longitudinal tensile strength
σ_T^t	transverse tensile strength
σ_L^c	longitudinal compressive strength
σ_T^c	transverse compressive strength
σ_{45}^t	diagonal tensile strength
σ_{45}^c	diagonal compressive stength
τ_{LT}	principal shear strength
τ_{45}^+	positive diagonal shear strength
τ_{45}^-	negative diagonal shear strength

The superscripts t and c refer to tension and compression respectively, the subscripts T, L and 45 refer to transverse axis of orthotropy T, longitudinal axis of orthotropy L and axis that makes an angle of 45° with the T axis. Shear strengths τ_{45}^+ and τ_{45}^- are defined in Fig. 2.
The superscript + indicates that shear stress causes tension along L axis and compression along T axis. The superscripts - indicates compresson along L axis. Nine engineering strength constants have been listed but only six of them are independent (in a case of plane stress), since tensors Π_{ij} and Π_{ijkm} have two and four independent components respectively.

Components of the tensors Π_{ij} and Π_{ijkm} are related to engineering strengths by the following equations

$$\Pi_{TT} = \frac{1}{2}\left(\frac{1}{\sigma_T^t} - \frac{1}{\sigma_T^c}\right)$$

$$\Pi_{LL} = \frac{1}{2}\left(\frac{1}{\sigma_L^t} - \frac{1}{\sigma_L^c}\right)$$

$$\Pi_{TL} = \Pi_{LT} = 0$$

$$\Pi_{TTTT} = \frac{1}{4}\left(\frac{1}{\sigma_T^t} + \frac{1}{\sigma_T^c}\right)^2 \qquad (7)$$

$$\Pi_{LLLL} = \frac{1}{4}\left(\frac{1}{\sigma_L^t} + \frac{1}{\sigma_L^c}\right)^2$$

$$\Pi_{TLTL} = \frac{1}{4\,\tau_{LT}^2}$$

$$\Pi_{TTLL} = \frac{1}{8}\left[\left(\frac{1}{\sigma_T^t} + \frac{1}{\sigma_T^c}\right)^2 + \left(\frac{1}{\sigma_L^t} + \frac{1}{\sigma_L^c}\right)^2 - \left(\frac{1}{\tau_{45}^+} + \frac{1}{\tau_{45}^-}\right)^2\right.$$

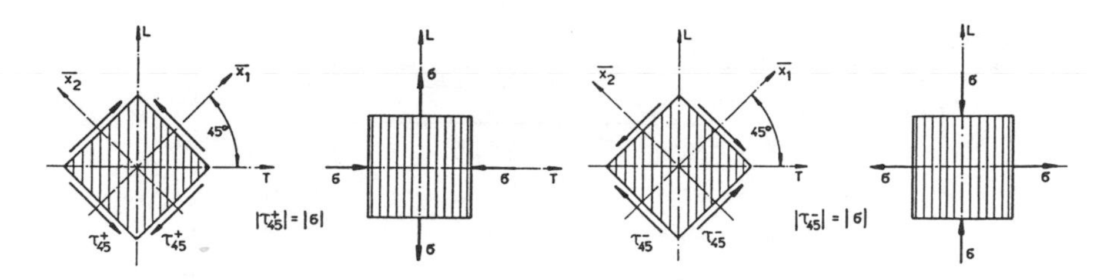

Fig. 2.

Instead of using τ^{+}_{45} and τ^{-}_{45}, σ^{t}_{45} of σ^{c}_{45} can be used. In that case Π_{TTLL} is given by

$$\Pi_{TTLL} = \frac{1}{2}\left[\left(\frac{2}{\sigma^{t}_{45}} - \Pi_{LL} - \Pi_{TT}\right)^{2} + \left(\Pi_{TTTT} + \Pi_{LLLL} + 4\,\Pi_{TLTL}\right)\right]$$

$$\Pi_{TTLL} = \frac{1}{2}\left[\left(\frac{2}{\sigma^{c}_{45}} + \Pi_{LL} + \Pi_{TT}\right)^{2} - \left(\Pi_{TTTT} + \Pi_{LLLL} + 4\,\Pi_{TLTL}\right)\right] \tag{8}$$

where Π_{TT}, Π_{LL}, Π_{TTTT}, Π_{LLLL}, Π_{TLTL}, Π_{TTLL} are principal components of the tensor Π_{ij} and Π_{ijkm} refering to the axes of orthotropy T, L.

When all the principal components of tensors Π_{ij} and Π_{ijkm} are determined, components of tensors in any other coordinate axes can be easily determined using tensor transformation law. Now from the components of the strength tensors, engineering strengths can be obtained. When this is done, the following equations are obtained for tensile and compressive strengths

$$\frac{1}{\sigma^{t}_{x}} = \Pi_{TT}\cos^{2}\varphi + \Pi_{LL}\sin^{2}\varphi + \sqrt{\Pi_{TTTT}\cos^{4}\varphi + \Pi_{LLLL}\sin^{4}\varphi + \left(\Pi_{TLTL} + \tfrac{1}{2}\Pi_{TTLL}\right)\sin^{2}\varphi} \tag{9}$$

$$\frac{1}{\sigma^{c}_{x}} = -\,\Pi_{TT}\cos^{2}\varphi - \Pi_{LL}\sin^{2}\varphi + \sqrt{\Pi_{TTTT}\cos^{4}\varphi + \Pi_{LLLL}\sin^{4}\varphi + \left(\Pi_{TLTL} + \tfrac{1}{2}\Pi_{TTLL}\right)\sin^{2}\varphi} \tag{10}$$

where σ^{t}_{x} and σ^{c}_{x} are tensile and compressive strengths along the x axis that makes an angle φ with T axis. If a material exhibits the same strength properties in tension and in compression eqs (9) and (10) take the simpler form

$$\sigma^{t}_{x} = \sigma^{c}_{x} = \sigma^{t}_{T}\left\{\cos^{2}\varphi + \left(\frac{\sigma^{t}_{T}}{\sigma^{t}_{L}}\right)^{2} + 2\left[2\left(\frac{\sigma^{t}_{T}}{\sigma^{t}_{45}}\right) - \left(\frac{\sigma^{t}_{T}}{\sigma^{t}_{L}}\right) - 1\right]\sin^{2}\varphi\,\cos^{2}\varphi\right\}^{1/2} \tag{11}$$

TABLE 1.

Elastic constant	Unit	Material A	Material B
E_T	MPa	9820	28100
E_L	MPa	45000	28100
G_{TL}	MPa	4300	4300
ν_{TL}	-	0,27	0,07

Two kinds of orthotropic material were used: angle-ply consisting of 32 unidirectional plys that make an angle of $\pm$ 5° with longitudinal axis L, and cross-ply consisting of 32 alternating plys half of them aligned along L axis and half parallel to T axis. The former material is designated as material A and the latter as material B. Elastic and strength properties of these materials are given in Table 1. and Table 2. Using values from this table and formulae (4), (9), (10) and (11) diagrams showing variation of modulus of elasticity and tensile and compressive strengths are constructed. These diagrams are shown in Fig. 3. and Fig. 4. respectively.

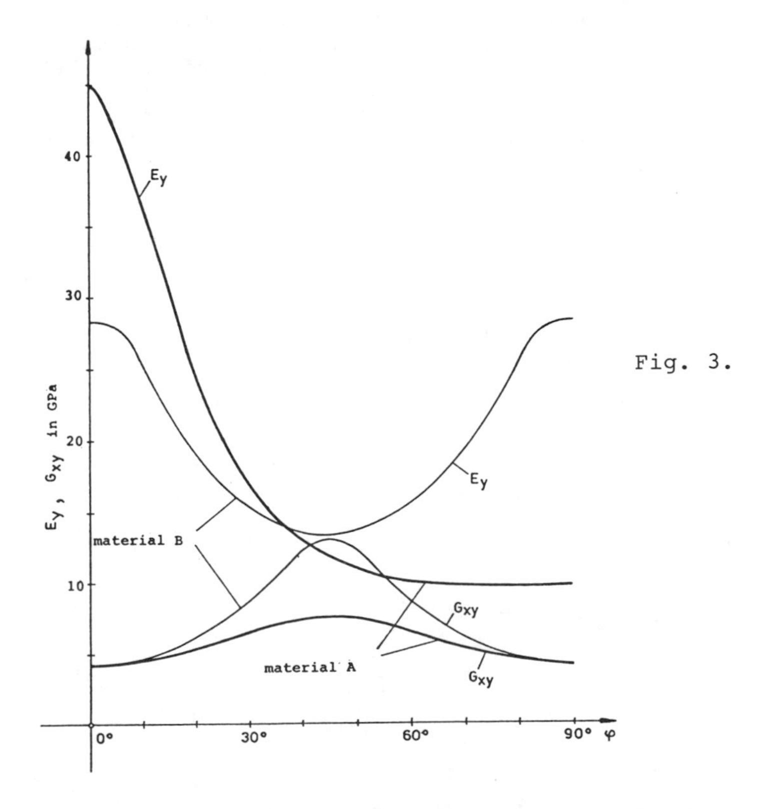

Fig. 3.

TABLE 2.

Strength	Unit	Material A	Material B
σ_T^t	MPa	21	525
σ_L^t	MPa	775	525
σ_{45}^t	MPa	34	154
σ_T^c	MPa	140	525
σ_L^c	MPa	540	525
σ_{45}^c	MPa	175	162
τ_{TL}	MPa	31	29

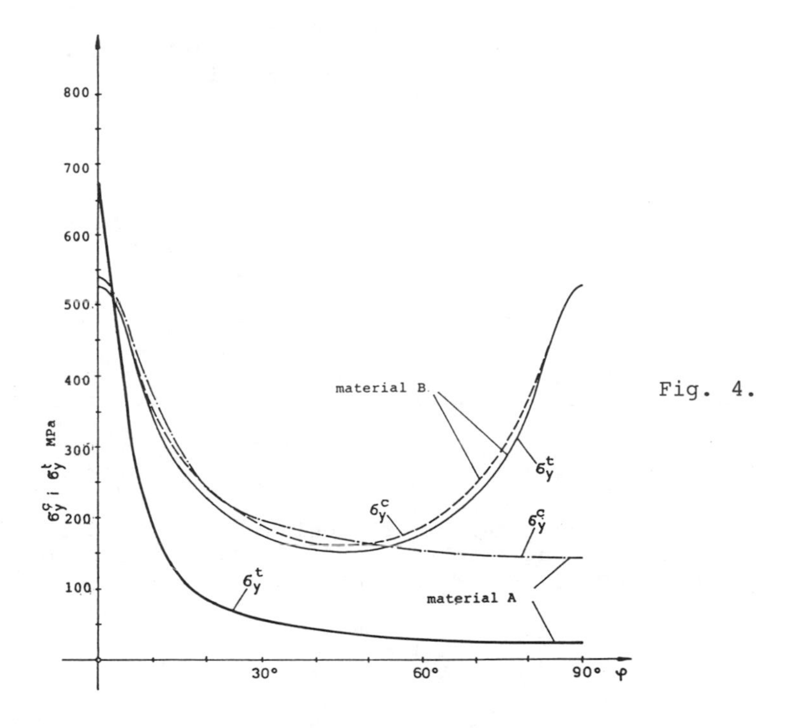

Fig. 4.

APPLICATION OF BIREFRINGENT COATINGS TO ORTHOTROPIC SPECIMENS. Geometry of the investigated models is defined in Fig. 1 and Fig. 5. The angle γ is the angle between the transverse axis T of the model and the principal axis of the ellipse. The angle γ was selected to be 0°, 30°, 45°, 60° and 90°, so that totally ten models were investigated, five for each material.

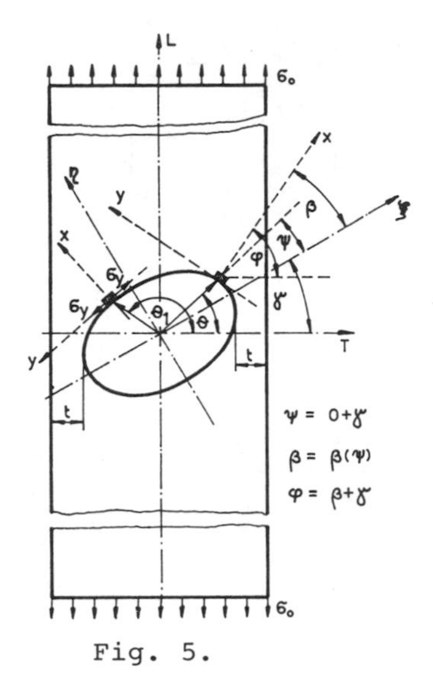

Fig. 5.

When the specimens were produced, birefringent coating was cemented to them and after machining off the exces of the coating, models were loaded and investigated in reflexion polariscope. Photographs were taken in dark and light fields to improve the accuracy of the method. A basic assumption in birefringent coating method is that surface strains of the specimen or structure wholly transmit to the coating and that in the coating exists plane stress. This is not always the case due to different disturbing effects: reinforcing effect, strain gradient effect, change of the sarface curvature effect and Poisson's ratio mismatch effect. All the effects were eliminated or accounted for. The basic assumption gives

$$\varepsilon_x^c = \varepsilon_x^c, \qquad \varepsilon_y^c = \varepsilon_y^s, \qquad \gamma_{xy}^c = \gamma_{xy}^s \tag{12}$$

where the superscripts c and s refer to coating and specimen respecitvely. Because of Poisson's ratio mismatch effect explained in [1], at the free boundary we have

$$\varepsilon_x^c = -\nu_c \cdot \varepsilon_y^c = -\nu_c \varepsilon_y^s, \qquad \varepsilon_y^c = \varepsilon_y^s, \qquad \gamma_{xy}^c = 0$$

so that ε_x^c, ε_y^c are principal strains. Thus

$$\varepsilon_1^c - \varepsilon_2^c = \varepsilon_y^c - \varepsilon_x^c = \frac{N\ f_\varepsilon}{2\ h_c} \tag{13}$$

where N is fringe order, f_ε photoelastic constant in terms of strains, h_c coating thickness and ν_c Poisson's ratio of the coating. Since $\varepsilon_x^c = -\nu_c\ \varepsilon_y^c$ it follows

$$\varepsilon_y^c\ (1 + \nu_c) = \frac{N\ f_\varepsilon}{2\ h_c}$$

$$\varepsilon_y^c = \varepsilon_y^s = \frac{N\ f_\varepsilon}{2\ h_c\ (1 + \nu_c)} \tag{14}$$

At the free boundary $\sigma_x = \tau_{xy} = 0$ and Hooke's law (3) gives $\sigma_y \equiv \sigma_y^s = \varepsilon_y^s\ E_y$. Thus

$$\sigma_y = \frac{N\ f_\varepsilon}{2\ h_c\ E_y\ (1 + \nu_c)} \tag{15}$$

where σ_y is the stress in the specimen tangential to the boundary. The stress ratio σ_y/σ_o is given by

$$\frac{\sigma_y}{\sigma_o} = \frac{f_\varepsilon}{2\ h_c\ (1 - \nu_c)\ \sigma_o}\ N\ E_y \tag{16}$$

where σ_o is uniform stress in cross-section far from geometrical discontinuity. Fringe order N and modulus of elasticity E_y are functions of θ. Thus

$$\sigma_y/\sigma_o = C\ N\ (\theta)\ E_y\ (\theta) \tag{17}$$

where $C = f_\varepsilon\ /\ [2\ h_c(1+\ \nu_c)\ \sigma_o]$ is a constant. The fringe order N is taken from photographs of the models and plotted in diagramms shown in Fig. 6 and Fig. 7. In the same figures the diagramms of stress ratio $\sigma_y/\ \sigma_o$ is also given.

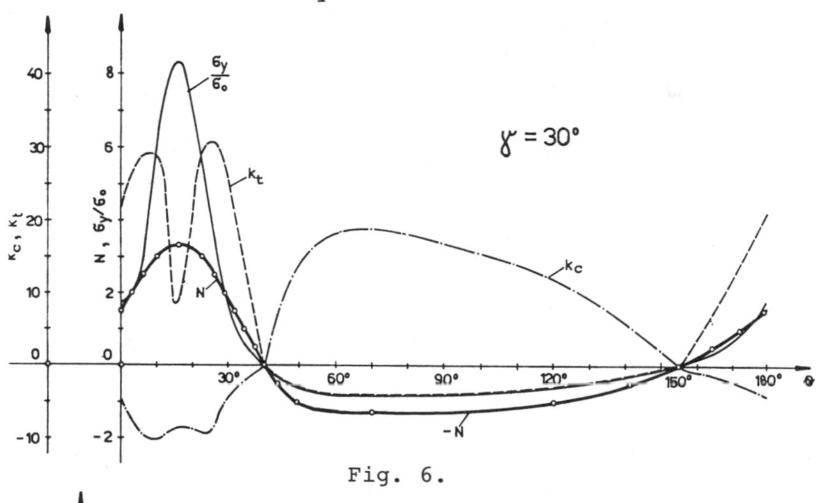

Fig. 6.

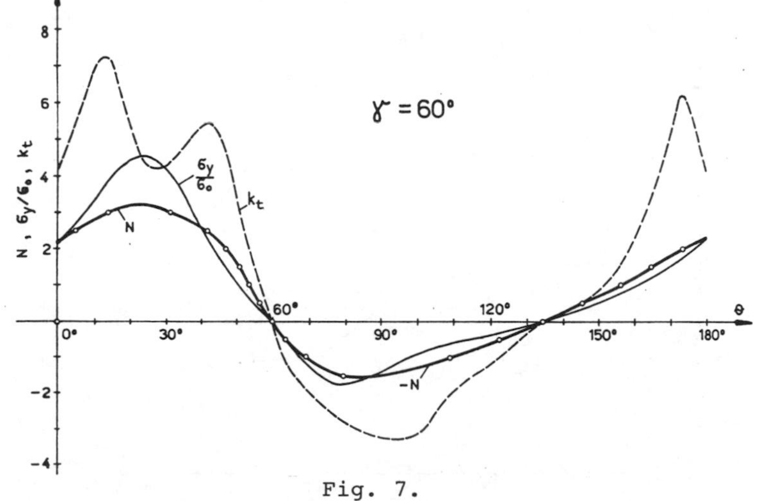

Fig. 7.

DIFINITION OF THE REDUCED STRESS CONCENTRATION FACTORS The weakest point at a free boundary of an orthotropic model is a point where the ratio of local stress to local strength is a maximum rather then at the point of maximum stress. In order to find the weakest point the following stress ratios are defined: $k_t(\theta)$ if the model is loaded in tension and $k_c(\theta)$ if model is loaded in compression

$$
\begin{aligned}
k_t &= \frac{\sigma_y/\sigma_y^t}{\sigma_o/\sigma_L^t} && \text{if} \quad \sigma_y > 0 \\
k_t &= \frac{\sigma_y/\sigma_y^c}{\sigma_o/\sigma_L^t} && \text{if} \quad \sigma_y < 0 \\
k_c &= \frac{\sigma_y/\sigma_y^c}{\sigma_o/\sigma_L^c} && \text{if} \quad \sigma_y < 0 \\
k_c &= \frac{\sigma_y/\sigma_y^t}{\sigma_y/\sigma_L^c} && \text{if} \quad \sigma_y > 0
\end{aligned}
\tag{18}
$$

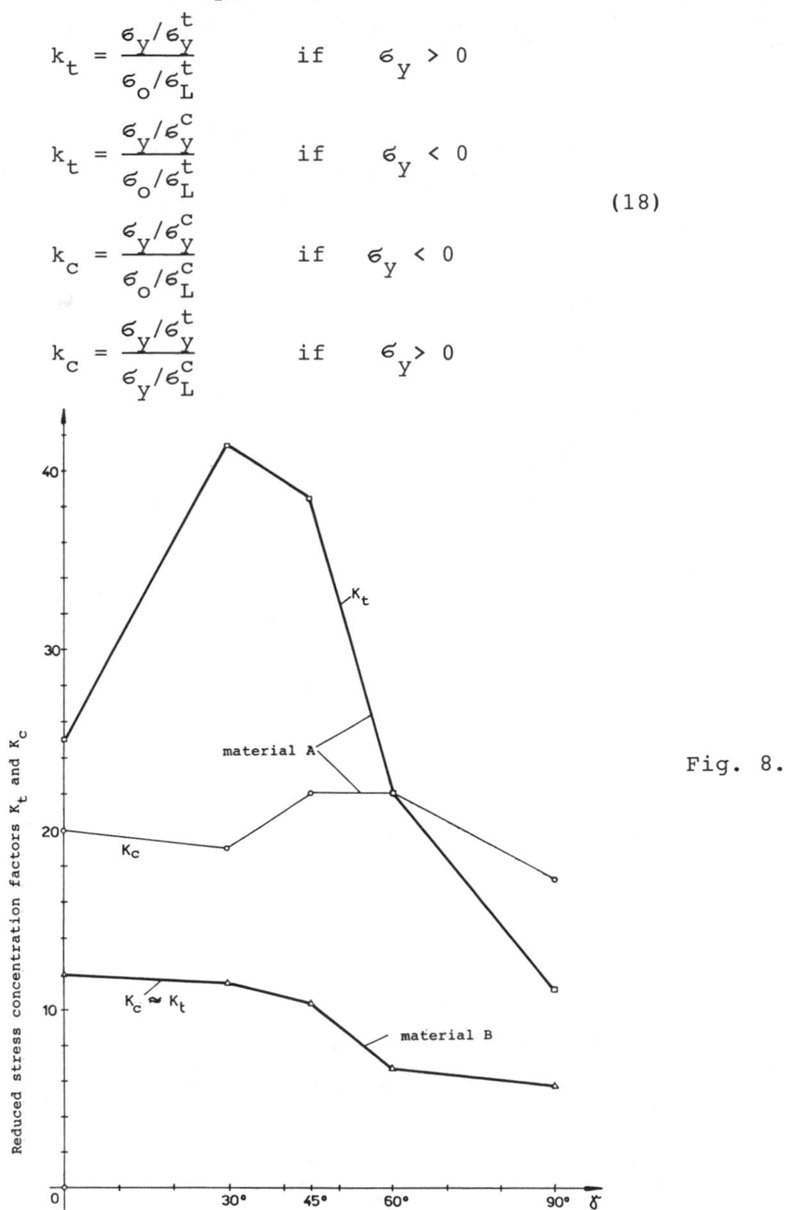

Fig. 8.

The reduced stress concentration factors are defined by

$$K_t = (k_t)_{max}$$
$$K_c = (k_c)_{max} \tag{19}$$

where K_t and K_c are reduced stress concentration factors in tension and compression respectively. The ordinary stress concentration factor K and K_n are defined by

$$K = \left(\frac{\sigma_y}{\sigma_o}\right)_{max}, \qquad K_n = \left(\frac{\sigma_y}{\sigma_n}\right)_{max} \tag{20}$$

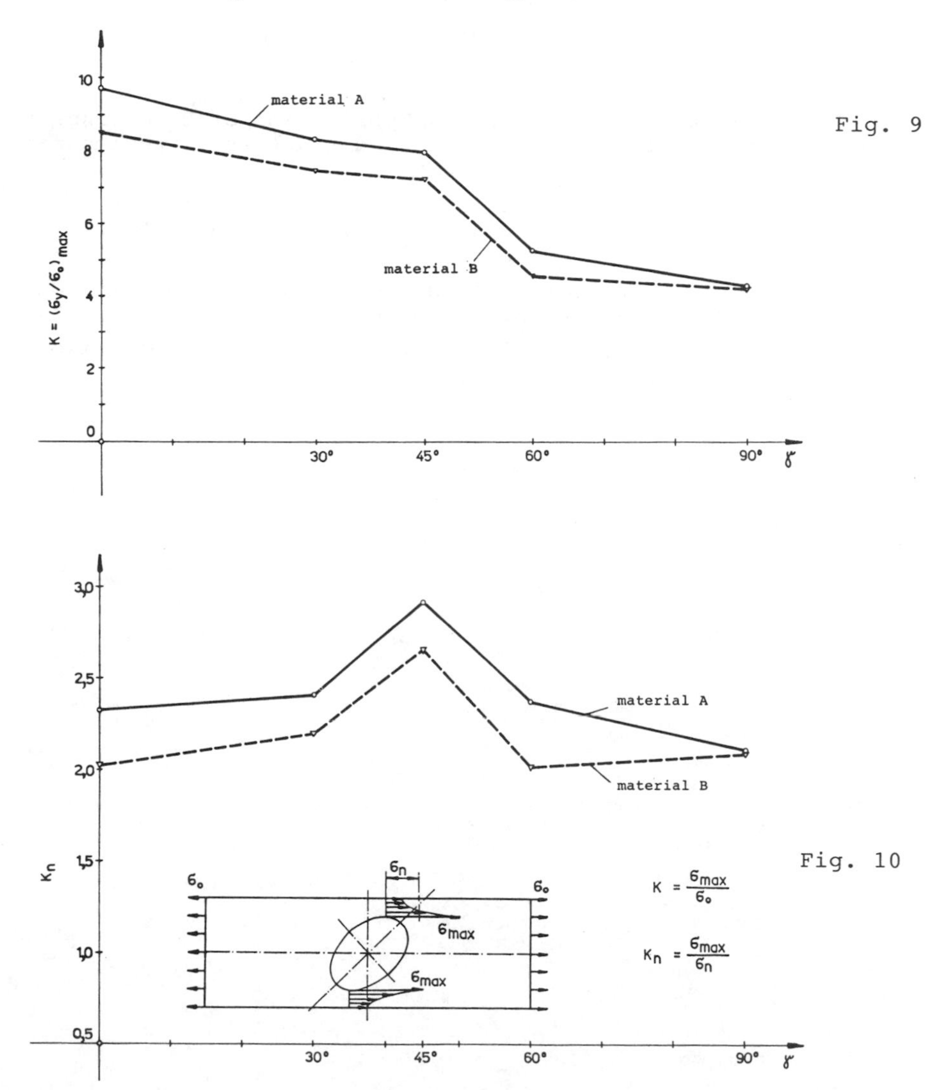

Fig. 9

Fig. 10

where σ_n is average stress across the weakest cross-section. The reduced stress concentration factors for different orientation of the ellipse are shown in Fig. 8. Stress concentration factors K and K_n are shown in Fig. 9 and Fig. 10 respectively.

CONCLUSION The stress concentration factors for non-symmetric (materially or geometrically) models defined in the usual way are of little use for a designer of an orthotropic structure. They do not indicate either position of the first crack or the load at which it will occur. The reduced stress concentration factors can be used for prediction of the load at which failure of the model or structure will start.

References

1. Dally, J. W., Alfirevich, I., Application of Birefringent Coatings to Glass-fiber-reinforced Plastics, Experimental Mechanics 9 (1969) pp 97-102.
2. Alfirević, I., Photoelastic Determination of the Stress Concentration Factors in Anisotropic Materials, Acta Technica ČSAV 2 (1969) pp 129-141.
3. Alfirević, I., Faktori Koncentracije naprezanja u anizotropnim materijalima, Strojarstvo 16 (1974) 1/2 pp 5-15.
4. Gol'denblat, I. I. et al., Soprotivlenie stekloplastikov, Mashinostroenie, Moskva 1968.

APPROXIMATE DETERMINATION OF STRENGTH TENSORS COMPONENTS FOR ORTHOTROPIC MATERIALS

Andrzej P.Wilczynski
Warsaw Technical University
POLAND

ABSTRACT

An approximate strength function for orthotropic materials is taken to consist of both linear and quadratic terms in the stress components. The requirements that this function should yield a positive safety factor for all admissible stress fields gives rise to relations between the components of the tensors characterising the strength function. These relations are employed to allow the calculation of approximate values of the tensor components in plane stress states using the results of five experiments in simple tension.

Introduction

Establishment of a strength theory by Malmeister [1] in 1966, developed further by several authors, has given an useful tool for prediction of strength of a orthotropic material in a combined stress state. At the some time this strength theory has introduced strength tensors the value of which is perhaps underestimated. These tensors are very useful when investigating ultimate stresses in directions not necessarily coinciding with the principal material directions. General requirements of the strength theory introduce additional requirements on tensorial coordinates, which may lead to their limiting values, useful in some calculations.

Another problem concerned with experimental investigations can be readily solved. The laboratory techniques for experiments in simple tension are well established and in general results obtained in this way are quite reliable. But experiments in compresion, shear and especialy in biaxial stresses have no such advantage and in certain cases may lead to erroneous results. This paper gives an approximate solution for such cases.

The strength theory for orthotropic media

The proposal of Malmeister [1] has been developed to an approximate form by Tsai and Wu [2] and further investigated by Wu [3] so that a form

$$F(\sigma) = F_1(\sigma) + F_2(\sigma) = \Pi_{ik}\sigma_{ik} + \Pi_{iklm}\sigma_{ik}\sigma_{lm} \qquad /1/$$

was proved to be admissible and further to cover all well defined strength theories. This form is an approximation in the sens of truncation of higher orders of tensorial coefficients. For a planar case the equation /1/ when written in the principal directions of the material has an explicit form

$$F(\sigma) = \Pi_{11}\sigma_{11} + \Pi_{22}\sigma_{22} + \Pi_{1111}\sigma_{11}^2 + 2\Pi_{1122}\sigma_{11}\sigma_{22} + \Pi_{2222}\sigma_{22}^2 + 4\Pi_{1212}\sigma_{12}^2 \leqslant 1 \qquad /2/$$

General requirements for the forms /1/ and /2/ have not been perhaps investigated to the limit. These additional requirements may be listed as follows:

- The strength function /1/ should be convex in the six-dimensional stresses space.
- The function F should give for all admissible stress fields a possitive safety factor.
- The function F should represent an elipse in the two dimensional stress space.
- The strength function should be symmetric and invariant to the choice of principal directions of the material.

For the present purpose most of valuable knowlege can be obtained investigating the safety factor value. The safety factor n, for a given stress field σ, can be introduced as a ratio of the ultimate value of this stress field σ_0 to its actual value

$$n = \sigma_0 / \sigma \qquad /3/$$

The ultimate value σ_0 leads to an equation

$$F_1(\sigma_0) + F_2(\sigma_0) = 1 \qquad /4/$$

As both the linear and the quadratic part of equation /4/ are homogeneous, using /3/ it is possible to obtain a relation

$$n^2 F_2(\sigma) + n F_1(\sigma) = 1 \qquad /5/$$

leading to a positive value of the safety factor n

$$n = 2/\left(F_1(\sigma) + \sqrt{F_1^2(\sigma) + 4F_2(\sigma)}\right) \qquad /6/$$

provided that

$$F_2(\sigma) \geqslant 0 \qquad /7/$$

for all admissible stress fields.
This usefull statement can be utilised further. The function $F_2(\sigma)$ for a three dimensional case has in the principal material directions a form

$$\begin{aligned} F_2(\sigma) = {} & \Pi_{1111}\sigma_{11}^2 + \Pi_{2222}\sigma_{22}^2 + \Pi_{3333}\sigma_{33}^2 + \\ & + 2\Pi_{1122}\sigma_{11}\sigma_{22} + 2\Pi_{2233}\sigma_{22}\sigma_{33} + 2\Pi_{3311}\sigma_{33}\sigma_{11} + \\ & + 4\Pi_{1212}\sigma_{12}^2 + 4\Pi_{2323}\sigma_{23}^2 + 4\Pi_{3131}\sigma_{31}^2 \end{aligned} \qquad /8/$$

If for all stress fields $F_2(\sigma) \geqslant 0$ then the relation /8/ should be written as a sum of squares as follows

$$\begin{aligned} F_2(\sigma) = {} & (1-2\beta)\left(\sqrt{\Pi_{1111}}\,\sigma_{11} + \sqrt{\Pi_{2222}}\,\sigma_{22} + \sqrt{\Pi_{3333}}\,\sigma_{33}\right)^2 + \\ & + \beta\Big[\left(\sqrt{\Pi_{1111}}\,\sigma_{11} - \sqrt{\Pi_{2222}}\,\sigma_{22}\right)^2 + \left(\sqrt{\Pi_{2222}}\,\sigma_{22} - \sqrt{\Pi_{3333}}\,\sigma_{33}\right)^2 + \\ & + \left(\sqrt{\Pi_{3333}}\,\sigma_{33} - \sqrt{\Pi_{1111}}\,\sigma_{11}\right)^2 + 2\sigma_{11}\sigma_{22}\left[\Pi_{1122} - (1-3\beta)\sqrt{\Pi_{1111}\Pi_{2222}}\right] + \\ & + 2\sigma_{22}\sigma_{33}\left[\Pi_{2233} - (1-3\beta)\sqrt{\Pi_{2222}\Pi_{3333}}\right] + 2\sigma_{33}\sigma_{11}\Big[\Pi_{3311} - \\ & - (1-3\beta)\sqrt{\Pi_{3333}\Pi_{1111}}\Big] + 4\left(\Pi_{1212}\sigma_{12}^2 + \Pi_{2323}\sigma_{23}^2 + \right. \\ & \left. + \Pi_{3131}\sigma_{31}^2\right) \end{aligned} \qquad /9/$$

where β is a dimensionless unknown factor.

The function $F_2(\sigma)$ will be non negative, provided that

$$
\begin{aligned}
&1-2\beta \geq 0 \; : \; \beta \geq 0 \\
&\Pi_{1122} = (1-3\beta)\sqrt{\Pi_{1111}\Pi_{2222}} \\
&\Pi_{2233} = (1-3\beta)\sqrt{\Pi_{2222}\Pi_{3333}} \\
&\Pi_{3311} = (1-3\beta)\sqrt{\Pi_{3333}\Pi_{1111}}
\end{aligned}
\qquad /10/
$$

In addition, if $1 - 3\beta \leq 0$ as the strength theory demands, then the factor β is limitted to a value

$$1/3 \leq \beta \leq 1/2 \qquad /11/$$

with an mean square value

$$\beta = 0.423 \qquad /12/$$

when the possible error does not exceed 20%.

In a case of approximate calculations however using value /12/ may lead to seriously erroneus results. It is necessary to estimate the whole value of $1 - 3\beta$ coefficient. Accordingly to previous statements

$$-1/2 \leq (1-3\beta) \leq 0 \qquad /13/$$

with a mean square value

$$1-3\beta = -0.354 \qquad /14/$$

from which $\beta = 0.451$. It seems that the real value of β should be close to 0.44. The above estimations were performed using the mean square values with no good reason for it but it apears that such a method fits the experimental results best. This enables to postpone the experimental determination of the stresses interaction coefficients and calculate their values from tension and compression tests with errors not greater than 20%. It also can be shown, that the coefficient $\beta = 0.423$ ensures fulfilment of other necessary requirements mentioned earlier.

Experimental Evaluation

A paper by Ru-Yu-Wu and Stachurski [4] gives experimental values of strength in tension and compression for several materials as well as the experimentally determined interaction coefficients value. In Table 1 these values are compared with proposals from [4], [5] and the present paper. Besides the interaction coefficient an angle of roatation of the strength elipse is also calculated using a known relation

$$\tan 2\theta = 2\Pi_{1122} / (\Pi_{1111} - \Pi_{2222}) \qquad /15/$$

Experimental Procedure

Having **established** an approximate relation for the normal **stresses** interaction tensorial coordinates of the type

$$\Pi_{1122} = -0.354\sqrt{\Pi_{1111}\Pi_{2222}} \qquad /16/$$

Table 1

Material	Experimental					Theoretical			Exptl.	Theoretical		
	T_{11}	C_{11}	T_{22}	C_{22}	$-\Pi_{1122} \times 10^{4}$	$-\Pi_{1122} \times 10^{4}$ [4]	$-\Pi_{1122} \times 10^{4}$ [5]	$-\Pi_{1122} \times 10^{4}$ current paper	θ^{o}	θ^{o} [4]	θ^{o} [5]	θ^{o} current paper
Paper	7.3	2.4	3.5	1.19	400	460	585	410	11,8	13.4	16.3	16.30
Paper	6.9	2.26	2.7	0.93	500	550	810	560	8.06	8.89	12.54	12.78
Thermo-plastic	77	59	50	65.5	1.1	1.27	1.3	0.91	34.5	35.7	35.95	35.9
Graphite Epoxy	185.6	127.1	7.52	33.8	1.62	0.42	2.04	1.43	2.37	0.62	3.0	2.99
Graphite Epoxy	150	100	6	17	2.45	0.7	4.04	2.83	1.47	0.41	2.37	2.37
Graphite Epoxy	149	103	6.3	18.2	2	0.65	3.8	2.63	1.32	0.43	2.51	2.49
Glass Epoxy	120,8	88.1	3.25	13.57	6.39	0.94	7.3	5.11	1.62	0.24	1.85	1.85

it is possible to obtain approximate values of all tensorial coefficients for a plane stress strength theory from five simple tension experiments, performed in five different directions in relation to principal directions of the material in question.

Table 2 gives values of stresses in the principal directions of the material for specimens in simple tension in five different directions as shown in figure 1. Using values from Table 2 and the strength theory in the form /2/ it is possible to obtain five linear equations for determination of tensorial coefficients.

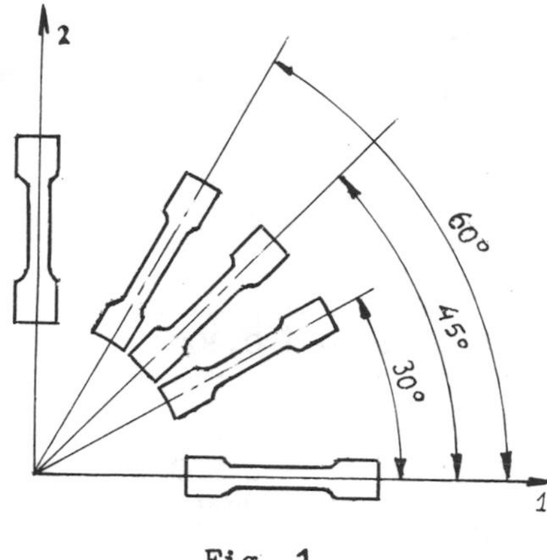

Fig. 1

Table 2

Angle	σ_{11}	σ_{22}	σ_{12}
0°	σ	0	0
30°	$\frac{3}{4}\sigma$	$\frac{1}{4}\sigma$	$-\frac{3}{4}\sigma$
45°	$\frac{1}{2}\sigma$	$\frac{1}{2}\sigma$	$-\frac{1}{2}\sigma$
60°	$\frac{1}{4}\sigma$	$\frac{3}{4}\sigma$	$\frac{3}{4}\sigma$
90°	0	σ	0

These equations can be written in a form

$$
\begin{aligned}
&T_1 \Pi_{11} + T_1^2 \Pi_{1111} = 1 \\
&4T_2 (3\Pi_{11} + \Pi_{22}) + T_2^2 (9\Pi_{1111} + \Pi_{2222} + 6\Pi_{1122} + 12\Pi_{1212}) = 16 \\
&6T_3 (\Pi_{11} + \Pi_{22}) + T_3^2 (3\Pi_{1111} + 3\Pi_{2222} + 6\Pi_{1122} + 12\Pi_{1212}) = 12 \\
&4T_4 (\Pi_{11} + 3\Pi_{22}) + T_4^2 (\Pi_{1111} + 9\Pi_{2222} + 6\Pi_{1122} + 12\Pi_{1212}) = 16 \\
&T_5 \Pi_{22} + T_5^2 \Pi_{2222} = 1
\end{aligned}
\qquad /17/
$$

where T_i is the ultimate stress in the direction i, accordingly to Table 2. The set of equations /17/ can be treated numericaly obtaining as results Π_{11} , Π_{22} , Π_{1111} , Π_{2222} , $\Pi_{1122} + 2\Pi_{1212}$. Employing equation /16/ it is possible to find the value of Π_{1122} and finaly the value of Π_{1212} . As the tensorial coefficients are given by relations [2]:

$$
\begin{aligned}
&\Pi_{11} = 1/T_{11} - 1/C_{11} : \Pi_{1111} = 1/(T_{11} C_{11}) \\
&\Pi_{22} = 1/T_{22} - 1/C_{22} : \Pi_{2222} = 1/(T_{22} C_{22}) \\
&4\Pi_{1212} = 1/S_{12}^2
\end{aligned}
\qquad /18/
$$

where S_{12} is the ultimate shear stress in the principal directions of the material and C represents ultimate compression strength. A numeric solution of the set of equations /18/ gives finaly the approximate values of the ultimate stresses T_{11}, C_{11}, T_{22}, C_{22}, S_{12} in the principal directions of the material.

From the set of equations /17/ can be seen that separation of tensorial coefficients in results of any simple tension experiments is not possible without additional relations such as the approximation /16/.

Numerical Simulation

To investigate the errors resulting from introduction of a constant value of $1 - 3\beta = -0{,}354$ a numerical procedure was prepared, simulating the process of calculation of results with $1 - 3\beta$ regarded as a variable. Table 3 gives final results of calculation

Table 3

$1-3\beta$	T_{11}	C_{11}	T_{22}	C_{22}	S_{12}
—	5.000	1.000	0.500	0.100	0.200
- 0.20	5.000	1.000	0.500	0.100	0.202
- 0.30	5.000	1.000	0.500	0.100	0.200
- 0.40	5.000	1.000	0.500	0.100	0.199

while the values inserted into the procedure are given in the first row. It can be seen that variating the value $1 - 3\beta$ by 33% gives differences in results not exceeding 1% and only in calculation of the ultimate value of shear stress S_{12}. This proves that the true value of the coefficient $1 - 3\beta$ for experimental purposes is not vital and does not introduce serious errors. It also leads to an assumption, that the presented "constant coefficient" method may be introduced into experimental practice. Another problem however was encountered. The solution of the set of equations /17/, due to experimental errors had to be performed using the least squares method, somewhat "rounding up" the experimental findings.

References

[1] Malmeister, A.K., Geometrija Teorii Pročnosti, Mekh.Polim., 4,p.519, /1966/

[2] Tsai, T.S., Wu, E.M., A General Theory of Strength for Anisotropic Materials, J.Comp.Mat., vol. 5, p.58 /1971/

[3] Wu, E.M., Phenomenological Criterion of Destruction for Anisotropic Materials. In: Mechanics of Composite Materials, Ed. G.P. Sendeckyj, Academic Press, /1974/

[4] Ru-Yu-Wu, Stachurski, Z., Evaluation of Normal Stress Interaction Parameter in the Tensor Polynomial Strength Theory for Anisotropic Materials, J.Comp.Mat., vol. 18 /1984/

[5] Tsai, S.W., Han, H.T., Introduction to Composite Materials, Technomic, /1980/

FLOQUET WAVES IN INEXTENSIBLE LAMINATES

E. Rhian Baylis

Department of Theoretical Mechanics,
The University, Nottingham, NG7 2RD,
England

The laminate is formed of a single fibre-reinforced material arranged in layers of uniform thickness, with the fibre direction in adjoining layers being at right angles to each other and in the plane of the layers. The dispersion equation is obtained for waves propagating along a fibre direction and this defines a surface in the frequency-wave number space. The regions corresponding to the passing bands and stopping bands are identified and particular attention is devoted to the bounding curves of the Brillouin zones.

INTRODUCTION

We consider the propagation of harmonic waves in an unbounded, periodically layered elastic solid. Each layer consists of the same transversely isotropic material which is inextensible in the direction of transverse isotropy. The layers are so arranged that the axes of transverse isotropy in adjoining layers are orthogonal. We shall examine waves propagating in the plane of the layers and for simplicity we restrict attention to waves whose direction of propagation is parallel to an axis of transverse isotropy.

The inextensible transversely isotropic material is intended to model a composite material consisting of a single family of parallel strong fibres embedded in an elastic matrix. Spencer [1] has pointed out that for many static stress analysis problems the idealized material which is inextensible in the direction of transverse isotropy can provide an adequate simple model of such a composite. Baylis and Green [2,3] and Baylis [4] have shown that for three-ply and two-ply laminated plates the idealized material can give an acceptable approximation to the short wavelength (high frequency) behaviour of the composite material, with a considerable simplification in the analysis involved.

Delph et al. [5,6] have considered wave propagation in a similar medium where the two alternating layers are composed of different homogeneous isotropic materials. They have applied Floquet's theory of ordinary differential equations with periodic coefficients to derive the dispersion equation, which is a relation between the frequency and the wave number. Our approach closely follows that of Delph et al. [5], but we also draw on the work of Gilbert [7] and Schoenberg [8] by making use of the propagator matrix technique. This technique allows the displacement and traction components at the upper surface of a layer to be expressed in terms of those at the lower surface by means of the lamina propagator matrix. For perfect bonding between the layers, these components on the lower surface of one lamina are identical to the corresponding components on the upper surface of the lamina immediately below it. We can therefore express these components at one interface in terms of those at another interface, n layers away, by compounding the n matrices for the individual layers. Green and Baylis [9] have previously applied this approach to obtain dispersion curves in a plate of finite depth made of the laminated material which is being considered here.

We apply the propagator method coupled with Floquet's theory to derive the dispersion equation which, in this case, defines a multi-valued surface in the frequency - wave number space. One of the most important features of this surface is the presence of passing and stopping bands. These are regions of the spectrum where harmonic waves in the direction normal to the layering are propagated or attenuated respectively. These passing and stopping bands have been shown to be a characteristic feature of wave propagation in periodic media (Brillouin [10]). The curves on the surface which define the boundary between the passing and stopping bands divide the surface into Brillouin zones. In examining the dispersion surface, these curves can be analyzed in detail, since the dispersion equation factorizes along the ends of the Brillouin zones. Each factor corresponds to a motion with a particular symmetry.

The long and short wavelength limits of the dispersion equation are examined analytically. Numerical results are obtained using data measured by Markham [11], for a carbon fibre - epoxy resin composite, and the paper closes with a discussion of these results.

A more detailed account of this work will be found in the Ph.D. thesis by Baylis [12]. In particular, this contains a full analysis of the limiting behaviour of the dispersion surfaces in the long wavelength region.

SOLUTIONS OF THE EQUATIONS OF MOTION (PROPAGATOR MATRICES)

We consider harmonic waves propagating through a periodically layered elastic body of unbounded extent. We set up a Cartesian system of axes $Ox_1x_2x_3$ so that Ox_1 is normal to the layering. The body consists of an infinite sequence of two alternating layers. One of these layers is of depth 2h and has its axis of transverse isotropy parallel to Ox_3. The other, of depth 2d, has its fibre-direction along Ox_2. For mathematical convenience, however, we shall interpret this periodic layering as follows. We define a unit cell to consist of three layers - an inner core of depth 2d and two outer layers, each of depth h. The body under consideration consists of an infinite number of these unit cells.

We consider waves propagating in the x_3-direction and with displacements polarized in the x_1x_3-plane so that in each layer the displacements have the form

$$u_1(x_i,t) = U(x_1)\cos k(x_3-vt)\ , \qquad u_2(x_i,t) = 0\ ,$$

$$u_3(x_i,t) = W(x_1)\sin k(x_3-vt)\ , \qquad (i = 1,2,3), \qquad (1)$$

where v is the speed of propagation and k the wave number. To a disturbance travelling in the x_3-direction, the layers of depth 2d appear isotropic and we shall refer to these as the isotropic layers, as opposed to the inextensible layers of depth h. The governing equations in each layer have been derived by Baylis and Green [2] and are quoted here without derivation.

(a) Inextensible layer

The inextensibility constraint gives $W(x_1) = 0$ throughout the layer, and the equations of motion reduce to

$$c_1^2 \frac{d^2U}{dx_1^2} + k^2(v^2 - c_3^2)U = 0 , \tag{2}$$

where

$$c_1^2 = (\lambda + 2\mu_T)/\rho , \qquad c_3^2 = \mu_L/\rho \qquad \text{and} \qquad c_2^2 = \mu_T/\rho . \tag{3}$$

ρ is the density and λ, μ_T and μ_L are elastic moduli. Because of the inextensibility constraint, the longitudinal stress component t_{33} is not determined by the constitutive equations, but is obtained directly as a reaction stress from the equation of motion in the x_3-direction. Baylis and Green [2] have pointed out that this reaction stress can be singular at the interfaces between the layers, allowing a discontinuity in the shear stress t_{13} across the singularity. The normal stress component t_{11} must be continuous, and this is given by the expression

$$t_{11} = T_{11}(x_1)\cos k(x_3 - vt) = \rho c_1^2 \frac{dU}{dx_1} \cos k(x_3 - vt) . \tag{4}$$

Writing $T(x_1) = T_{11}(x_1)/\rho c_1^2 k$ and defining the vector $\underset{\sim}{X}(x_1)$ by

$$\underset{\sim}{X}(x_1) = \{T(x_1), U(x_1)\}^T ,$$

the solution to equation (2) may be expressed in the form

$$\underset{\sim}{X}(x_1) = \underset{\sim}{M}_1(x_1)\underset{\sim}{X}(0) , \tag{5}$$

where $\underset{\sim}{M}_1(x_1)$ is the propagator matrix for the inextensible layer. In particular, for the layer of thickness h, we have

$$\underset{\sim}{X}(h) = \underset{\sim}{M}_1(h)\underset{\sim}{X}(0) . \tag{6}$$

$\underset{\sim}{M}_1(h)$ is defined by

$$\underset{\sim}{M}_1(h) = \begin{bmatrix} C & pS \\ S/p & C \end{bmatrix} , \tag{7}$$

where

$$p^2 = (c_3^2 - v^2)/c_1^2 , \qquad S = \sinh pkh , \qquad C = \cosh pkh . \tag{8}$$

That is, we can express the value of $\underset{\sim}{X}(x_1)$, $\underset{\sim}{X}^u$, say, at the upper surface of the layer of depth h in terms of its value, $\underset{\sim}{X}^\ell$ at the lower surface, in the form

$$\underset{\sim}{X}^{u} = \underset{\sim}{M}_1(h)\underset{\sim}{X}^{\ell} . \tag{9}$$

(b) Isotropic layer

The equations of motion in this layer are

$$\begin{aligned} c_1^2 \frac{d^2U}{dx_1^2} + k^2(v^2-c_3^2)U + (c_1^2-c_2^2)k\frac{dW}{dx_1} &= 0 , \\ -(c_1^2-c_2^2)k\frac{dU}{dx_1} + c_2^2\frac{d^2W}{dx_1^2} + k^2(v^2-c_1^2)W &= 0 . \end{aligned} \tag{10}$$

The traction components $t_{11} = T_{11}(x_1)\cos k(x_3-vt)$ and $t_{13} = T_{13}(x_1)\sin k(x_3-vt)$ on any plane x_1 = constant are given by

$$T_{11} = \rho c_1^2\frac{dU}{dx_1} + \rho(c_1^2-2c_2^2)kW , \qquad T_{13} = \rho c_2^2\left(\frac{dW}{dx_1} - kU\right) . \tag{11}$$

Writing $T(x_1) = T_{11}/\rho c_1^2 k$ and $S(x_1) = T_{13}/\rho c_1^2 k$, the solution of equations (10) give an expression for the vector $\underset{\sim}{Y}(x_1) = \{T(x_1),S(x_1),U(x_1),W(x_1)\}^T$ in terms of its value at $x_1 = 0$, say, in the form

$$\underset{\sim}{Y}(x_1) = \underset{\sim}{P}(x_1)\underset{\sim}{Y}(0) . \tag{12}$$

In particular, we can write, for a layer of thickness d, that

$$\underset{\sim}{Y}(d) = \underset{\sim}{P}(d)\underset{\sim}{Y}(0) , \tag{13}$$

where

$$\underset{\sim}{P}(d) = \frac{c_1^2}{v^2}\begin{pmatrix} -\alpha C_1 + \beta C_2 & -\frac{\alpha}{q_1}S_1 + \beta q_2 S_2 & -\frac{\alpha^2}{q_1} + \beta^2 q_2 S_2 & -\alpha\beta(C_1-C_2) \\ \beta q_1 S_1 - \frac{\alpha}{q_2}S_2 & \beta C_1 - \alpha C_2 & \alpha\beta(C_1-C_2) & \beta^2 q_1 S_1 - \frac{\alpha^2}{q_2}S_2 \\ -q_1 S_1 + \frac{S_2}{q_2} & -C_1 + C_2 & -\alpha C_1 + \beta C_2 & -\beta q_1 S_1 + \frac{\alpha}{q_2}S_2 \\ C_1 - C_2 & \frac{S_1}{q_1} - q_2 S_2 & \frac{\alpha S_1}{q_1} - \beta q_2 S_2 & \beta C_1 - \alpha C_2 \end{pmatrix} . \tag{14}$$

The terms appearing in the matrix are defined by

$$S_1 = \sinh q_1 kd , \quad S_2 = \sinh q_2 kd , \quad \alpha = (2c_2^2-v^2)/c_1^2 , \quad q_1^2 = 1 - v^2/c_1^2 ,$$

$$C_1 = \cosh q_1 kd , \quad C_2 = \cosh q_2 kd , \quad \beta = 2c_2^2/c_1^2 , \quad q_2^2 = 1 - v^2/c_2^2 .$$

$\underset{\sim}{P}(2d)$ is the propagator matrix for the isotropic layer, but it is convenient to write this as the product $\underset{\sim}{P}(d)\underset{\sim}{P}(d)$. This isotropic layer is bonded on both upper and lower surfaces to an inextensible layer, and it is therefore subject to the constraint $W = 0$ on these surfaces. Using these conditions gives an expression for $\underset{\sim}{X}$ at the upper surface, $\underset{\sim}{X}^u$ in terms of its value at the lower surface $\underset{\sim}{X}^\ell$, given by

$$\underset{\sim}{X}^u = \underset{\sim}{M}_2 \underset{\sim}{X}^\ell = \frac{\underset{\sim}{R}\cdot\hat{\underset{\sim}{R}}}{|\underset{\sim}{R}|}\cdot\underset{\sim}{X}^\ell \tag{15}$$

where

$$\underset{\sim}{R} = \begin{pmatrix} r_{11} & r_{12} \\ r_{21} & r_{22} \end{pmatrix}, \qquad \hat{\underset{\sim}{R}} = \begin{pmatrix} r_{22} & r_{12} \\ r_{21} & r_{11} \end{pmatrix},$$

and the elements r_{ij} are related to the elements p_{ij} of the propagator $\underset{\sim}{P}(d)$ by the expressions

$$r_{11} = p_{11} - \frac{p_{14}p_{41}}{p_{44}}, \qquad r_{12} = p_{13} - \frac{p_{12}p_{43}}{p_{42}},$$

$$r_{21} = p_{31} - \frac{p_{34}p_{41}}{p_{44}}, \qquad r_{22} = p_{33} - \frac{p_{32}p_{43}}{p_{42}}.$$

(c) Unit cell

In order to obtain the transfer matrix for the unit cell, we must make use of the interface conditions between the isotropic layer and the two inextensible layers bonded to it. These are that U and T must be continuous across the interfaces and that $W = 0$ at the interfaces. The value of S can be discontinuous in view of the possible singularity in the reaction stress at the boundary in the inextensible layers.

The vector $\underset{\sim}{X}^u$ at the top of the unit cell is then given in terms of the vector $\underset{\sim}{X}^\ell$ at the bottom by the expression

$$\underset{\sim}{X}^u = \underset{\sim}{M}\underset{\sim}{X}^\ell = \underset{\sim}{M}_1\underset{\sim}{M}_2\underset{\sim}{M}_1\underset{\sim}{X}^\ell = \frac{\underset{\sim}{N}\cdot\hat{\underset{\sim}{N}}}{|\underset{\sim}{N}|}\underset{\sim}{X}^\ell, \tag{16}$$

where $\underset{\sim}{N} = \underset{\sim}{M}_1\underset{\sim}{R}$, $\hat{\underset{\sim}{N}} = \hat{\underset{\sim}{R}}\underset{\sim}{M}_1$ and $|\underset{\sim}{N}| = |\underset{\sim}{M}_1\underset{\sim}{R}| = |\underset{\sim}{R}|$.

We can also rewrite the matrix $\underset{\sim}{M}$ in the form

$$\underset{\sim}{M} = \frac{\underset{\sim}{Q}\cdot\hat{\underset{\sim}{Q}}}{|\underset{\sim}{Q}|}, \tag{17}$$

where the elements q_{ij} of the matrix $\underset{\sim}{Q}$ are related to those of $\underset{\sim}{N}$ and are given by

$$q_{11} = \frac{v^2}{c_1^2}CC_1C_2 - pS\left\{q_1S_1C_2 - \frac{S_2}{q_2}C_1\right\},$$

$$q_{12} = -\frac{v^2}{c_1^2}\frac{q_2}{q_1}CS_1S_2 - pS\left\{\frac{S_1}{q_1}C_2 - q_2S_2C_1\right\}, \tag{18}$$

$$q_{21} = \frac{v^2}{c_1^2}\frac{S}{p}C_1C_2 - C\left\{q_1S_1C_2 - \frac{S_2}{q_2}C_1\right\},$$

$$q_{22} = -\frac{v^2}{c_1^2}\frac{q_2}{q_1}\frac{S}{p}S_1S_2 - C\left(\frac{S_1}{q_1}C_2 - q_2S_2C_1\right) .$$

The elements m_{ij} of the matrix $\underset{\sim}{M}$ are given by

$$\begin{aligned} m_{11} &= m_{22} = (q_{11}q_{22}+q_{12}q_{21})/|\underset{\sim}{Q}| , \\ m_{12} &= 2q_{11}q_{12}/|\underset{\sim}{Q}| , \qquad m_{21} = 2q_{21}q_{22}/|\underset{\sim}{Q}| . \end{aligned} \tag{19}$$

Note that

$$|\underset{\sim}{M}| = m_{11}^2 - m_{12}m_{21} = 1 . \tag{20}$$

FLOQUET THEORY

The equation of motion for the laminated body takes the form of a partial differential equation with coefficients which are periodic in the x_1-direction, with period $\ell = 2(h+d)$. This follows from the fact that the density and elastic moduli are piecewise constant in each layer and have periodic variation from cell to cell. We can thus apply Floquet's theorem which states that a differential equation with periodic coefficients admits a solution of the form

$$u(x_1) = g(x_1)e^{irx_1} ,$$

where r is a constant and $g(x_1)$ is periodic with the same periodicity as the coefficients of the differential equation. The equations of motion thus admit the solution

$$u_i(x_1,x_2,x_3,t) = g_i(x_1,x_2,x_3,t)e^{irx_1} , \tag{21}$$

where $g_i(x_1+\ell,x_2,x_3,t) = g_i(x_1,x_2,x_3,t)$. We can then deduce that

$$u_i(x_1+\ell,x_2,x_3,t) = u_i(x_1,x_2,x_3,t)e^{ir\ell} . \tag{22}$$

From equations (1) and (22) we can write

$$U(x_1+\ell) = U(x_1)e^{ir\ell} \qquad \text{and also} \qquad T(x_1+\ell) = T(x_1)e^{ir\ell} ,$$

since $T(x_1)$ is a linear combination of u_i and their first derivatives. In terms of the vector $\underset{\sim}{X}$ this gives

$$\underset{\sim}{X}^u = e^{ir\ell}\underset{\sim}{X}^\ell \tag{23}$$

From equation (21) we see that r has the character of a wave number and is known as the wave number of the Floquet wave.

DISPERSION EQUATION

Equations (16) and (23) together give the equations

$$\underset{\sim}{M}\underset{\sim}{X}^\ell = e^{ir\ell}\underset{\sim}{X}^\ell$$

which have non-trivial solutions for $\underset{\sim}{X}^\ell$ provided

$$|\underset{\sim}{M}-e^{ir\ell}\underset{\sim}{I}| = 0 \ , \tag{24}$$

where $\underset{\sim}{I}$ is the unit matrix. By using equation (20), this can be expressed as

$$e^{2ir\ell} - 2m_{11}e^{ir\ell} + 1 = 0 \ .$$

Therefore $e^{ir\ell} = m_{11} \pm \sqrt{m_{11}^2-1}$ and

$$m_{11} = \cos r\ell \ . \tag{25}$$

The element m_{11} appearing in equation (25) is defined through equations (19) and (18) and is a function of the reduced wave number $k\ell$, the ratio $\nu = d/h$ and the phase velocity v. Writing $v = \omega/k$, where ω is the angular frequency of the wave motion, equation (25) is then seen to be a relationship between the frequency ω and the independent wave numbers k and r. This is the dispersion equation. It defines a surface, known as the dispersion surface, in frequency - wave number space. From equation (25) we see that the surface is real valued in the range $-1 \leqslant m_{11} \leqslant 1$. The surface is in general discontinuous at the planes $r\ell = n\pi$, and these planes divide the surface into Brillouin zones. For $|m_{11}| > 1$, the surface will be complex-valued. The existence of real and complex portions of the surface may be interpreted in terms of passing and stopping bands. On the passing bands, the wave number r is real and harmonic waves are propagated along Ox_3 such that the amplitude normal to the layering varies periodically and is unattenuated. However, the stopping bands correspond to complex or imaginary values of r, and by virtue of the Floquet solution (21) the amplitude of harmonic waves is attenuated along the normal to the layering.

We now investigate the form of the dispersion equation on the end points of the Brillouin zones. Here, $m_{11} = \pm 1$ and $m_{12}m_{21} = 0$. From equation (19), we find that either $q_{11} = 0$ or $q_{12} = 0$ or $q_{21} = 0$ or $q_{22} = 0$. In particular, $q_{11} = 0$ and $q_{22} = 0$ correspond to $e^{ir\ell} = -1$, $r\ell = (2n+1)\pi$, whilst $q_{12} = 0$ and $q_{21} = 0$ correspond to $e^{ir\ell} = +1$, $r\ell = 2n\pi$. These four curves give the edges of the dispersion surface. Numerically we find two surfaces and we plot them in $k\ell$, $r\ell$, $\omega\ell/c_1$ space.

We can show that $q_{11} = 0$ corresponds to a motion which is symmetric relative to the mid-plane of the isotropic layer and antisymmetric relative to the mid-plane of the inextensible layer. Applying the notation of Delph et al. [5], we designate this as the SA mode. Similarly, $q_{12} = 0$ represents an AA mode, $q_{21} = 0$ an SS mode and $q_{22} = 0$ an AS mode.

We shall present some numerical results later, but it is possible to examine the long and short wavelength behaviour analytically.

(a) Long wave limit

The only edge curve with a finite long wave limiting speed is $q_{12} = 0$, with

$$v^2 = \frac{c_2^2+\nu c_3^2}{1+\nu} \ ,$$

where $\nu = d/h$.

When we examine the limiting behaviour under the assumption that v is infinite, we replace v by ω/k and apply the approximations $kp \simeq i\omega/c_1$, $kq_1 \simeq i\omega/c_1$ and $kq_2 \simeq i\omega/c_2$. We then find that unless

$$\frac{\omega\ell}{c_1} = (1+\nu)\frac{c_2}{c_1}n\pi \tag{26}$$

the long wave limiting behaviour is given by

$$r\ell = \pm\frac{\omega\ell}{c_1} + 2n\pi \ . \tag{27}$$

When equation (26) is satisfied, m_{11} is indeterminate. The numerical results indicate that here $r\ell$ takes all values from $(n-1)\pi$ to $n\pi$. This corresponds to a straight line when we plot $\omega\ell/c_1$ against $r\ell$.

(b) Short wave limit

When p, q_1 and q_2 remain real as $k\ell \to \infty$, we replace the hyperbolic sines and cosines by one half the relevant positive exponentials. However, it is easy to show that there are no real solutions to equation (25) in this case.

We re-examine this limiting behaviour with p and q_1 real and q_2 imaginary. The hyperbolic functions of q_2 have to be replaced by circular functions. It turns out that $v \to c_2$ as $k\ell \to \infty$ and the limiting behaviour of the surface is given by

$$\omega\ell/c_1 = c_2k\ell/c_1 \ .$$

This is independent of $r\ell$ and indicates that the surface becomes horizontal as $k\ell \to \infty$. The short wave behaviour is a shear disturbance confined to the isotropic layers.

NUMERICAL RESULTS

Figure 1 shows results for the first four Brillouin zones for sections of the first surface lying in planes of constant $k\ell$ ($k\ell$ = 0.001, 1, 2.5, 5 and 10). It can be seen that the surface is discontinuous at $r\ell = n\pi$. Figure 2 shows corresponding results for the second surface. The existence of these two distinct surfaces is established by the numerical solution of equation (25). For very small values of $k\ell$, $\cos r\ell$ decreases from +1 at $\omega\ell/c_1 \simeq 0$, through 0 at $\omega\ell/c_1 \simeq \frac{1}{2}\pi$ and reaches -1 at $\omega\ell/c_1 \simeq 2c_2\pi/c_1$. $\cos r\ell$ then very rapidly tends to $-\infty$, comes back through $+\infty$ and reaches +1 almost instantaneously. But this implies that there is a definite gap between the surfaces. The long-wave limiting behaviour of the first surface is given by equation (27) for values of $r\ell$ satisfying $(n-1)\pi \leqslant r\ell \leqslant 2nc_2\pi/c_1$ and by equation (26) otherwise. The limiting behaviour of the second surface is given by the equation (26) for $(n-1)\pi \leqslant r\ell \leqslant 2nc_2\pi/c_1$ and by equation (27) otherwise.

Figure 3 shows the fundamental mode and first three harmonics for each of the four curves $q_{11} = 0$ (solid), $q_{12} = 0$ (dot), $q_{21} = 0$ (dash) and $q_{22} = 0$ (dot/dash). They have all been projected onto the plane $r\ell = 0$. Because of this, there appear to be more intersections than is, in fact, the case. For example, dash 1 and dot/dash 1 (numbering from $\omega\ell/c_1 = 0$ upwards) appear to intersect. But since the dash curves represent $q_{21} = 0$, dash 1 lies in the plane $r\ell = 0$ whereas dot/dash 1, representing $q_{22} = 0$, lies in the $r\ell = \pi$

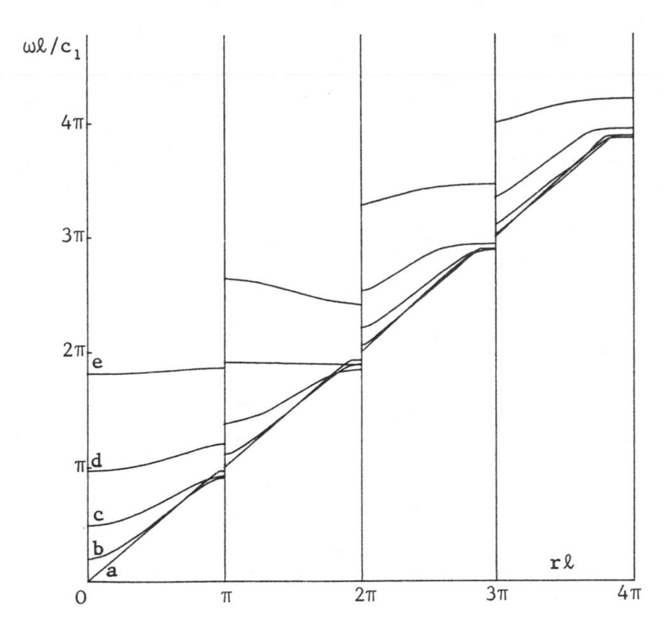

FIGURE 1. Curves of constant $k\ell$ on the dispersion surface 1. (a) $k\ell = 0.001$; (b) $k\ell = 1$; (c) $k\ell = 2.5$; (d) $k\ell = 5$; (e) $k\ell = 10$

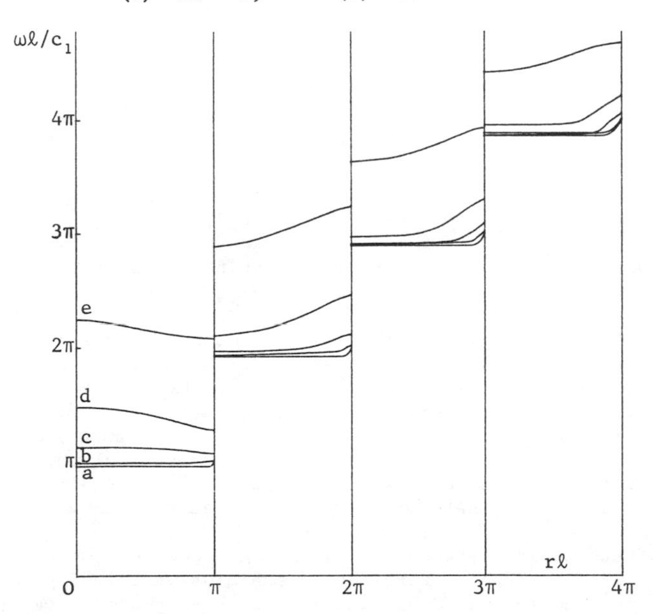

FIGURE 2. Curves of constant $k\ell$ on the dispersion surface 2. (a) $k\ell = 0.001$; (b) $k\ell = 1$; (c) $k\ell = 2.5$; (d) $k\ell = 5$; (e) $k\ell = 10$

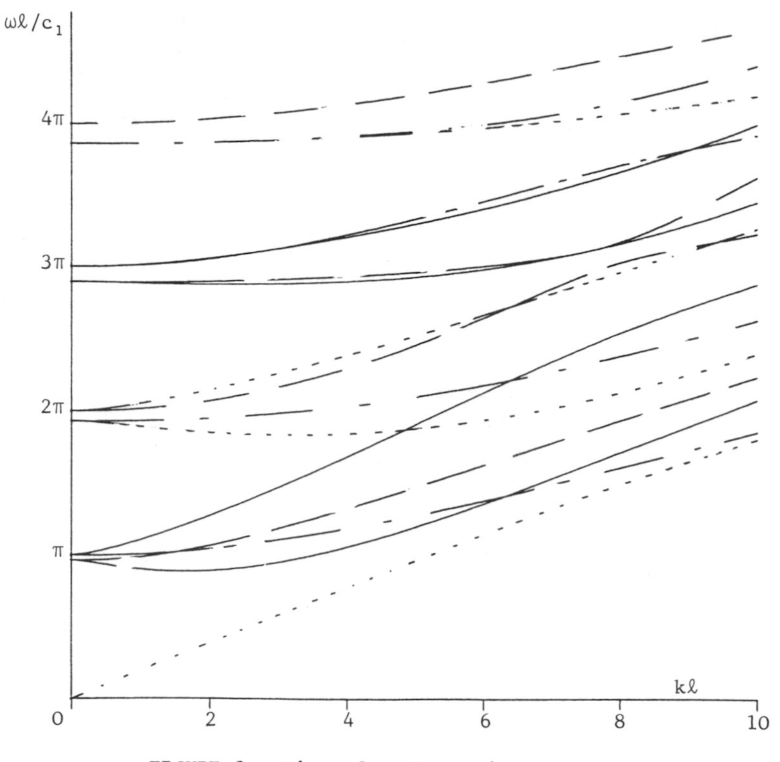

FIGURE 3. First four harmonics of the curves $q_{11} = 0$ (solid), $q_{12} = 0$ (dot), $q_{21} = 0$ (dash), $q_{22} = 0$ (dot/dash)

plane. However, the intersections of dot/dash 1 with solid 1 are genuine since both curves lie in the $r\ell = \pi$ plane. The points of intersection are called conical points, and were also met by Delph et al. [5].

We interpret the surfaces as follows. Zone 1 of surface 1 lies between dot 1 and the lower of the curves solid 1 and dot/dash 1. Consider a point on this surface lying close to $k\ell = 0$ on solid 1. By moving along the zone boundary (solid 1) in the direction of increasing $k\ell$, we pass through a conical point. In order to remain on the same surface, we must then pass onto dot/dash 1. We have passed from a curve of one symmetry SA to one of the opposite symmetry AS. This will occur at each conical point since the intersecting curves are either $q_{11} = 0$ (SA) and $q_{22} = 0$ (AS), at $r\ell = (2n+1)\pi$, or $q_{12} = 0$ (AA) and $q_{21} = 0$ (SS), at $r\ell = 2n\pi$.

Zone 1 of surface 2 lies between dash 1 and the upper of the curves solid 1 and dot/dash 1. The shape of this region explains why, in Figure 2, the curves corresponding to $k\ell = 0.001$ and $k\ell = 1$ slope upwards from $r\ell = 0$ to $r\ell = \pi$, whereas for $k\ell = 2.5$, 5 and 10, the curves slope downwards. Zone 2 of surface 1 lies between solid 2 and dot 2, and zone 2 of surface 2 starts from dot/dash 2 and ends on the lower of dash 2 and dot 3. We can continue to find all the zones in this manner.

Delph et al. [6] have previously established the existence of two dispersion surfaces for a laminated body consisting of two different isotropic materials. They were able to distinguish between the two surfaces by considering the long wavelength limiting behaviour. Here we are dealing with a transversely isotropic material and the constraint of inextensibility gives rise to a discrepancy between the long wave limiting behaviour and the corresponding solution for waves propagating normal to the layering. Distinguishing between the two surfaces in our case is therefore not straightforward, and we have relied heavily on the numerical results. Further details on the explanation for this discrepancy may be found in the author's thesis [12].

ACKNOWLEDGEMENTS

This work has been carried out under a research grant from the Science and Engineering Research Council and I am grateful for their support. I am indebted to Dr. W. A. Green for many helpful discussions during the preparation of this paper.

REFERENCES

[1] A. J. M. Spencer, *Deformations of Fibre-reinforced Materials*, Oxford (1972)

[2] E. R. Baylis & W. A. Green, *J. Sound Vibration* 110 (1986) (to appear)

[3] E. R. Baylis & W. A. Green, *J. Sound Vibration* 111 (1986) (to appear)

[4] E. R. Baylis, *Acta Mech.* (1986) (to appear)

[5] T. J. Delph, G. Herrmann & R. K. Kaul, *J. Appl. Mech.* 45 (1978), 343

[6] T. J. Delph, G. Herrmann & R. K. Kaul, *J. Appl. Mech.* 46 (1979), 113

[7] K. E. Gilbert, *J. Acoust. Soc. Am.* 73 (1983), 137

[8] M. Schoenberg, *Geophysical Prospecting* 31 (1983), 265

[9] W. A. Green & E. R. Baylis, *Arch. Mech.* 38 (1986) (to appear)

[10] L. Brillouin, *Wave Propagation in Periodic Structures*, Dover Publications, New York (1953)

[11] M. F. Markham, *Composites* 1 (1970), 145

[12] E. R. Baylis, *Wave Propagation in Fibre-reinforced Laminates*, Ph.D. thesis, University of Nottingham (1986)

On the non-existence of interfacial waves in an inextensible elastic composite with distinct fibre directions

P. Chadwick and V. S. Captain

(School of Mathematics and Physics, University of East Anglia, Norwich, England)

Summary

It is shown that interfacial waves cannot propagate along a plane of welded contact between dissimilar transversely isotropic elastic bodies when each material is inextensible along the axis of symmetry and the axial directions on the two sides of the interface do not coincide.

The secular equation: derivation and preliminary discussion

The composite body studied in this paper consists of two semi-infinite, homogeneous elastic bodies B and B' welded together at their plane interface I. We denote by $\underset{\sim}{m}$ the unit normal to I, directed away from B, and by $\underset{\sim}{x}$ the position vector of an arbitrary point relative to an origin in I. The interface is thus given by $\underset{\sim}{m}.\underset{\sim}{x} = 0$ and the regions occupied by B and B' by $\underset{\sim}{m}.\underset{\sim}{x} \leq 0$ and $\underset{\sim}{m}.\underset{\sim}{x} \geq 0$ respectively. The different materials forming B and B' are assumed to be unidirectionally inextensible and transversely isotropic in the direction of inextensibility with densities ρ and ρ' and elastic constants λ,μ_L,μ_T and λ',μ_L',μ_T', unprimed symbols referring throughout to B and primed symbols to B'. The directions of inextensibility in B and B' (described for brevity as the fibre directions) are defined by unit vectors $\underset{\sim}{e}$ and $\underset{\sim}{e}'$. These directions are taken to be distinct, so that $\underset{\sim}{E} = \underset{\sim}{e}\wedge\underset{\sim}{e}' \neq \underset{\sim}{0}$, and we suppose, in this section and the next, that neither fibre direction is parallel to I. External body forces are assumed absent.

We consider the possibility of transmitting through the composite $B \cup B'$ an interfacial wave, travelling in the direction of a unit vector $\underset{\sim}{n}$ orthogonal to $\underset{\sim}{m}$. For consistency with previous accounts of surface-wave propagation in inextensible elastic bodies [1], [2], we denote by (ℓ,m,n) and (ℓ',m',n') the components of $\underset{\sim}{e}$ and $\underset{\sim}{e}'$ relative to the orthonormal basis $\{\underset{\sim}{\ell},\underset{\sim}{m},\underset{\sim}{n}\}$, with $\underset{\sim}{\ell} = \underset{\sim}{m}\wedge\underset{\sim}{n}$, and utilize the dimensionless quantities

$$\Theta = \mu_L/\mu_T, \quad \Lambda = \mu_L/(\lambda + 2\mu_T), \quad \gamma = \rho v^2/\mu_T ,$$

and their primed counterparts.[+] A harmonic interfacial wave in $B \cup B'$ gives rise to a displacement field having the same form in each sub-body as a harmonic surface wave. Thus

---

[+] We note in passing that our results hold in full when one or both of B and B' is composed of isotropic material, that is when $\Theta = 1$ or $\Theta' = 1$ or $\Theta = 1 = \Theta'$.

$$\left.\begin{aligned} \underset{\sim}{u} &= \exp\{i\varkappa(\underset{\sim}{n}.\underset{\sim}{x} - vt)\}\Sigma\, c_\alpha \exp(i\varkappa p_\alpha \underset{\sim}{m}.\underset{\sim}{x})\underset{\sim}{a}_\alpha \quad \text{in} \quad B\,, \\ \underset{\sim}{u}' &= \exp\{i\varkappa(\underset{\sim}{n}.\underset{\sim}{x} - vt)\}\Sigma c'_\alpha \exp(i\varkappa p'_\alpha \underset{\sim}{m}.\underset{\sim}{x})\underset{\sim}{a}'_\alpha \quad \text{in} \quad B'\,, \end{aligned}\right\} \tag{1}$$

where the wave number $\varkappa$ and the speed of propagation v are real and positive and the summation is over $\alpha = 1,2$. The complex coefficients c_α and c'_α are ultimately determined by conditions at I: in common with the complex numbers p_α and p'_α and the complex vectors $\underset{\sim}{a}$ and $\underset{\sim}{a}'$ they depend on v. From [1,§§2,3] we have

$$p_\alpha = s_\alpha - iq_\alpha, \quad \alpha = 1,2, \tag{2}$$

with

$$s_1 = (1 - \Theta)M_1^{-1}mn, \qquad s_2 = (1 - \Lambda)M_2^{-1}mn, \tag{3}$$

$$q_1 = M_1^{-1}\{\ell^2 + \Theta(1 - \ell^2) - M_1\gamma\}^{\frac{1}{2}}, \quad q_2 = M_2^{-1}\{\ell^2 + \Lambda(1 - \ell^2) - \Theta^{-1}\Lambda M_2\gamma\}^{\frac{1}{2}}, \tag{4}$$

$$M_1 = 1 - m^2 + \Theta m^2, \qquad M_2 = 1 - m^2 + \Lambda m^2, \tag{5}$$

and

$$\left.\begin{aligned} \underset{\sim}{a}_1 &= (np_1 - m)\underset{\sim}{\ell} + \ell\underset{\sim}{m} - \ell p_1\underset{\sim}{n}, \\ \underset{\sim}{a}_2 &= -\ell(mp_2 + n)\underset{\sim}{\ell} + \{(1 - m^2)p_2 - mn\}\underset{\sim}{m} + (1 - n^2 - mnp_2)\underset{\sim}{n}. \end{aligned}\right\} \tag{6}$$

The definitions (3) to (6) and subsequent results relating solely to B have direct analogues for B', obtained by interchanging unprimed and primed quantities. However, in order that the displacement shall tend to zero with increasing distance from I, $\operatorname{Im} p_\alpha < 0$ and $\operatorname{Im} p'_\alpha > 0$. The counterpart of (2) is thus

$$p'_\alpha = s'_\alpha + iq'_\alpha\,, \quad \alpha = 1,2, \tag{7}$$

and corresponding changes of sign must be made when transferring between B and B'.

The traction produced by the interfacial wave (1) on the planes parallel to I is

$$\left.\begin{aligned} \underset{\sim}{t} &= i\varkappa\exp\{i\varkappa(\underset{\sim}{n}.\underset{\sim}{x} - vt)\}\Sigma^+ c_\alpha \exp(i\varkappa p_\alpha \underset{\sim}{m}.\underset{\sim}{x})\underset{\sim}{t}_\alpha \text{ in } B, \\ \underset{\sim}{t}' &= i\varkappa\exp\{i\varkappa(\underset{\sim}{n}.\underset{\sim}{x} - vt)\}\Sigma^+ c'_\alpha \exp(i\varkappa p'_\alpha \underset{\sim}{m}.\underset{\sim}{x})\underset{\sim}{t}'_\alpha \text{ in } B'. \end{aligned}\right\} \tag{8}$$

Here Σ^+ indicates summation from $\alpha = 1$ to $\alpha = 3$ and

$$p_3 = - n/m,$$

$$\left.\begin{aligned} \mu_T^{-1}\underset{\sim}{t}_1 &= - \{\Theta(mp_1 + n) - n\gamma\}\underset{\sim}{\ell} + 2\ell p_1\underset{\sim}{m} + \ell(2 - \gamma)\underset{\sim}{n} \\ &\qquad - 2\ell(1 - \Theta)(mp_1 + n)\underset{\sim}{e}, \\ \mu_T^{-1}\underset{\sim}{t}_2 &= - \ell(2 - \Theta)(np_2 - m)\underset{\sim}{\ell} - \{2\ell^2 + \Theta(1 - \ell^2) - \gamma\}\underset{\sim}{m} \\ &\qquad + \{2\ell^2 + \Theta(1 - \ell^2)\}p_2\underset{\sim}{n} - m\gamma\underset{\sim}{e}\,. \end{aligned}\right\} \tag{9}$$

Without loss of generality we can set

$$\underset{\sim}{t}_3 = \underset{\sim}{e}, \quad \underset{\sim}{t}'_3 = \underset{\sim}{e}' \tag{10}$$

[1,§§2,3].

The continuity at I of the displacement and traction leads, via (1) and (8), to the conditions

$$\Sigma c_\alpha \underset{\sim}{a}_\alpha = \Sigma c'_\alpha \underset{\sim}{a}'_\alpha, \quad \Sigma^+ c_\alpha \underset{\sim}{t}_\alpha = \Sigma^+ c'_\alpha \underset{\sim}{t}'_\alpha \ . \tag{11}$$

For $\alpha = 1,2$, $\underset{\sim}{a}_\alpha$ and $\underset{\sim}{a}'_\alpha$ are orthogonal to $\underset{\sim}{e}$ and $\underset{\sim}{e}'$ respectively [1,§3], whence, from $(11)_1$,

$$\Sigma c_\alpha \underset{\sim}{a}_\alpha . \underset{\sim}{e}' = 0, \quad \Sigma c'_\alpha \underset{\sim}{a}'_\alpha . \underset{\sim}{e} = 0 \ . \tag{12}$$

Forming scalar products in (11) with $\underset{\sim}{E}$ and appealing to (10) yields the further relations

$$\Sigma c_\alpha a_\alpha = - \Sigma c'_\alpha a'_\alpha, \quad \Sigma c_\alpha t_\alpha = - \Sigma c'_\alpha t'_\alpha \ , \tag{13}$$

in which

$$a_\alpha = \underset{\sim}{a}_\alpha . \underset{\sim}{E}, \quad a'_\alpha = - \underset{\sim}{a}'_\alpha . \underset{\sim}{E}, \quad t_\alpha = \underset{\sim}{t}_\alpha . \underset{\sim}{E}, \quad t'_\alpha = - \underset{\sim}{t}'_\alpha . \underset{\sim}{E}, \quad \alpha = 1,2. \tag{14}$$

Equations (12) and (13) are satisfied by values of c_α and c'_α not all zero if and only if the 4×4 determinant of coefficients is zero. This requirement can be expressed as

$$A'T - AT' = 0 \ , \tag{15}$$

where

$$\left. \begin{aligned} A &= (a_1 \underset{\sim}{a}_2 - a_2 \underset{\sim}{a}_1) . \underset{\sim}{e}', \quad A' = (a'_2 \underset{\sim}{a}'_1 - a'_1 \underset{\sim}{a}'_2) . \underset{\sim}{e} \ , \\ T &= (t_1 \underset{\sim}{a}_2 - t_2 \underset{\sim}{a}_1) . \underset{\sim}{e}', \quad T' = (t'_2 \underset{\sim}{a}'_1 - t'_1 \underset{\sim}{a}'_2) . \underset{\sim}{e} \ . \end{aligned} \right\} \tag{16}$$

The <u>secular equation</u> (15) determines, in principle, the speed of propagation v of the interfacial wave.

On calculating A and T from equations $(16)_{1,3}$ and (14), with the use of (6) and (9), we find that

$$A = - \{(1 - m^2)p_1 p_2 - mn(p_1 + p_2) + 1 - n^2\}\sin^2\varphi, \tag{17}$$

$$\begin{aligned} \mu_T^{-1} T = &- \Theta(MN + QR)p_1 p_2 + \{\Theta(M^2 - R^2) - M^2\gamma\}p_1 \\ &+ \{\Theta(Q^2 - N^2) - Q^2\gamma\}p_2 + (\Theta - \gamma)(MN + QR), \end{aligned} \tag{18}$$

where

$$\left. \begin{aligned} M &= \underset{\sim}{m} . \underset{\sim}{E} = n\ell' - \ell n', \quad N = \underset{\sim}{n} . \underset{\sim}{E} = \ell m' - m\ell', \\ Q &= [\underset{\sim}{m}, \underset{\sim}{e}, \underset{\sim}{E}] = m\cos\varphi - m', \quad R = [\underset{\sim}{n}, \underset{\sim}{e}, \underset{\sim}{E}] = n\cos\varphi - n', \end{aligned} \right\} \tag{19}$$

and $\varphi = \cos^{-1}(\underset{\sim}{e} . \underset{\sim}{e}')$ is the angle between the fibre directions.

It is evident from (2), (7), (17) and (18) that the secular equation

(15) is complex. If its real and imaginary parts are equivalent an interfacial wave of the form (1) exists when the real equation to which (15) is thereby reduced determines a positive real value of v in the subsonic interval $S: 0 < v < \hat{v}$, $\hat{v}$ being the least value of v for which q_α and q'_α are all real and positive. The considerations relating to the slowness surface of an inextensible, transversely isotropic elastic material set out in [1,§2] imply that

$$\hat{v} = \min([\{\mu_T \ell^2 + \mu_L(1 - \ell^2)\}/\rho M_1]^{\frac{1}{2}}, \ [\{\mu'_T \ell'^2 + \mu'_L(1 - \ell'^2)\}/\rho' M'_1]^{\frac{1}{2}}).$$

If it is not possible to put the secular equation into real form by removing a complex factor (as in [1,§3]), v satisfies simultaneously the independent conditions provided by the real and imaginary parts of (15) and only in highly exceptional circumstances could an interfacial wave then exist. We show in the next section that when mn and $m'n'$ are not both zero equation (15) has no real equivalent, and in the final section that, while (15) is real when $mn = 0 = m'n'$, the displacement field associated, through (1), with any real zero in S is identically zero. Except, perhaps, under artificial conditions, requiring special combinations of $\underset{\sim}{e}$, $\underset{\sim}{e}'$ and the material constants, it is therefore impossible for an interfacial wave to propagate in the composite $B \cup B'$ when the fibre directions are distinct.

Non-existence of interfacial waves when $|mn| + |m'n'| \neq 0$

With the aid of equations (2) to (5) and the relations

$$p_\alpha^2 = 2s_\alpha p_\alpha - s_\alpha^2 - q_\alpha^2, \quad \alpha = 1,2, \tag{20}$$

supplied by (2), the expression (17) for A can be written as the product of the real and complex factors

$$A_R = (\Theta - \Lambda)^{-1}(\Lambda M_1 q_1 + \Theta M_2 q_2)\sin^2\varphi \quad \text{and} \quad A_C = i(p_1 - p_2). \tag{21}$$

In deducing from (21) the corresponding representation $A'_R A'_C$ of A' the factor -1 must be inserted in A'_R as explained above in connection with (2) and (7). The secular equation (15) now becomes

$$A'_R A'_C T - A_R A_C T' = 0 \, . \tag{22}$$

Bearing in mind the definitions (4) and (5), the fact that (ℓ,m,n) are the components of a <u>unit</u> vector and the inequalities

$$0 < \Lambda < \Theta \tag{23}$$

[1,§2], we see from $(21)_1$ that A_R is positive and A'_R negative for all $v \in S$. The left side of (22) can therefore have a real factor with at least one real zero in S only if T and T' have the decompositions $T_R T_C$ and $T'_R T'_C$ into real and complex factors and, moreover, $A'_C T_C$ and $A_C T'_C$ are real multiples of one another. Since a real multiplier can be absorbed into T_R or T'_R, the latter requirement can be simplified to $A'_C T_C = A_C T'_C$, and it follows that, for some τ, $T = \tau T_R A_C$, $T' = \tau T'_R A'_C$.

It is clear, however, from (18) and (18)',[+] that there is no function of v which is a common factor of T and T'. We can accordingly set $\tau = 1$ and state

$$T = T_R A_C, \quad T' = T'_R A'_C \tag{24}$$

as necessary conditions for the real and imaginary parts of the secular equation (15) to be equivalent, the reduced real form then being

$$A'_R T_R - A_R T'_R = 0 .$$

If T is to have the representation $(24)_1$, equations $(21)_2$, (18) and (2) indicate that $\mu_T^{-1} T_R$ must be a polynomial in q_1 and q_2. On account of (20),

$$i(p_1 - p_2)(aq_1 + bq_2) = (a - b)p_1p_2 - (as_1 - bs_2)(p_1 + p_2)$$
$$+ a(s_1^2 + q_1^2) - b(s_2^2 + q_2^2) , \tag{25}$$

and the right sides of (18) and (25) agree if

$$\left.\begin{aligned} a - b &= - \theta(MN + QR) , \\ as_1 - bs_2 &= - \theta(M^2 - R^2) + M^2\gamma = - \theta(Q^2 - N^2) + Q^2\gamma , \\ a(s_1^2 + q_1^2) - b(s_2^2 + q_2^2) &= (\theta - \gamma)(MN + QR) . \end{aligned}\right\} \tag{26}$$

Since s_1 and s_2 are independent of γ (see (3)), equations $(26)_{2,3}$ show that a and b must be real linear functions of the form

$$a = c + e\gamma, \quad b = d + e\gamma. \tag{27}$$

The requirement that equations (26) hold identically in γ then yields relations between the constants c, d and e which are easily proved to be inconsistent. Equating coefficients of γ^2 in $(26)_4$ gives

$$0 = - e(M_1^{-1} - \theta^{-1}\Lambda M_2^{-1}) = - e(\theta M_1 M_2)^{-1}(\theta - \Lambda)(1 - m^2),$$

use being made of (27), (4) and (5). In view of the inequalities (23) we infer that either $e = 0$ or $m^2 = 1$. If $e = 0$ the coefficients of γ in $(26)_{2,3}$ agree only if $M = 0 = Q$, and $(26)_1$ is then compatible with the result of equating coefficients of γ in $(26)_4$ only if $m^2 = 1$. If $m^2 = 1$ we have $\ell = 0 = n$, $\underset{\sim}{e} = \pm \underset{\sim}{m}$ and, from $(19)_{1,3}$ and (3), M, Q, s_1 and s_2 are all zero. The equality of the constant terms in $(26)_{2,3}$ thereupon gives $N = 0 = R$, whence, from $(19)_{2,4}$, $\ell' = 0 = n'$, $\underset{\sim}{e}' = \pm \underset{\sim}{m}$. But this means that $\underset{\sim}{E} = \underset{\sim}{0}$ and we reach a contradiction.

In order to conclude from this reasoning that when p_α and p'_α are complex there is no T_R satisfying $(24)_1$, and hence no real equivalent of the secular equation (15), we have to confirm that the real factor $aq_1 + bq_2$ in (25) is of the most general form possible. Were this factor to contain a real multiple of q_1q_2 or a real polynomial in γ, the

---

[+]For economy we cite as (*)' the result of interchanging unprimed and primed quantities in the equation numbered (*).

polynomial in p_1 and p_2 on the right of (25) would have complex coefficients, contrary to (18). By the same token any term of even degree in q_1 and q_2 is inadmissible. Because of (4) the inclusion of terms of odd degree ≥ 3 in q_1 and q_2 would be tantamount to allowing a and b to be of degree ≥ 2 in γ. But then, from $(26)_1$, the coefficients in a and $\bar{b}$ of the highest power of γ would have to be equal and the absence of this power from the right side of $(26)_4$ would force m^2 to be unity, leading to inconsistency as above. The argument is now complete.

Special cases

When $|mn| + |m'n'| = 0$ either (a) $m = 0 = m'$, (b) $n = 0 = n'$, or (c) $m = 0 = n'$, the possibility $n = 0 = m'$ being equivalent to (c) due to the interchangeability of the sub-bodies B and B'. We consider the cases (a), (b) and (c) in turn, observing from (2), (3) and (7) that $p_\alpha = -iq_\alpha$, $p'_\alpha = iq'_\alpha$ in each instance.

$m = 0 = m'$. When the fibre directions are parallel to I the degenerate waves responsible for the $\alpha = 3$ terms in (8) are absent and a tangential discontinuity in traction appears on I (cf. [1,§§5(b), 6]). However, $(13)_2$, the only condition on the traction entering into the derivation of the secular equation, remains valid since $\underset{\sim}{E} = \pm \sin\varphi \underset{\sim}{m}$ and, from $(14)_{3,4}$, t_α and t'_α are accordingly normal components. We are therefore justified in making the substitutions $m = 0 = m'$ in (15) and (4).

With the help of (17) to (19) and noting that $M^2 = \sin^2\varphi$, $R^2 = \ell^2\sin^2\varphi$, we find that

$$\mu_T q_1(\Theta n^2 - \gamma)(q'_1 q'_2 - \ell'^2) + \mu'_T q'_1(\Theta' n'^2 - \gamma')(q_1 q_2 - \ell^2) = 0 \tag{28}$$

and

$$q_1 = (\ell^2 + \Theta n^2 - \gamma)^{\frac{1}{2}}, \quad q_2 = \{\ell^2 + \Theta^{-1}\Lambda(\Theta n^2 - \gamma)\}^{\frac{1}{2}}. \tag{29}$$

There follow from (29) the identities

$$\left.\begin{aligned} q_1 q_2 - \ell^2 &= (\Theta - \Lambda)^{-1}(q_1 - q_2)(\Lambda q_1 + \Theta q_2), \\ \Theta n^2 - \gamma &= \Theta(\Theta - \Lambda)^{-1}(q_1 - q_2)(q_1 + q_2) , \end{aligned}\right\} \tag{30}$$

which, with (30)', enable us to rewrite equation (28) as

$$(q_1 - q_2)(q'_1 - q'_2)\{\mu_T\Theta q_1(q_1 + q_2)(\Lambda' q'_1 + \Theta' q'_2)$$

$$+ \mu'_T\Theta' q'_1(q'_1 + q'_2)(\Lambda q_1 + \Theta q_2)\} = 0 . \tag{31}$$

The terms in curly brackets in (31) are clearly positive for all $v \in S$, so the secular equation has precisely two roots in this interval, given by $q_1 = q_2 \Rightarrow \gamma = \Theta n^2$ and $q'_1 = q'_2 \Rightarrow \gamma' = \Theta' n'^2$. We now show that, for each of these roots, the displacement vanishes identically in $B \cup B'$.

In the present special case we obtain from (6)

$$c_1\underset{\sim}{a}_1 + c_2\underset{\sim}{a}_2 = (iq_1c_1 + \ell c_2)(-n\underset{\sim}{\ell} + \ell\underset{\sim}{n}) + (\ell c_1 - iq_2c_2)\underset{\sim}{m}, \tag{32}$$

and the displacement conditions (12) and $(13)_1$ take the form

$$iq_1c_1 + \ell c_2 = 0, \quad iq_1'c_1' - \ell' c_2' = 0, \quad \ell c_1 - iq_2c_2 = \ell' c_1' + iq_2'c_2'. \tag{33}$$

When $q_1 = q_2$ the identities $(30)_1$ and

$$iq_2(iq_1c_1 + \ell c_2) + \ell(\ell c_1 - iq_2c_2) = -(q_1q_2 - \ell^2)c_1,$$

together with $(33)_1$, imply that either $\ell = 0$ ($\Rightarrow q_1 = q_2 = 0$) or $\ell c_1 - iq_2c_2 = 0$. For each alternative the right side of (32) is zero and, since $p_1 = p_2$, we see from $(1)_1$ that $\underset{\sim}{u} = \underset{\sim}{0}$ in B. If $q_1' \neq q_2'$, $(33)_2$ and $(33)_3$, now reduced to $\ell' c_1' + iq_2'c_2' = 0$, admit only the zero solution $c_1' = 0 = c_2'$, whence, from $(1)_2$, $\underset{\sim}{u}' = \underset{\sim}{0}$ in B'. If $q_1' = q_2'$ we arrive at the same conclusion from (32)' and $(33)_3$.

A parallel calculation shows that $\underset{\sim}{u} = \underset{\sim}{0} = \underset{\sim}{u}'$ when $q_1' = q_2'$.

<u>$n = 0 = n'$.</u> When the fibre directions are orthogonal to the wave normal, $\underset{\sim}{E} = \pm \sin\varphi\underset{\sim}{n}$ and the appropriate specializations of equations (15) and (24), again secured with the aid of (17) to (19), are

$$\mu_T q_2(\Theta m^2 + \ell^2\gamma)(\ell'^2 q_1'q_2' - 1) + \mu_T' q_2'(\Theta' m'^2 + \ell'^2\gamma')(\ell^2 q_1q_2 - 1) = 0 \tag{34}$$

and

$$q_1 = (\ell^2 + \Theta m^2)^{-\frac{1}{2}}(1 - \gamma)^{\frac{1}{2}}, \qquad q_2 = (\ell^2 + \Lambda m^2)^{-\frac{1}{2}}(1 - \Theta^{-1}\Lambda\gamma)^{\frac{1}{2}}. \tag{35}$$

The relation

$$\ell^2 q_1q_2 - 1 = -\{\Theta M_1M_2(q_1 + q_2)\}^{-1}(\Theta m^2 + \ell^2\gamma)(\Lambda M_1q_1 + \Theta M_2q_2), \tag{36}$$

furnished by (35), and its primed counterpart (36)' reveal that the left side of (34) is positive for all $v \in S$, thus prohibiting the existence of an interfacial wave.

<u>$m = 0 = n'$.</u> In this case $M = \ell' n$, $N = \ell m'$, $Q = -m'$, $R = \ell\ell' n$, and the secular equation is derived straightforwardly from (15), (17) and (18). With the application of the identities (30) and (36)' it is brought into the form

$$(q_1 - q_2)\{\mu_T\Theta(q_1 + q_2)(\ell'^2n^2q_1 + m'^2q_2)(\Lambda' M_1'q_1' + \Theta' M_2'q_2')$$

$$+ \mu_T'\Theta' M_1'M_2'(q_1' + q_2')(\Lambda q_1 + \Theta q_2)(n^2q_1' + \ell^2m'^2)q_2'\} = 0, \tag{37}$$

q_α being defined by (29) and q_α' by (35)'. The terms in curly brackets in (37) are plainly positive for all $v \in S$, so the only root is given by $q_1 = q_2$. As in the first special case the displacement then turns out to be identically zero.

Equation (32) continues to hold, but in place of (32)' we have, from

(6)',

$$c_1'\mathbf{a}_1' + c_2'\mathbf{a}_2' = (c_1' + i\ell' q_2' c_2')(-\,m'\boldsymbol{\ell} + \ell'\mathbf{m}) - (i\ell' q_1' c_1' - c_2')\mathbf{n}\ .$$

The displacement conditions (12) and $(13)_1$ thus read

$$\left.\begin{aligned}
&(\ell m' - i\ell' n q_1)c_1 - (\ell\ell' n + i m' q_2)c_2 = 0,\\
&(\ell m' + i\ell' n q_1')\, c_1' - (n - i\ell\ell' m' q_2')c_2' = 0,\\
&(\ell\ell' n + i m' q_1)c_1 + (\ell m' - i\ell' n q_2)c_2\\
&\qquad = (n - i\ell\ell' m' q_1')c_1' + (\ell m' + i\ell' n q_2')c_2'.
\end{aligned}\right\} \qquad (38)$$

When $q_1 = q_2$, $(30)_1$ gives $q_1 = q_2 = |\ell|$, and if $\ell \neq 0$ $(38)_1$ becomes $i|\ell|c_1 + \ell c_2 = 0$. It then follows from (32) and $(38)_3$ that

$$c_1\mathbf{a}_1 + c_2\mathbf{a}_2 = \mathbf{0} \qquad (39)$$

and

$$(n - i\ell\ell' m' q_1')c_1' + (\ell m' + i\ell' n q_2')c_2' = 0. \qquad (40)$$

Equations (39) and (40) also apply if $\ell = 0$ as $q_1 = q_2 = 0$ and the right side of (32) and the left side of $(38)_3$ vanish. Since $p_1 = p_2$ we deduce from (39) and $(1)_1$ that $\mathbf{u} = \mathbf{0}$ in B. The determinant of the coefficients of c_α' in $(38)_2$ and (40) is $-(\ell^2 m'^2 + n^2)(\ell'^2 q_1' q_2' - 1)$, which is non-zero by (36)'. Hence $c_1' = 0 = c_2'$ and, from $(1)_2$, $\mathbf{u}' = \mathbf{0}$ in B'.

We remark in conclusion that interfacial-wave propagation in $B \cup B'$ is no longer completely excluded by the constraint of inextensibility when the fibre directions in B and B' coincide. Details of the behaviour of interfacial waves in this situation will be published elsewhere.

References

1. V. S. Captain and P. Chadwick. Surface waves in an inextensible, transversely isotropic elastic body. Quart. J. Mech. Appl. Math. 39 (1986) 327 - 342.

2. A. M. Whitworth and P. Chadwick. The effect of inextensibility on elastic surface waves. Wave Motion 6 (1984) 289 - 302.

APPROXIMATE PHASE VELOCITY FOR BENDING WAVES IN ELASTIC PLATES REINFORCED BY TWO FAMILIES OF STRONG FIBRES

Dragan Milosavljević

Faculty of Mechanical Engineering ,University "Svetozar Marković", Kragujevac, Yugoslavia

ABSTRACT

Exact solutions are obtained for bending wave propagation in an elastic plate reinforced by two families of inextensible fibres. When the fibres are almost inextensible approximate solutions are derived using a singular perturbation technique for which the first term in the outer expansion is the solution of the inextensible problem. The perturbation scheme involves both a geometrical and a material small parameter and the inner expansion is obtained in terms of the ratio of these two quantities. Numerical results show that the matched composite solution gives excellent agreement with exact solutions over the entire range of wavelengths.

INTRODUCTION AND GOVERNING EQUATIONS

In this paper we shall consider wave propagation in an infinite plate reinforced by two families of straight mechanically equivalent fibres wich lie in the planes parallel to stress-free boundaries. The kind of composite material that we have in minds is one in which a matrix material is reinforced by strong stiff fibres which are systematically arranged in the matrix. Thus we condiser a laminated plate built up from a large number of thin unidirectionally reinforced laminae stacked alternately with the fibres aligned in two diferent directions. On the macroscopic scale such laminate will have two preferred directions, and so will have orthoropic symmetry. We suppose that the only anisotropic properties of the material are those which are due to the presence of the fibres which will be characterized by unit vectors $\underset{\sim}{a}$ and $\underset{\sim}{b}$. All vector and tensor components will be referred to a system of rectangular cartesian components x_i (i=1,2, 3). Components of the infinitesimal displacement vector $\underset{\sim}{u}$ are denoted by u_i, and components of infinitesimal strain tensor $\underset{\sim}{e}$ by e_{ij} so that $e_{ij} = (u_{i,j} + u_{j,i})/2$ where commas denote partial derivatives. The Cauchy stress tensor $\underset{\sim}{\sigma}$ has components σ_{ij}, and the fibre direction vectors $\underset{\sim}{a}$

and $\underset{\sim}{b}$ have components, a_i and b_i, respectively. The constitutive equation, derived by Spencer (1981 a,b), is then

$$\sigma_{ij} =\{\lambda e_{rr}+\gamma_3(a_r a_s e_{rs}+b_r b_s e_{rs})+\gamma_4 a_r b_s e_{rs}\cos 2\phi\}\delta_{ij}+2\mu e_{ij}$$
$$+\{\gamma_3 e_{rr}+2\gamma_1 a_r a_s e_{rs}+\gamma_6 b_r b_s e_{rs}+\gamma_5 a_r b_s e_{rs}\cos 2\phi\}a_i a_j$$
$$+\{\gamma_3 e_{rr}+\gamma_6 a_r a_s e_{rs}+2\gamma_1 b_r b_s e_{rs}+\gamma_5 a_r b_s e_{rs}\cos 2\phi\}b_i b_j \qquad (1)$$
$$+\frac{1}{2}\{\gamma_4 e_{rr}\cos 2\phi+\gamma_5(a_r a_s e_{rs}+b_r b_s e_{rs})\cos 2\phi+2\gamma_2 a_r b_s e_{rs}\}(a_i b_j+a_j b_i)$$
$$+\gamma_7\{a_r(e_{ri}a_j+e_{rj}a_i)+b_r(e_{ri}b_j+e_{rj}b_i)\}\ ,$$

where $\lambda,\mu,\gamma_1,\ldots,\gamma_7$ are even functions of $\cos 2\phi$ and 2ϕ is the angle between two families of the fibres. Indices take the velues 1, 2 and 3 and summation convention is employed. If we introduce the assumption of inextensibility in the fibre directions $\underset{\sim}{a}$ and $\underset{\sim}{b}$ then material has two kinematic constraints

$$a_i a_j e_{ij} = 0 \quad \text{and} \quad b_i b_j e_{ij} = 0, \qquad (2)$$

which give rise to reaction stresses T_a and T_b whose cartesian components are $T_a a_i a_j$ and $T_b b_i b_j$, respectively. The constitutive equation is then given by

$$\sigma_{ij}=(\lambda e_{rr}+\gamma_4 a_r b_s e_{rs}\cos 2\phi)\delta_{ij}+2\mu e_{ij}+(\tfrac{1}{2}\gamma_4 e_{rr}\cos 2\phi+\gamma_2 a_r b_s e_{rs})(a_i b_j+a_j b_i)$$
$$+\gamma_7\{a_r(e_{ri}a_j+e_{rj}a_i)+b_r(e_{ri}b_j+e_{rj}b_i)\}+T_a a_i a_j+T_b b_i b_j\ . \qquad (3)$$

Choosing cartesian coordinates Ox_1 perpendicular to the plate, Ox_2 and Ox_3 in the midle plane of the plate along bisectors of two families of the fibres, the boundary surfaces of the plate are given as $x_1=\pm h$, where $2h$ is the plate thickness. The equations of motion with no body forces are

$$\sigma_{ij,j} = \rho\ddot{u}_i,\ (i,j=1,2,3), \qquad (4)$$

where ρ is density of the material and dots denote diferentiation with respect to time. Equations (4) are to be solved subject to traction-free boundary conditions expressed as

$$\sigma_{1i}=0, \quad (i=1,2,3), \quad \text{at} \quad x_1=\pm h. \qquad (5)$$

INEXTENSIBLE FIBRES

Suppose that a plane wave propagates with a phase velocity v in the direction which makes an angle α with x_3 - axis. Let us assume the displacements and reaction stresses in the form

$$u_1 = U(x_1)\cos\psi,\; u_2 = V(x_1)\sin\psi,\; u_3 = W(x_1)\sin\psi, \tag{6}$$

and

$$T_a = T_A(x_1)\cos\psi,\quad T_b = T_B(x_1)\cos\psi\;, \tag{7}$$

respectively, where $\psi = k(sx_2+cx_3-vt)$, $s\equiv\sin\alpha$, $c\equiv\cos\alpha$ and k is a wave number. The wave number k is related to the wavelength Λ by the expression $k=2\pi/\Lambda$ and the limits $kh\to\infty$ and $kh\to 0$ correspond to waves of vanishingly small and infinitely large wavelengths,respectively. Substituting (6) and (7) into (3) and then into (4) the equations of motion become

$$\begin{aligned}
&a_{11}U''+ks(a_{12}+a_{66})V'+kc(a_{13}+a_{55})W'+k^2[\rho v^2-(s^2a_{66}+c^2a_{55})]U = 0,\\
&a_{66}V''-ks(a_{12}+a_{66})U'+k^2[\rho v^2-(s^2a_{22}+c^2a_{44})]V-k^2sc(a_{23}+a_{44})W\\
&\qquad -k\sin\phi\{(s\sin\phi+c\cos\phi)T_A+(s\sin\phi-c\cos\phi)T_B\}= 0,\\
&a_{55}W''-kc(a_{13}+a_{55})U'-k^2sc(a_{23}+a_{44})V+k^2[\rho v^2-(s^2a_{44}+c^2a_{33})]W\\
&\qquad -k\cos\phi\{(s\sin\phi+c\cos\phi)T_A-(s\sin\phi-c\cos\phi)T_B\}= 0,
\end{aligned} \tag{8}$$

where ϕ is angle between one family of the fibres and x_3-axis and material constants a_{ij} are given by

$$\begin{aligned}
&a_{11}=\lambda+2\mu,\; a_{12}=\lambda-\gamma_4\cos2\phi\sin^2\phi,\; a_{13}=\lambda+\gamma_4\cos2\phi\cos^2\phi,\\
&a_{22}=\lambda+2\mu-2\gamma_4\cos2\phi\sin^2\phi+2\gamma_2\sin^4\phi+4\gamma_7\sin^2\phi,\\
&a_{23}=\lambda+\gamma_4\cos^2 2\phi-\tfrac{1}{2}\gamma_2\sin^2 2\phi,\\
&a_{33}=\lambda+2\mu+2\gamma_4\cos2\phi\cos^2\phi+2\gamma_2\cos^4\phi+4\gamma_7\cos^2\phi,\\
&a_{44}=\mu+\gamma_7,\; a_{55}=\mu+\gamma_7\cos^2\phi,\; a_{66}=\mu+\gamma_7\sin^2\phi.
\end{aligned} \tag{9}$$

The boundary conditions (5) can be expressed as

$$\begin{aligned}
&a_{11}U' + ksa_{12}V + kca_{13}W = 0,\\
&W' - kcU = 0,\quad V' - ksU = 0,\quad \text{at } x_1 = \pm h
\end{aligned} \tag{10}$$

and the constraint conditions (2) lead to

$$k\{sV\sin^2\phi \pm (cV + sW) + cW\cos^2\phi\} = 0. \tag{11}$$

In order to get non-travial solutions of equations of motion (8) subject to the boundary conditions (10) it is necessary to postulate existence of singular layers at the upper and lower surfaces of the plate as it has been done for plates reinforced by one family of the fibres by Green (1982, 1984) and Green, Milosavljević (1984,1985). In the singular layers the reaction stresses T_a and T_b become infinite, corresponding to finite loads in the upper and lower surfaces. Across the singular layers, the shear stresses σ_{12} and σ_{13} are discontinuous, being zero at $x_1=\pm h$ but having a non-zero limits as $x_1 \to \pm h$ from the interior of the plate. With above assumptions and constraint conditions (11), for asymmetric deformations due to bending, we get solutions

$$u_1 = A\cos\psi\ ,\ u_2 = u_3 = 0,\ \ \rho v^2 = c^2 a_{55} + s^2 a_{66}, \tag{12}$$

for arbitrary direction of propagation exept when direction of propagation is perpendicular to one of the families of the fibres, when we have kinematically admissiable deformations according the constraint conditions (11), that is $s^2\sin^2\phi - c^2\cos^2\phi = 0$. Kinematically admissiable deformations imply $V/W=-s/c$ and from $(8)_{2,3}$ we obtain

$$dW'' - kc(c^2b_2 - s^2b_3)U' - k^2(s^4a_{22} + c^4a_{33} - 2s^2c^2a_{23} - \rho v^2)W = 0, \tag{13}$$

where

$$d = c^2a_{55} + s^2a_{66},\quad b_2 = a_{13} + a_{55},\qquad b_3 = a_{12} + a_{66}. \tag{14}$$

Equations $(8)_1$ and (13) lead to general solutions

$$\begin{aligned} U &= A_1\cosh kp_1x_1 + A_2\cosh kp_2x_1, \\ W &= -\frac{ca_{11}}{c^2b_2 - s^2b_3}\{(p_1^2 - a_1)A_1\frac{\sinh kp_1x_1}{p_1} - (p_2^2 - a_1)A_2\frac{\sinh kp_2x_1}{p_2}\}\ , \end{aligned} \tag{15}$$

where $a_1 = (d-\rho v^2)/a_{11}$, A_1 and A_2 are arbitrary constants and $p_1^2 < p_2^2$ are solutions of equation

$$\begin{aligned} &dp^4 - \{da_1 + s^4a_{22} + c^4a_{33} - 2s^2c^2a_{23} - (c^2b_2 - s^2b_3)^2/a_{11} - \rho v^2\}p^2 \\ &\quad + a_1(s^4a_{22} + c^4a_{33} - 2s^2c^2a_{23} - \rho v^2) = 0. \end{aligned} \tag{16}$$

Again if we assume the existence of the singular fibres at the upper and lower boundary it is easy to prove that instead of $(10)_{2,3}$ it must be satisfied

$$(s^2a_{66} - c^2a_{55})kcU + dW' = 0,\quad \text{at } x_1 = \pm h. \tag{17}$$

The boundary conditions $(10)_1$ and (17) are satisfied by the solutions (15) provided

$$\begin{aligned} &\{dep_1^2p_2^2 + e[e(c^2b_2 - s^2b_3)/a_{11} - da_1]p_1^2 + da_1(c^2a_{13} - s^2a_{12})\,p_2^2 \\ &\qquad + a_1(c^2a_{13} - s^2a_{12})[e(c^2b_2 - s^2b_3)/a_{11} - da_1]\}\frac{\tanh khp_1}{p_1} \\ &- \{\,dep_1^2p_2^2 + e[e(c^2b_2 - s^2b_3)/a_{11} - da_1]p_2^2 + da_1(c^2a_{13} - s^2a_{12})p_1^2 \\ &\qquad + a_1(c^2a_{13} - s^2a_{12})[e(c^2b_2 - s^2b_3)/a_{11} - da_1]\}\frac{\tanh khp_2}{p_2} = 0, \end{aligned} \tag{18}$$

where $e = c^2a_{55}-s^2a_{66}$. The dispersion equation (18) leads in long wave limit, when $kh\to o$, to squared phase velocity

$$\rho v^2 = d-e^2/d = 4s^2c^2c_{55}c_{66}/d \ , \qquad (19)$$

and in short wave limit, when $kh\to\infty$, to Rayleigh wave speed. General consideration of surface waves in anisotropic elastic materials has been given by Chadwick and Smith (1977) and for problems considered here by Milosavljević (1986 a,b).

MATCHED COMPOSITE SOLUTIONS FOR STRONGLY ANISOTROPIC MATERIALS

Material that has two families of extensible but strong fibres we shall call strongly anisotropic material. A material with inextensible fibres can be regarded as limiting case of the strongly anisotropic material when $\gamma_1\to\infty$ and $\gamma_6\to\infty$ with $a_ia_je_{ij}\to 0$, $b_ib_je_{ij}\to 0$ in such a way that relations

$$\begin{aligned} 2\gamma_1a_ra_se_{rs}+\gamma_6b_rb_se_{rs} &\to T_a-\gamma_3e_{rr}-\gamma_5a_rb_se_{rs}\cos 2\phi \ , \\ \gamma_6a_ra_se_{rs}+2\gamma_1b_rb_se_{rs} &\to T_b-\gamma_3e_{rr}-\gamma_5a_rb_se_{rs}\cos 2\phi , \end{aligned} \qquad (20)$$

are satisfied. From equations (20) it is easy to conclude that constants γ_3 and γ_5 may be incorporated, in limiting process, into reaction stresses. Similarly γ_2 and γ_6 may be incorporated into large constant γ_1 and we can set $\gamma_2=\gamma_3=\gamma_5=\gamma_6=0$ without loosing of any important information in limiting process which transforms the strongly anisotropic material into the inextensible material.

When we introduce dimensionless coordinate $x=x_1/h$ the equations of motions (4) for extensible fibres, with displacements assumed in form (6), become

$$\begin{aligned} &c_{11}U'' -k^2h^2(d-\rho v^2)U+khsb_3V' + khcb_2W' = 0, \\ &-khsb_3U'+c_{66}V''-k^2h^2(g_2-\rho v^2)V-k^2h^2scb_1W = 0, \\ &-khcb_2U'-k^2h^2scb_1V+c_{55}W'' - k^2h^2(g_3-\rho v^2)\ W = 0, \end{aligned} \qquad (21)$$

where prime denote diferentiation with respect to x and

$$\begin{aligned} &c_{11}=a_{11}, \quad c_{12}=a_{12}, \quad c_{13}=a_{13}, \quad c_{22}=a_{22}+4\gamma_1\sin^4\phi, \\ &c_{23}=a_{23}+4\gamma_1\sin^2\phi\cos^2\phi, \quad c_{33}=a_{33}+4\gamma_1\cos^4\phi, \\ &c_{44}=a_{44}+4\gamma_1\sin^2\phi\cos^2\phi, \quad c_{55}=a_{55}, \quad c_{66}=a_{66}, \\ &g_2=s^2c_{22}+c^2c_{44}, \quad g_3=s^2c_{44}+c^2c_{33}, \quad b_1=c_{23}+c_{44}. \end{aligned} \qquad (22)$$

The boundary conditions (5) become

$$c_{11}U'+khsc_{12}V+khc\ c_{13}W = 0,$$
$$-\ khcU + W'=0,\ -khsU+V' = 0,\ \text{at } x = \pm 1. \tag{23}$$

To describe large constant γ_1 we shall introduce a small parameter $\varepsilon<<1$ in following way

$$4\gamma_1 = d/\varepsilon^2, \tag{24}$$

and from (22) we can write

$$g_2=\bar{g}_2/\varepsilon^2+\hat{g}_2,\ g_3=\bar{g}_3/\varepsilon^2+\hat{g}_3,\ \ b_1=\bar{b}_1/\varepsilon^2+\hat{b}_1\ , \tag{25}$$

where bar and hat quantities are O(1). For kh=O(1) equations of motion $(21)_{2,3}$ become singular becouse the higest derivatives of V and W are multiplied by ε^2. Therefore there is a boundary layer when $\varepsilon<<1$ and we have to develop a perturbation sheme based on inner and outer expansion and then to match these in overlap region to obtain matched composite solutions. In that case, however, we are restricted on kh=O(1) becouse when kh become comparable to ε diferential equations (21) become regular. To overcome this dificulty we can introduce another parameter m which represents ratio of geometrical parameter kh and material parameter ε that is $m=kh/\varepsilon$ and solutions, based on parameter ε, will involve m as a parameter representing inner solutions. Therefore if we write formal series inner solutions in powers of ε in the form

$$\{U_{in},\ V_{in},\ W_{in},\ v^2_{in}\} = \sum_{\beta=0}^{\infty}\{U_{2\beta},m\varepsilon V_{2\beta},\ m\varepsilon W_{2\beta},\ v^2_{2\beta}\}\varepsilon^{2\beta}, \tag{26}$$

and substitute them into equations (21) then these lead to an infinite system of equations and may be solved successively starting with the lowest order terms. Imposing the boundary conditions (23) on these solutions then yields the expression for inner speed of propagation in terms of m and ε. The lowest order solutions of displacements U_{in}, V_{in} and W_{in} and squared phase velocity v^2_{in} are

$$U_o=A_o,\ \ V_o = \frac{-sA_o}{\bar{p}_3^2-\bar{p}_2^2}\sum_{\alpha=2}^{3}(-1)^{\alpha}\left(\bar{p}_\alpha^2-\frac{\bar{g}_2+c^2\bar{b}_1}{c_{66}}\right)\frac{\sinh m\bar{p}_\alpha x}{m\bar{p}_\alpha\cosh m\bar{p}_\alpha},$$
$$W_o = \frac{sA_o}{\bar{p}_3^2-\bar{p}_2^2}\sum_{\alpha=2}^{3}(-1)^{\alpha}\frac{c_{66}}{c_{55}}\ \frac{c_{55}\bar{p}_\alpha^2-\bar{g}_3}{sc\bar{b}_1}\left(\bar{p}_\alpha^2-\frac{\bar{g}_2+c^2\bar{b}_1}{c_{66}}\right)\frac{\sinh m\bar{p}_\alpha x}{m\bar{p}_\alpha\cosh m\bar{p}_\alpha}\ , \tag{27}$$
$$\rho v_o^2 = \frac{-1}{\bar{p}_3^2-\bar{p}_2^2}\sum_{\alpha=2}^{3}(-1)^{\alpha}\left(d\bar{p}_\alpha^2-s^2\bar{g}_3\frac{c_{66}}{c_{55}}-c^2\bar{g}_2\frac{c_{55}}{c_{66}}+2s^2c^2\bar{b}_1\right)\left(1-\frac{\tanh m\bar{p}_\alpha}{m\bar{p}_\alpha}\right),$$

respectively. Here $\bar{p}_2^2 < \bar{p}_3^2$ are solutions of the equation

$$c_{55}c_{66}\bar{p}^4 - (c_{55}\bar{g}_2 + c_{66}\bar{g}_3)\bar{p}^2 + \bar{g}_2\bar{g}_3 - s^2c^2\bar{b}_1^2 = 0. \tag{28}$$

Specifically if direction of propagation is perpendicular to one of the families of the fibres, that is $s^2\sin^2\phi - c^2\cos^2\phi = 0$, then $\bar{g}_2\bar{g}_3 - s^2c^2\bar{b}_1^2 = 0$ and equation (28) has solutions

$$\bar{p}_2 = 0, \qquad \bar{p}_3^2 = \bar{g}_2/c_{66} + \bar{g}_3/c_{55} = 2d^2s^2c^2/c_{55}c_{66}. \tag{29}$$

Same procedure as that leading to (27) or limit of (27), when $\bar{p}_2 \to 0$, leads, at that particular angle, to displacements and squared phase velocity in following form

$$U_o = A_o,\ V_o = \frac{sA_o}{d}\left(-ex + 2c_{55}c^2 \frac{\sinh m\bar{p}_3 x}{m\bar{p}_3 \cosh m\bar{p}_3}\right),$$

$$W_o = \frac{cA_o}{d}\left(ex + 2c_{66}s^2 \frac{\sinh m\bar{p}_3 x}{m\bar{p}_3 \cosh m\bar{p}_3}\right), \tag{30}$$

$$\rho v_o^2 = 4s^2c^2 \frac{c_{55}c_{66}}{d}\left(1 - \frac{\tanh m\bar{p}_3}{m\bar{p}_3}\right).$$

The boundary layer, as we have seen, occurs at $kh = O(1)$ and to obtain outer solutions from the equations (21) and the boundary conditions (23) we assume displacements and phase velocity in the following form

$$\{U_{ou}, V_{ou}, W_{ou}, v_{ou}^2\} = \sum_{\beta=o}^{\infty} \{U_{2\beta}, V_{2\beta}, W_{2\beta}, v_{2\beta}^2\}\varepsilon^{2\beta}. \tag{31}$$

It is easy to show that for any direction of propagation, except when it is perpendicular to one of the families of the fibres, solutions are

$$U_{ou} = \bar{A},\ V_{ou} = W_{ou} = 0,\ \ \rho v_{ou}^2 = d. \tag{32}$$

If direction of propagation is perpendicular to one of the families of the fibres then the lowest order terms in equations $(21)_{2,3}$ allow nontrivial solutions for displacements V_o and W_o in form

$$W_o = -\bar{g}_2 V_o/(sc\bar{b}_1) = -cV_o/s, \tag{33}$$

where, for asymmetric deformations, V_o and, therefore, W_o can be any odd function of x. It is obvious that the lowes order terms of the boundary conditions $(23)_{2,3}$ cannot be imposed on (33), becouse the lowest order terms of $(21)_{2,3}$ were not diferential but algebriac equations, and, therefore, have to be satisfied by matching with inner solutions. Using $(21)_1$ and $(23)_1$ we obtain

$$cc_{11}U_o'' - ck^2h^2(d-\rho v_o^2)U_o + kh(c^2b_2-s^2b_3)W_o' = 0, \tag{34}$$

and

$$cc_{11}U_o' + kh(c^2c_{13}-s^2c_{12})\,W_o = 0, \text{ at } x = \pm 1, \tag{35}$$

respectively. By considering next order terms in the equations of motion we come to concludion that $(21)_2$ and $(21)_3$ are consistent if and only if following equation

$$-khc(c^2b_2-s^2b_3)U_o' + dW_o'' - k^2h^2(c^2\hat{g}_3+s^2\hat{g}_2-2s^2c^2\hat{b}_1-\rho v_o^2)V_o = 0, \tag{36}$$

is satisfied. The boundary conditions $(23)_{2,3}$ have to be replaced by matching conditions

$$-khc\,U_o + W_o' = -khc\,U_m + W_m',\ -khsU_o + V_o' = -khsU_m + V_m',\ \text{at } x = \pm 1, \tag{37}$$

where V_m and W_m can be obtained as outer limit of inner solutions (or inner limit of outer solutions) as follows

$$V_m = \lim_{\substack{\varepsilon\to 0\\ m\to\infty\\ m\varepsilon\to kh}} V_{in} = -skhA_o x\,\frac{e}{d}\ ,\ W_m = \lim_{\substack{\varepsilon\to 0\\ m\to\infty\\ m\varepsilon\to kh}} W_{in} = ckhA_o x\,\frac{e}{d}\ . \tag{38}$$

Substitution of (38) into (37) and elimination of arbitrary constant A_o leads to condition

$$-khceU_o + dW_o' = 0, \qquad \text{at } x = \pm 1. \tag{39}$$

Thus the lowest order outer solutions of discplacements can be obtained from diferential equations (34) and (36) subject to boundary conditions (35) and (39) and that is equivalent to equations $(8)_1$, (13), (10), and (17), respectively, that we had in theory with inextensible fibres. Therefore inextensible solutions are the lowest order outer solutions of the extensible theory. If v_{ou} is phase velocity obtained in inextensible theory, then matched approximate solution of phase velocity is represented by

$$\rho v^2 = \rho v_{in}^2 + \rho v_{ou}^2 - \rho v_m^2 + O(\varepsilon^2)\ , \tag{40}$$

where matched phase velocity v_m represents either inner limit of outer solution or outer limit of inner solution. For arbitrary angle of propagation we obtain $\rho v_m^2 = \rho v_{ou}^2 = d$, whereas for direction of propagation perpendicular to one of the families of the fibres we have

$$\rho v_m^2 = \lim_{kh\to o} \rho v_{ou}^2 = \lim_{\substack{\varepsilon\to o \\ m\to\infty \\ m\varepsilon\to kh}} \rho v_{in}^2 = 4s^2c^2 \frac{c_{55}c_{66}}{d} \quad . \tag{41}$$

Therefore matched approximate solutions of squared phase velocity become

$$\rho v^2 = \rho v_{in}^2 + O(\varepsilon^2), \tag{42}$$

for arbitrary angle of propagation and

$$\rho v^2 = \rho v_{ou}^2 - 4s^2c^2 \frac{c_{55}c_{66}}{d} \frac{\tanh m\bar{p}_3}{m\bar{p}_3} + O(\varepsilon^2), \tag{43}$$

for direction of propagation perpendicular to one of the families of the fibres.

NUMERICAL RESULTS

For a numerical example we employ carbon fibres-epoxy resin composite for which nonzero material constants are as follows

$$\lambda = 5.65 \text{ GPa}, \quad \mu = 2.46 \text{ GPa}, \quad \gamma_4 = -1.28 \text{ GPa},$$
$$\gamma_7 = 3.20 \text{ GPa}, \quad 4\gamma_1 = 220.90 \text{ GPa}. \tag{44}$$

These constants are choosen in such a way that if two families of the fibres coincide, that is $\phi = 0$, then material has same elastic constants as

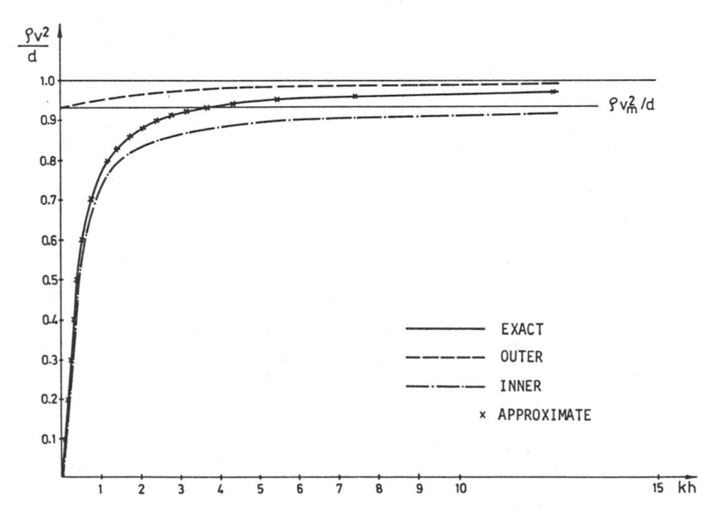

Fig.1. Variation of phase velocity with wavelength for $\phi=33^0 12'39''$ and $\alpha = 56^0 47'21''$

those determined experimentally by Markham (1970) for composite reinforced by one family of the fibres. Matched approximate solution of squared phase velocity scaled by d/ρ as a function of kh, according to equation (40), is shown on Figure 1. This Figure corresponds to angle between two families of the fibres given as $\sin^2\phi=0.3$ and direction of propagation given as $s^2\equiv\sin^2\alpha=0.7$ for which material parameter ε has value 0.131. Approximate solution is compared with exact solution, obtained by Milosavljević (1986 a), represented with solid line. Agreement between matched approximate solution and exact solution is remarcable in entire range of wavelengths. Such agreement can be shown for any angle between two families of the fibres and any direction of propagation.

ACKNOWLEDGEMENT

Results presented here are part of my Ph.D. thesis supervised by Dr. W.A. Green. I would like to express my thanks to Dr. W.A. Green for his invaluable guidance and encouradgement to present this paper here.

REFERENCES

Chadwick, P., Smith, G.D., (1977), Foundations of the Theory of surface Waves in Anisotropic Elastic Materials, in *Advances in Applied Mechanics*, Vol. 17, Academic Press, New York.

Green, W.A., (1982) , *Quart. J. Mech. Appl. Maths*, Vol.35.

Green, W.A., (1984) , *Report on Research Grant GR/C 39552 from SERC*, Nottingham.

Green, W.A., Milosavljević, D., (1984), Bending Waves in Strongly Anisotropic Plates and Cylinders, presented at *XVI Yugoslav Kongress of Th. and Appl. Mechanics*, Bečići.

Green, W.A., Milosavljević, D., (1985), *Int. J.of Solids Structures*, Vol. 21, No.4.

Markham, M.F., (1970), *Composites*, Vol.1.

Milosavljević,D., (1986 a) Wave Propagation in Elastic Plates reinforced by two families of fibres, *Ph.D. thesis submited at The Belgrade University*, Belgrade

Milosavljević, D., (1986 b), Surface Waves in orthotropic elastic materials, presented at *XVII Yugoslav Kongress of Th. and Appl. Mechanics*, Zadar.

Spencer, A. J.M., (1981 a), Continuum models of fibre-reinforce materials, in *Mechanics of Structured Media* (Ed. by Selvadurai, A.P.S.), Elsevier, Amsterdam.

Spencer, A.J.M., (1981 b), Constitutive theory for strongly anisotropic Solids, *Lectures in Fibre Reinforced Materials*, Udine.

THE EFFECT OF INTERFACE CONDITIONS ON WAVE PROPAGATION IN INEXTENSIBLE LAMINATES

W. A. Green

Department of Theoretical Mechanics,
The University, Nottingham, NG7 2RD,
England

This talk is concerned with a 2N-ply laminate formed from an inextensible fibre-reinforced material with the fibre directions in adjoining laminae being orthogonal. The effect of the interfacial bonding layer is modelled by allowing a discontinuity in tangential displacement between adjoining laminae, with the shear stress being taken proportional to this discontinuity. Results are obtained for both purely transverse (SH) waves and for coupled longitudinal and transverse (P and SV) waves.

INTRODUCTION

Laminated plates of fibre-reinforced material are often fabricated from prepreg tapes, laid up according to some specified arrangement of fibre orientations and then bonded together. Typical of these prepreg tapes is the material PEEK, manufactured by ICI and consisting of a single family of straight parallel carbon fibres embedded in a thermohardening polymer resin. A single sheet of this tape has a thickness of 125μm and contains 60% by volume of fibres whose diameter is approximately 6μm. The process of fabricating a multilayered plate of this material gives rise to a laminate in which the plies are separated either by a resin rich layer containing no fibres or by a layer of bonding material. In either case this thin interface layer has elastic characteristics which differ from those of the fibre-reinforced composite forming the prepreg. This paper examines the effect of such a resin rich layer or other bonding material on the propagation of plane harmonic waves in the laminate.

The particular laminate to be considered is formed by alternating the plies so that the fibre families in adjacent laminae are orthogonal to each other. The plies are modelled using a continuum approach whereby the prepreg composite is treated as a transversely isotropic elastic material which is inextensible in the fibre direction. This constraint of inextensibility is an idealization of the property that the extensional modulus of the composite in the fibre direction is very much greater than that in any direction at right angles to the fibre and has previously been adopted to consider waves in multi-ply laminates by Green and Baylis [1]. The propagation of Floquet waves in an unbounded medium of this idealized material is considered in the paper by Baylis [2] contained in this volume.

The results obtained using this idealization are compared with those obtained using a continuum model which is transversely isotropic but not inextensible, in papers by Baylis and Green [3], [4] and Baylis [5]. These show that the idealized model can give a good approximation to the behaviour of the composite in the intermediate and shortwavelength regions but that it fails to account adequately for the motion associated with longwavelength disturbances. The results reported here, using the inextensible model, may likewise by regarded as providing an approximation to the propagation of intermediate and shortwavelength disturbances in a layered fibre composite and in this interpretation the longwavelength results are not significant. On the other hand, the results for the whole range of wavelengths have a validity in an exact theory of elastic wave propagation in inextensible materials.

In all the papers referred to above, it is assumed that there is perfect bonding at the interface between each of the plies. Here the interest is in examining the effect of the bonding layer which has a thickness of the order of a fibre diameter or less. Since the continuum model is based on the assumption that the waves are long compared with the fibre diameter, the approach adopted here is to neglect this thickness. The presence of the shear layer is then taken into account by allowing the tangential displacement at the interface to be discontinuous with the discontinuity being determined by the interfacial shear traction. A similar condition has been employed by Schoenberg [6] to model a homogeneous isotropic medium with periodically spaced parallel slip interfaces. He uses the propagator matrix method to obtain reflection and transmission coefficients for a half space of such a material both for an incident SH wave and for incident P and SV waves. In addition, Schoenberg derived the variation of wave speed with direction for long wave propagation in an infinite expanse of the medium and he shows that wave speeds are either decreased or unchanged but never increased due to the elastic tangential slip condition.

Here also the propagator matrix method is employed to derive dispersion curves for plane harmonic waves propagating in a laminate of finite depth but infinite lateral extent. Attention will be restricted to waves propagating in the plane of the laminate and parallel to the fibre directions in one set of plies but the techniques employed here may equally be applied to examine waves travelling in the plane at any angle to the fibre direction. Consideration is given to two distinct problems. One is concerned with shear waves polarized parallel to the plane of the laminae (SH waves) corresponding to an antiplane motion. The second problem relates to coupled longitudinal and shear waves (P and SV waves) which give rise to a motion satisfying plane strain conditions. For the former, results for a multi-layered two ply laminate with perfect bonding are contained in the paper by Herrmann, Beaupre and Auld [7] whilst the case of the coupled P and SV wave with perfect bonding is treated by Green and Baylis [1].

BASIC CELL

The motion is referred to a Cartesian coordinate system of axes $Ox_1x_2x_3$ with Ox_1 normal to the plane of the laminae and with Ox_2 and Ox_3 parallel to the fibre directions in adjacent layers. The basic unit cell is then taken to be formed of an inner core of depth 2h of material with fibre direction along Ox_2, bounded above and below by a layer of depth h of the same material but with fibre direction along Ox_3. The displacement components $u_i(x_k,t)$ of the material at the point x_k at time t are assumed to have the form

$$u_1(x_k,t) = U_\alpha(x_1)\cos(kx_3-\omega t)\ , \qquad u_2(x_k,t) = V_\alpha(x_1)\sin(kx_3-\omega t)\ ,$$

$$u_3(x_k,t) = W_\alpha(x_1)\sin(kx_3-\omega t)\ , \tag{1}$$

where $\alpha = 1$ for the outer layers and $\alpha = 2$ in the core material. These displacements correspond to plane waves of angular frequency ω and wave number k propagating parallel to the x_3 axis with phase velocity $v = \omega/k$. The wave number k is related to the wavelength Λ by the expression $k = 2\pi/\Lambda$.

When the displacements given by equations (1) are substituted into the stress-strain relations for each of the two materials and thence into the equations of motion, there results a system of coupled ordinary differential equations for the amplitude functions $U_\alpha(x_1)$, $V_\alpha(x_1)$ and $W_\alpha(x_1)$ in each material. These equations have been derived for waves propagating in any direction in the x_2x_3 plane in [3]. For propagation along Ox_3 the equations in the outer layers have the form

$$c_1^2U_1'' + k^2(v^2-c_3^2)U_1 = 0\ ,$$

$$c_2^2V_1'' + k^2(v^2-c_3^2)V_1 = 0\ ,$$

$$W_1 = 0\ , \tag{2}$$

and in the core material they become

$$c_1^2U_2'' + k^2(v^2-c_2^2)U_2 + (c_1^2-c_2^2)kW_2' = 0\ ,$$

$$c_3^2V_2'' + k^2(v^2-c_3^2)V_2 = 0\ ,$$

$$-(c_1^2-c_2^2)U_2' + c_2^2W_2'' + k^2(v^2-c_1^2)W_2 = 0\ . \tag{3}$$

In equations (2) and (3) the primes denote differentiation with respect to x_1, c_1 and c_2 are speed of longitudinal and shear waves respectively propagating in a direction normal to the fibres and c_3 is the speed of shear waves travelling along the fibre direction. The third of equations (2) arises from the inextensibility constraint which is active in the outer layers by virtue of the fact that the waves travel in the fibre direction whereas since the direction of propagation is normal to the fibres in the core, the inextensibility constraint there is inactive.

The interface conditions to be applied to the solutions of equations (2) and (3) at the core boundaries $x_1 = \pm h$ are

$$U_1 = U_2\ , \qquad c_1^2U_1' = c_1^2U_2' + (c_1^2-2c_2^2)kW_2\ , \qquad c_2^2V_1' = c_3^2V_2'\ ,$$

$$V_1' = MG(V_1-V_2)/h\ , \qquad W_2' - kU_2 = -MGW_2/h\ . \tag{4}$$

The first three of equations (4) arise from continuity of normal displacement, normal traction and the transverse shear stress component (t_{12}) and the last two equations relate the transverse shear stress component (t_{12}) and the shear stress component in the core along the propagation direction (t_{13}) to the displacement discontinuity at the interface. In these last two equations the parameter G is a measure of the elastic response of the bonding layer and M is +1 at $x_1 = h$ and -1 at $x_1 = -h$. Note that because of the inextensibility constraint there is no continuity of the shear stress component t_{13} across the interfaces, the discontinuity being equilibrated by a singular reaction stress t_{33} at each interface in the outer material (see Green and Baylis [1] for details).

It may be seen from equations (2), (3) and (4) that the antiplane (SH) motion associated with the displacement $u_2(x_k,t)$ separates out from the plane strain motion (coupled P and SV) associated with the displacements $u_1(x_k,t)$ and $u_3(x_k,t)$. The antiplane strain motion is examined in detail in the next section and the following section is devoted to the plane strain disturbances.

We shall consider a multi-ply laminate formed of N of these unit cells bonded together with perfect bonding. This is equivalent to a symmetric plate consisting of (2N-1) layers each of depth 2h but bounded at the top and bottom by layers of depth h.

ANTIPLANE STRAIN MOTION

It is convenient to express the solutions of the differential equations for V_1 and V_2 in (2) and (3) by means of propagator matrices. To do this, the shear stress component $t_{12}^{(\alpha)}$ in layer α is written in the form

$$t_{12}^{(\alpha)} = \rho c_2^2 k R_\alpha(x_1)\sin(kx_3-\omega t) .$$

From the constitutive equations it may be shown that the amplitudes $R_\alpha(x_1)$ are related to the displacement amplitudes in the outer layer and in the core by the expressions

$$R_1(x_1) = V_1'(x_1)/k , \qquad R_2(x_1) = m^2 V_2'(x_1)/k ,$$

where $m = c_3/c_2$. The only other non-zero stress components in this problem are $t_{23}(x_k,t)$ which are given in both materials by the expression

$$t_{23}^{(\alpha)}(x_k,t) = \rho c_3^2 k V_\alpha(x_1)\cos(kx_3-\omega t) , \qquad \alpha = 1,2.$$

The propagator matrix $\underset{\sim}{P}_\alpha(x_1,\bar{x}_1)$ is obtained by solving the differential equation for $V_\alpha(x_1)$ and then relating the vector $\underset{\sim}{Z}_\alpha(x_1) = \{R_\alpha(x_1),V_\alpha(x_1)\}^T$ at x_1 and at a fixed point $\bar{x}_1$, through the matrix equation

$$\underset{\sim}{Z}_\alpha(x_1) = \underset{\sim}{P}_\alpha(x_1,\bar{x}_1)\underset{\sim}{Z}_\alpha(\bar{x}_1) . \tag{5}$$

Doing this, the propagators in the two materials may be written as

$$\underset{\sim}{P}_1 = \begin{pmatrix} C_1 & -mrS_1 \\ \dfrac{S_1}{mr} & C_1 \end{pmatrix} , \qquad \underset{\sim}{P}_2 = \begin{pmatrix} C_2 & -m^2 rS_2 \\ \dfrac{S_2}{m^2 r} & C_2 \end{pmatrix} , \tag{6}$$

where $C_1 = \cos kmr(x_1-\bar{x}_1)$, $S_1 = \sin kmr(x_1-\bar{x}_1)$, $C_2 = \cos kr(x_1-\bar{x}_1)$, $S_2 = \sin kr(x_1-\bar{x}_1)$ and $r^2 = v^2/c_3^2 - 1$. It may be seen from these expressions that $\underset{\sim}{P}_\alpha(x_1,\bar{x}_1)$ is a function of $(x_1-\bar{x}_1)$ only and these matrices will be denoted by $\underset{\sim}{P}_\alpha(x_1-\bar{x}_1)$. At the interfaces between the two materials the boundary conditions contained in equations (4) lead to relations between $\underset{\sim}{Z}_1$ and $\underset{\sim}{Z}_2$ on either side of the interface, in the form

$$\underset{\sim}{Z}_1 = \underset{\sim}{K}\underset{\sim}{Z}_2 \quad \text{at} \quad x_1 = h , \qquad \underset{\sim}{Z}_2 = \underset{\sim}{K}\underset{\sim}{Z}_1 \quad \text{at} \quad x_1 = -h , \tag{7}$$

where

$$\underset{\sim}{K} = \begin{pmatrix} 1 & 0 \\ kh/G & 1 \end{pmatrix} .$$

Choosing the origin to lie in the mid-plane of the basic cell, the vector $\underset{\sim}{Z}_1(2h)$ at the upper surface may then be expressed in terms of the vector $\underset{\sim}{Z}_1(-2h)$ at the bottom surface by repeated application of equations (5) and equations (7), the expression being

$$\underset{\sim}{Z}_1(2h) = \underset{\sim}{P}_1(h)\underset{\sim}{K}\underset{\sim}{P}_2(h)\underset{\sim}{P}_2(h)\underset{\sim}{K}\underset{\sim}{P}_1(h)\underset{\sim}{Z}_1(-2h) = \underset{\sim}{L}\underset{\sim}{Z}_1(-2h) . \tag{8}$$

For a multi-ply plate composed of N such unit cells with perfect bonding between the cells, the vector $\underset{\sim}{Z}^u$ at the upper surface is given in terms of the vector $\underset{\sim}{Z}^\ell$ at the lower surface by a repeated application of the relation (8) so that

$$\underset{\sim}{Z}^u = \underset{\sim}{L}^N\underset{\sim}{Z}^\ell = \underset{\sim}{N}\underset{\sim}{Z}^\ell .$$

Using the Cayley-Hamilton theorem, the matrix $\underset{\sim}{N}$ can be written in the form

$$\underset{\sim}{N} = \frac{\lambda_1^N-\lambda_2^N}{\lambda_1-\lambda_2}\underset{\sim}{L} - \lambda_1\lambda_2\frac{\lambda_1^{N-1}-\lambda_2^{N-1}}{\lambda_1-\lambda_2}\underset{\sim}{I} , \tag{9}$$

where $\underset{\sim}{I}$ is the unit matrix and λ_1 and λ_2 are the eigenvalues of $\underset{\sim}{L}$. Setting the traction R^ℓ on the lower surface to zero so that $\underset{\sim}{Z}^\ell = (0,V^\ell)^T$, the traction at the upper surface vanishes provided $n_{12} = 0$ which by virtue of equation (9) will be satisfied if $\ell_{12} = 0$. This gives the dispersion equation for SH waves in the multi-ply laminate and is clearly independent of N - a result derived for the case of perfect bonding by Herrmann et al. [7]. The dispersion equation $\ell_{12} = 0$ factorises into two branches given by

$$mC_1C_2 - S_1S_2 - \frac{m^2rkh}{G}S_1C_2 = 0 \quad \text{and} \quad mC_1S_2 + S_1C_2 - \frac{m^2rkh}{G}S_1S_2 = 0 . \tag{10}$$

The first of these corresponds to a motion which is antisymmetric with respect to the middle surface of the basic cell whilst the second is associated with a motion which is symmetric with respect to that surface. The fundamental mode for SH waves is given by the solution of the second of equations (10) corresponding to $r = 0$ for which $S_1 = S_2 = 0$. This gives a wave propagating with velocity $v = c_3$ for which $V_1 = V_2 = A$, where A is a constant. This motion is independent of G since the only non-zero shear stress associated with it is t_{23} and there is no interfacial shear stress. The higher harmonics are all dependent on the parameter G and in the limit as $1/G \to 0$ equations (10) are equivalent to those obtained by Herrmann et al. [7] for two different isotropic elastic materials.

It is a straightforward matter to obtain the dispersion curves from equations (10) for any values of the parameters. Thus writing $rkh = y$ the equations (10) may be written in the form

$$\tan y = m\left(\cot my - \frac{my}{G}\right) , \qquad -\cot y = m\left(\cot my - \frac{my}{G}\right) .$$

The phase velocity is then given in terms of a root $\bar{y}$ of one of these equations by the expression

$$v^2 = c_3^2(1+\bar{y}^2/k^2h^2)$$

and it is simple to show that at any kh the wave speed given by this expression is less than the corresponding speed for the case of perfect bonding ($1/G = 0$). For all roots other than the root $\bar{y} = 0$ of the equation for symmetric motion, the phase velocity becomes unbounded in the long wavelength limit ($kh \to 0$) and the associated angular frequency ω tends to a finite cut-off limit ω_0 given by $\omega_0 = c_3\bar{y}/h$, which depends on the parameter G. In the short wavelength limit ($kh \to \infty$), the phase velocity tends to the limit c_3 for all values of G.

PLANE STRAIN MOTION

This motion involves the displacement amplitudes U and W only, for which the governing equations are the first and third of equations (2) and (3) together with the associated interface conditions from equations (4). The propagator matrix solutions for equations (2) and (3) have been derived by Green and Baylis [1]. They are contained in the paper by Baylis [2] in this volume and will not be reproduced here. For the outer material, governed by equations (2), the propagator matrix for a layer of depth h is the 2×2 matrix $\underset{\sim}{M}_1(h)$ defined in equation (7) of the paper by Baylis [2] whilst that for a layer of depth d of the inner core is the 4×4 matrix $\underset{\sim}{P}(d)$ defined in equation (14) of the same paper. In order to reduce this latter to a matrix of order 2, it is necessary to make use of the interface conditions (4) together with the constraint condition $W_1 = 0$ at the two interfaces. When this is done, the vector $\underset{\sim}{X}(2h)$ at the upper surface of the unit cell is related to its value $\underset{\sim}{X}(-2h)$ at the lower surface by the expressions

$$\underset{\sim}{X}(2h) = \frac{\underset{\sim}{M}_1(h)\underset{\sim}{S}(h)\hat{\underset{\sim}{S}}(h)\underset{\sim}{M}_1(h)\underset{\sim}{X}(-2h)}{|\underset{\sim}{S}(h)|} = \underset{\sim}{H}\underset{\sim}{X}(-2h) \ . \tag{11}$$

The components s_{ij} of the 2×2 matrix $\underset{\sim}{S}(h)$ in equation (11) are related to the components p_{ij} of the 4×4 matrix $\underset{\sim}{P}(h)$ by the expressions

$$s_{11} = p_{11} - \frac{khp_{21}+Gp_{41}}{khp_{24}+Gp_{44}}p_{14} \ , \qquad s_{12} = p_{13} - \frac{khp_{23}+Gp_{43}}{khp_{22}+Gp_{42}}p_{12} \ ,$$

$$s_{21} = p_{31} - \frac{khp_{21}+Gp_{41}}{khp_{24}+Gp_{44}}p_{34} \ , \qquad s_{22} = p_{33} - \frac{khp_{23}+Gp_{43}}{khp_{22}+Gp_{42}}p_{32} \ , \tag{12}$$

and

$$\hat{\underset{\sim}{S}} = \begin{pmatrix} s_{22} & s_{12} \\ s_{21} & s_{11} \end{pmatrix} .$$

Applying the same procedure as for the SH waves, the overall transfer matrix for a laminate formed of N unit cells is $\underset{\sim}{H}^N$ and this is expressible in the same form as $\underset{\sim}{L}^N$ in equation (9). The condition that the upper surface of the laminate be traction free when the lower surface is traction free then reduces to $h_{12} = 0$ and is independent of N. A similar result has been obtained by Green and Baylis [1] in the case of perfect bonding. The dispersion equation $h_{12} = 0$ factorises into two branches given by

$$Cs_{11} + pSs_{21} = 0 , \qquad Cs_{12} + pSs_{22} = 0 , \tag{13}$$

where p, C and S are as defined by Baylis [2]. The first of equations (13) corresponds to a motion which is symmetric with respect to the middle plane of the basic unit and gives rise to longitudinal waves whilst the second equation is associated with a motion that is antisymmetric with respect to the middle surface and may be interpreted as corresponding to flexural waves.

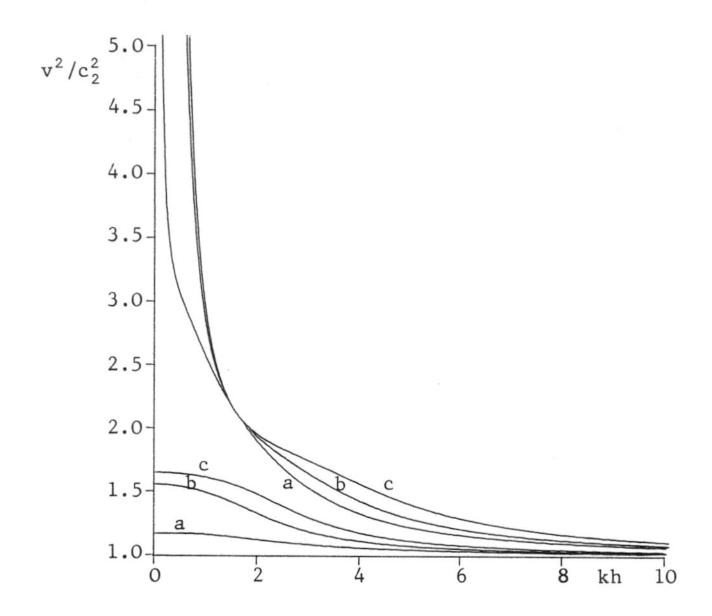

FIGURE 1. Dispersion curves for fundamental modes of each of the two branches for plane strain motion
(a) G = 0.01, (b) G = 1.0, (c) G = 100.
Lower curves are antisymmetric branch, upper curves symmetric

Equations (13) are readily solved numerically to give the dispersion curves relating the phase velocity v to the reduced wave number kh. Examples of these are shown in Figure 1 which contains the curves corresponding to the fundamental mode of each of the two branches at three different values of G (100, 1, 0.01). The curves relate to a material for which $c_1^2/c_2^2 = 4.297$ and $c_3^2/c_2^2 = 2.301$ and may be compared with corresponding curves given by Green and Baylis [1] (Figure 2, curves b) for the same material with perfect bonding ($1/G = 0$). The longwavelength and shortwavelength limiting solutions may again be obtained analytically for all values of the parameters. Letting $kh \to 0$ in the first of equations (13) it may be shown that there are no solutions for which v remains finite and each mode has a cut-off frequency given by $\omega h = \frac{1}{2}(p+\frac{1}{2})\pi c_1$, $p = 0,1,2,\ldots$, which is independent of the parameter G. The limiting velocity of the fundamental mode of equation (13_2) is given, as $kh \to 0$, by the expression

$$\frac{v^2}{c_2^2} = \frac{1}{2}\left\{\frac{c_3^2}{c_2^2} + \frac{c_1^2 G}{c_1^2 G + c_2^2}\right\} .$$

For $1/G = 0$, this agrees with the limiting velocity obtained by Green and Baylis [1] in the case of perfect bonding. The higher harmonics have cut-off frequency given by $\omega h = \frac{1}{2}m\pi c_1$, $m = 1,2,\ldots$, independent of G. The shortwave-length limit for both of equations (13) is obtained by letting $kh \to \infty$ and $q_2 = \sqrt{\{1-(v^2/c_2^2)\}} \to 0$ in such a way that $q_2 kh$ tends to a finite limit $i\bar{\omega}_2$, where $\bar{\omega}_2 = s\pi$ $(s = 1,2,\ldots)$ for the first of equations (13) and $\bar{\omega}_2 = (s-\frac{1}{2})\pi$ $(s = 1,2,\ldots)$ for the second of equations (13). Thus the limiting velocity is c_2 for both branches and $v \to c_2$ in a manner independent of G.

ACKNOWLEDGEMENTS

I am indebted to my colleague E. R. Baylis for computing the dispersion curves plotted in Figure 1. This work was carried out under a research grant from the Science and Engineering Research Council.

REFERENCES

[1] W. A. Green & E. R. Baylis, *Arch. Mech.* 38 (1986) (to appear)

[2] E. R. Baylis, *Proceedings of Euromech Colloquium 214*, Elsevier (1986)

[3] E. R. Baylis & W. A. Green, *J. Sound Vibration* 110 (1986) (to appear)

[4] E. R. Baylis & W. A. Green, *J. Sound Vibration* 111 (1986) (to appear)

[5] E. R. Baylis, *Acta Mech.* (1986) (to appear)

[6] M. Schoenberg, *Geophysical Prospecting* 31 (1983) 265

[7] G. Herrmann, G. S. Beaupre & B. A. Auld, *Mechanics Today* 5 (1980) 83

INTERFACE EFFECTS ON ATTENUATION AND PHASE VELOCITIES IN COMPOSITES

S.K. Datta
Department of Mechanical Engineering and CIRES
University of Colorado, Boulder, CO 80309

H.M. Ledbetter
Y. Shindo[1]
Fracture and Deformation Division
National Bureau of Standards, Boulder, CO 80303

A.H. Shah
Department of Civil Engineering
University of Manitoba, Winnipeg, Canada R3T2N2

ABSTRACT

Much current practical interest exists concerning wave propagation through a composite medium with a random distribution of inclusions: particles, flakes, long continuous or chopped fibers in a homogeneous matrix. Several theoretical studies report wave speeds and attenuation of coherent plane waves propagating through an elastic homogeneous medium containing reinforcing particles or fibers. All these studies assume that the interface between the matrix and the inclusion is sharply defined. Also, it is mostly assumed that the inclusion is perfectly bonded to the matrix.

In the present study, we analyze the problem of damping in metal-matrix composites when there is an interface layer through which the inclusion property changes continuously to that of the matrix.

INTRODUCTION

Determination of effective elastic moduli and damping properties of a heterogeneous material by using elastic waves (propagating or standing) is very effective. Several theoretical studies show that for long wavelengths one can calculate the effective wave speeds of plane longitudinal and shear waves through a composite material. At long wavelengths the wave speeds thus calculated are non-dispersive and hence provide the values for the static effective elastic properties. References to some of the recent theoretical and experimental works can be found in [1-12]. The scattering formulations developed in [1-8] provide a means to obtain not only the effective wave speeds but also the damping of wave amplitudes due to scattering.

In this paper we present results of some of our recent investigations of phase velocity and attenuation of plane longitudinal and shear waves in a medium with inclusions with interfacial layers. We use a wave-scattering approach together with Lax's quasicrystalline approximation to predict the phase velocity of either a longitudinal or a shear wave propagating through the medium. The scattering approach leads also to an estimation of attenuation.

---

1. On leave from Department of Mechanical Engineering II, Tohoku University, Sendai 980, Japan.

MULTIPLE SCATTERING BY A DISTRIBUTION OF SPHERICAL INCLUSIONS AT LOW FREQUENCIES

In this section we present expressions for the phase velocities and attenuation of longitudinal and shear waves in a medium containing a distribution of spherical inclusions with thin interface layers through which the elastic properties vary rapidly, but continuously, from those of the inclusions to those of the matrix. Such interface layers are often present due to processing [13,14].

In order to use the scattering approach it is necessary to calculate the scattered field due to a single spherical inclusion with an interface layer (Fig. 1).

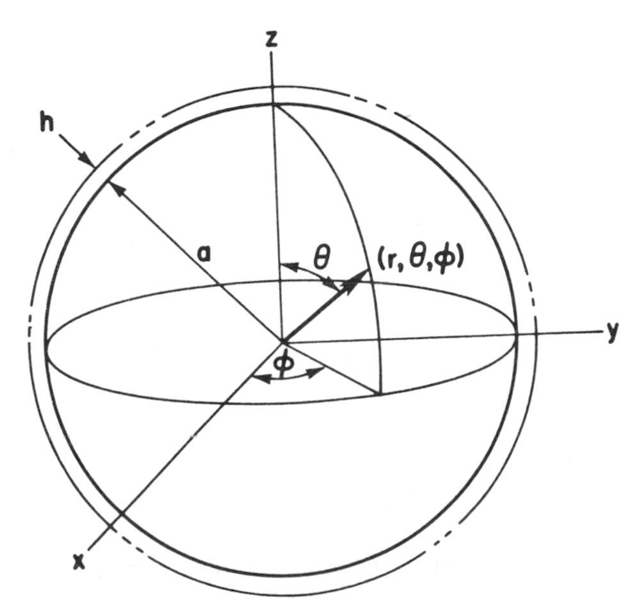

Fig. 1. Spherical inclusion with interface layer

Consider a spherical inclusion of radius a and elastic properties, λ_1, μ_1, ρ_1 embedded in an elastic matrix of material properties λ_2, μ_2 and ρ_2. Also, let the inclusion be separated from the matrix by a thin layer of uniform thickness ($h \ll a$) but variable properties $\lambda(r)$, $\mu(r)$, and $\rho(r)$. Here, λ, μ denote Lamé constants and ρ the density. Let $\lambda(r)$, $\mu(r)$ be expressed as

$$\lambda(r) + 2\mu(r) = (\lambda_1' + 2\mu_1')\, f(r), \quad a < r < a + h \tag{1}$$

$$\mu(r) = \mu_1'\, g(r), \quad a < r < a + h \tag{2}$$

f(r) and g(r) are general functions of r. A special case arises when

$$f(a) = \frac{\lambda_1 + 2\mu_1}{\lambda_1' + 2\mu_1'}\,, \quad g(a) = \frac{\mu_1}{\mu_1'}$$

$$f(a + h) = \frac{\lambda_2 + 2\mu_2}{\lambda_1' + 2\mu_1'}, \quad g(a + h) = \frac{\mu_2}{\mu_1'} \tag{3}$$

with the stipulation that f(r) and g(r) with their first derivatives are continuous in (a, a+h). Since h is assumed to be much smaller than a, it follows from (3) that f′(a) can be approximated by

$$f'(a) = \frac{(\lambda_2 + 2\mu_2) - (\lambda_1 + 2\mu_1)}{h(\lambda_1' + 2\mu_1')}$$

$$g'(a) = \frac{\mu_2 - \mu_1}{h\mu_1'} \tag{4}$$

Note that λ_1' and μ_1' are Lamé constants of the interface material at some value of r $(a<r<a+h)$, say at r=a+h/2.

Another special case would be that the interface material possesses constant properties. Then we have

$$f(r) = g(r) = 1, \quad a < r < a + h \tag{5}$$

We also make the assumption that h is very much smaller than the wavelength of the propagating wave. Then, to first order in h/λ, λ being the wavelength,

$$\tau^t_{rr} = \tau^s_{rr} + \tau^i_{rr}, \quad \tau^t_{r\theta} = \tau^s_{r\theta} + \tau^i_{r\theta}$$
$$\tau^t_{r\phi} = \tau^s_{r\phi} + \tau^i_{r\phi} \tag{6}$$

Here τ_{ij} is the stress tensor and superscripts t, s, and i denote the transmitted, scattered, and incident quantities, respectively. Note that τ^s_{ij}, τ^i_{ij}, and τ^t_{ij} appearing above are calculated at r=a. The spherical polar coordinates r, θ, ϕ are defined in Fig. 1. Boundary conditions (6) express the fact that, to first order in h/λ, the traction components do not suffer any jump across the layer. However, the displacement components suffer jumps. Two parameters that characterize interface are:

$$K_1 = \int_0^1 \frac{dx}{f(a+hx)}, \quad K_2 = \int_0^1 \frac{dx}{g(a+hx)} \tag{7}$$

Using equations (3) and (4) in (7), it can be shown that K_1 and K_2 are given approximately by

$$K_1 = \frac{\lambda_1'+2\mu_1'}{\lambda_2+2\mu_2-(\lambda_1+2\mu_1)} \ln\left(1 + \frac{\lambda_2+2\mu_2-(\lambda_1+2\mu_1)}{\lambda_1+2\mu_1}\right) \tag{8}$$

$$K_2 = \frac{\mu_1'}{\mu_2-\mu_1} \ln\left(1 + \frac{\mu_2 - \mu_1}{\mu_1}\right) \tag{9}$$

On the other hand, if eq. (5) is used, then

$$K_1 = K_2 = 1 \tag{10}$$

Mal and Bose [2] studied a problem similar to the one considered here. They assumed a thin viscous third layer between the sphere and the matrix and imposed the condition of continuity of radial displacement.

The incident wave will be assumed to be either a plane longitudinal wave propagating in the positive z-direction or a plane shear wave polarized in the x-direction and propagating in the positive z-direction. Thus,

$$\underset{\sim}{u}^i = e^{ik_1 z} \underset{\approx}{e}_z + e^{ik_2 z} \underset{\sim}{e}_x \tag{11}$$

where $k_1=\omega/c_1$ and $k_2=\omega/c_2$. ω denotes the circular frequency of the wave and c_1, c_2 denote the longitudinal and shear wave speeds in the matrix. The factor $e^{-i\omega t}$ has been suppressed.

The scattered and transmitted fields can be written as

$$\underset{\sim}{u}^{(s)} = \sum_{n=0}^{\infty} \sum_{m=-1}^{1} [A_{mn} \underset{\sim}{L}^{(3)}_{mn} \delta_{mo} + B_{mn} \underset{\sim}{M}^{(3)}_{mn} + C_{mn} \underset{\sim}{N}^{(3)}_{mn}] \tag{12}$$

$$\underset{\sim}{u}^{t} = \sum_{n=0}^{\infty} \sum_{m=-1}^{1} [A'_{mn} \underset{\sim}{L}^{(1)\prime}_{mn} \delta_{mo} + B'_{mn} \underset{\sim}{M}^{(1)\prime}_{mn} + C'_{mn} \underset{\sim}{N}^{(1)\prime}_{mn}] \tag{13}$$

where the prime denotes that k_1 and k_2 are to be replaced by k_1' $(=\omega/c_1')$, and k_2' $(=\omega/c_2')$, respectively. c_1' and c_2' are the wave speeds in the inclusion. $\underset{\sim}{L}^{(3)}$, $\underset{\sim}{M}^{(3)}$ and $\underset{\sim}{N}^{(3)}$ are defined in [15].

The constants A, B, C, A′, B′, C′ are found by the use of conditions (6). For details, see [15].

Once A_{mn}, C_{mn}, and B_{mn} are determined, the scattered field is then found from equation (12). Since the expressions for the field inside the inclusion will not be needed for the derivation of the dispersion equation governing the effective wave number of plane-wave propagation through the composite medium, we do not give these here. In the following we present equations governing propagation of effective plane waves through a medium composed of a random homogeneous distribution of identical spherical inclusions surrounded by the layers as discussed above.

To derive approximately the phase velocities of plane waves moving through the composite medium, we assume that wavelengths are long compared to the radius of each inclusion. In this long-wavelength limit it can be shown that, correct to $O(\varepsilon^3)$,

$$A_{mn} = i\varepsilon^3 [P_n \phi_{mn} + Q_n X_{mn}] \tag{14}$$

$$C_{mn} = i\tau^3 \varepsilon^3 [R_n \phi_{mn} + S_n X_{mn}] \tag{15}$$

Expressions for P_n, Q_n, R_n, and S_n are defined in [15], where ϕ_{mn} and X_{mn} are also given.

Once the scattered field due to a single inclusion is known, multiple scattering due to a number of inclusions can easily be calculated. In particular, following the steps discussed before by us [8] it can be shown that effective speeds of propagation of plane longitudinal and shear waves are given by

$$\frac{k_1^{*2}}{k_1^2} = \frac{(1 + 9cP_1)(1 + 3cP_o)\ \{1 + \frac{3c}{2} P_2\ (2 + 3\tau^2)\}}{1 - 15cP_2\ (1 + 3cP_o) + \frac{3}{2}cP_2\ (2 + 3\tau^2)} \tag{16}$$

$$\frac{k_2^{*2}}{k_2^2} = \frac{(1 + 9cP_1)\ \{1 + \frac{3}{2}cP_2\ (2 + 3\tau_2^{\ 2})\}}{1 + \frac{3}{4}cP_2\ (4 - 9\tau^2)} \tag{17}$$

$$k_1^* = \omega/c_1^*\ ,\quad k_2^* = \omega/c_2^*$$

c_1^* and c_2^* are the effective wave speeds of plane longitudinal and shear waves, respectively. c is the volume concentration of inclusions in the matrix.

The attenuation caused by scattering (to this order of approximation) can also be calculated using equations (12) and (15). We find that the attenuation coefficients α_p and α_s are given by

$$\frac{\alpha_p}{k_1} = 3c\varepsilon^3\ [P_o^2 + 3P_1^2\ (1 + 2\tau^3) + 5P_2^2(1 + 3\tau^5/2)] = Q_p^{-1} \tag{18}$$

$$\frac{\alpha_s}{k_2} = 3c\varepsilon^3\ [3P_1^2\ (1 + \frac{1}{2}\tau^3) + \frac{15}{4}\ \tau^2P_2^2\ (1 + 3\tau^5/2)] = Q_s^{-1} \tag{19}$$

It may be noted that good approximations to the attenuation coefficients are obtained if P_o, P_1 and P_2 are calculated after replacing the matrix properties by the effective composite properties given by Eqs. (16) and (17).

SCATTERING CROSS SECTIONS OF A SPHERICAL INCLUSION WITH INTERFACE LAYER AT ARBITRARY FREQUENCY

Since attenuation coefficients are directly related to the scattering cross sections, in this section we present some results for scattering cross sections at finite frequencies of a sphere with interface layer.

From equation (12) the scattered field at a large distance from the sphere is given by

$$\underset{\sim}{u}^{(s)} = \underset{\sim}{u}^p + \underset{\sim}{u}^s \tag{20}$$

where superscripts p and s denote longitudinal and shear wave components, respectively. It is easily shown that

$$\underset{\sim}{u}^p = g(\theta)\ \frac{e^{ik_1r}}{r}\ \underset{\sim}{e}_r \tag{21}$$

$$\underset{\sim}{u}^s = h(\theta)\ \frac{e^{ik_2r}}{r}\ \underset{\sim}{e}_\theta \tag{22}$$

where, for incident longitudinal wave corresponding to the first term in equation (11),

$$g(\theta) = \frac{1}{k_1} \sum_{n=0}^{\infty} (-i)^n A_{on}^P P_n(\cos\theta) \tag{23}$$

$$h(\theta) = \frac{1}{k_2} \sum_{n=0}^{\infty} (-i)^n C_{on}^P \frac{dP_n}{d\theta}$$

The scattering cross section is then (24)

$$\Sigma_p = \frac{4\pi}{k_1} \mathrm{Im}\{\frac{1}{k_1} \sum_{n=0}^{\infty} (-i)^n A_{on}^P\}$$

For incident shear wave given by the second term in Eq. (11) we get

$$\Sigma_s = \frac{4\pi}{k_2} \mathrm{Im}\{\frac{1}{k_2} \sum_{n=0}^{\infty} (-i)^n \frac{n(n+1)}{2} (C_n^s - iB_n^s)\} \tag{25}$$

In writing (25) we have rewritten (16) as

$$\underset{\sim}{u}^s = \sum_{n=0}^{\infty} [A_n^s \underset{\sim}{L}_{e1n}^{(3)} + B_n^s \underset{\sim}{M}_{01n}^{(3)} + C_n^s \underset{\sim}{N}_{e1n}^{(3)}] \tag{26}$$

Here subscripts e and o refer to even and odd spherical wave functions (see [16]).

The coefficients appearing in (23) and (26) were calculated when the interface layer was very small ($h/a \ll 1$), and K_1 and K_2 were given by Eqs. (8) and (9). Scattering cross sections then were calculated by using equations (26) and (27). These results are discussed in the next section.

NUMERICAL RESULTS AND DISCUSSION

Attenuation coefficients for two composite materials were calculated using the procedure outlined in the previous two sections. The first composite was a SiC/Al material which was studied in detail in [8]. The constituent properties are given in Table 1. The other material was a lead/epoxy composite discussed in [17]. Table 1 lists the properties of the constituents.

Table 1. Constituent Properties of SiC/Al and Lead/Epoxy Composites

Constituents	ρ (g/cm^3)	μ (GPa)	$\lambda+2\mu$ (GPa)
SiC	3.181	188.1	474.2
Al	2.706	26.7	110.5
Lead	11.300	8.35	55.46
Epoxy	1.202	1.71	8.36

Figures 2-5 show the variations α_p/k_1 and α_s/k_2 with volume concentration of inclusions at different frequencies. Also shown in these figures is the effect of a thin interface layer with K_1 and K_2 given by Eqs. (8) and (9). For SiC/Al composite Figs. 2 and 3 show that the effect of the interface is to decrease the attenuation. This is perhaps understandable because the interface is not a sharp discontinuity. However, as shown in the case of Lead/Epoxy composite this is not always the case. These figures also show that attenuation reaches a maximum at higher and higher concentrations as the frequency increases. Behavior of attenuation vs.

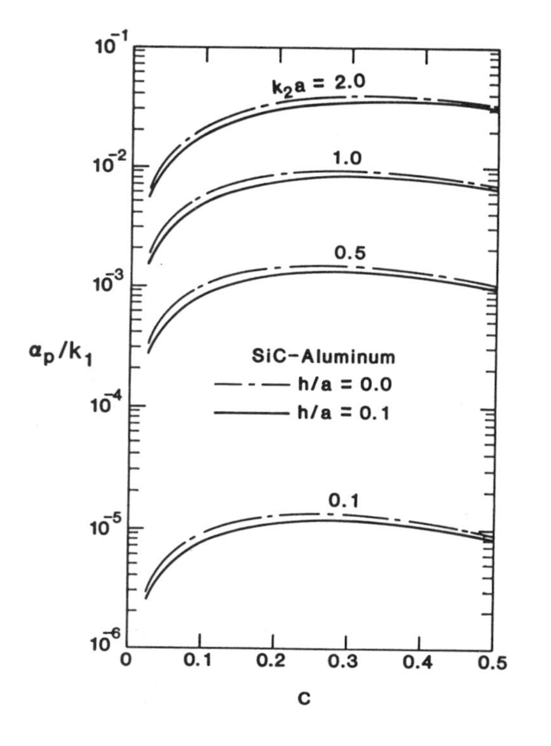

Fig. 2. Attenuation of P wave in SiC/Al Composite

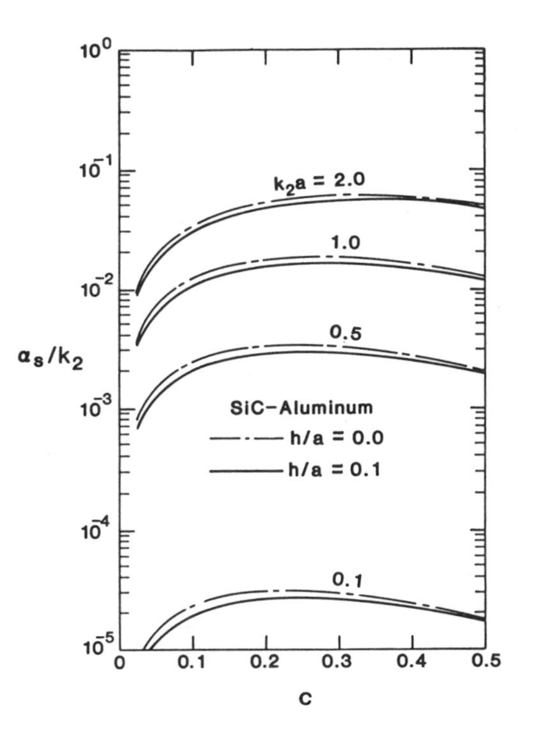

Fig. 3. Attenuation of S wave in SiC/Al Composite

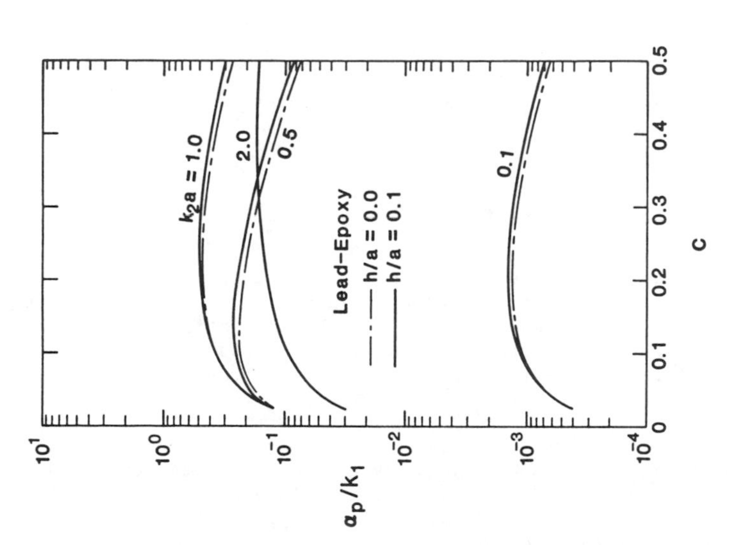

Fig. 4. Attenuation of P wave in Lead/Epoxy Composite

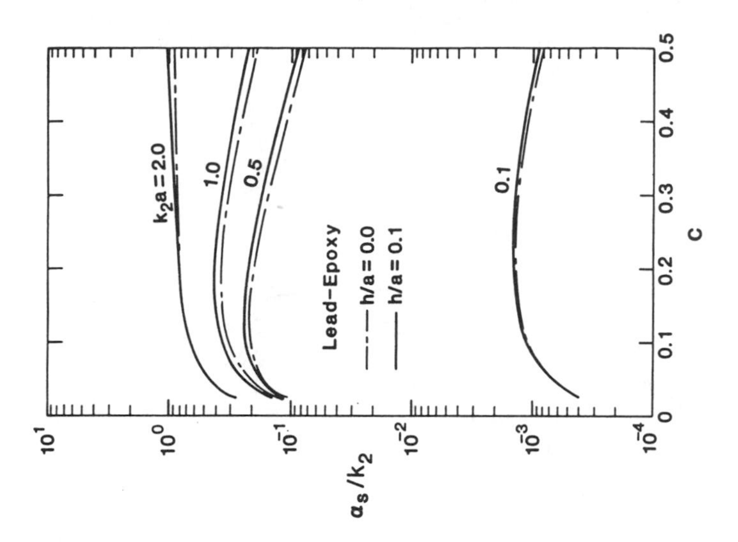

Fig. 5. Attenuation of S wave in Lead/Epoxy Composite

concentration in Lead/Epoxy composite is markedly different from the SiC/Al material. It is seen that the interface generally increases the attenuation. The peak attenuation is reached at concentrations that first increase with frequency, but then decrease and increase again at high frequencies. Figure 6 shows the P-wave attenuation vs. frequency at different concentrations. This agrees well with the results obtained in [7] using a different approach.

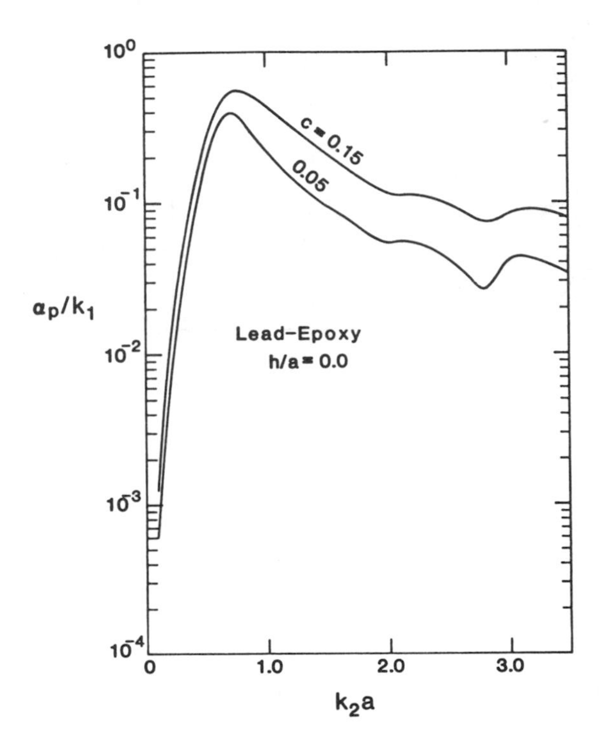

Fig. 6. Attenuation of P wave in Lead/Epoxy Composite.

In conclusion, it may be stated that attenuation in a particulate composite depends crucially on the constituents and the nature of interface layers. These layers can enhance or diminish attenuation. More model studies are needed to characterize attenuation in composites with thick interface layers. This work is in progress and will be reported later.

ACKNOWLEDGMENT

This work was supported in part by a grant from the Office of Naval Research (ONR 00014-86-K0280) and by a grant (A-7988) from National Science and Engineering Research Council of Canada. Support was also received from the Office of Nondestructive Evaluation, NBS.

REFERENCES

1. S.K. Bose and A.K. Mal, J. Mech. Phys. Solids, 22, 217 (1974).

2. A.K. Mal and S.K. Bose, Proc. Camb. Philos. Soc. 76, 587 (1974).

3. S.K. Datta, J. Appl. Mech. 44, 657 (1977).

4. A.J. Devaney, J. Math. Phys. 21, 2603 (1980).

5. J.G. Berryman. J. Acoust. Soc. Am. 68, 1809 (1980).

6. J.R. Willis, J. Mech. Phys. Solids 28, 307 (1980).

7. V.K. Varadan, Y. Ma and V.V. Varadan, J. Acoust. Soc. Am. 77, 375 (1985).

8. H.M. Ledbetter and S.K. Datta, J. Acoust. Soc. Am. 79, 239 (1986).

9. D.T. Read and H.M. Ledbetter, J. Appl. Phys. 48, 2827 (1977).

10. V.K. Kinra, M.S. Petraitis and S.K. Datta, Int. J. Solids Str. 16, 301 (1980).

11. S.K. Datta and H.M. Ledbetter, in "Mechanics of Nondestructive Testing," W.W. Stinchcomb, ed., Plenum Press, New York (1980).

12. H.M. Ledbetter and S.K. Datta, Mat. Sci. Eng. 67, 25 (1984).

13. C.G. Rhodes and R.A. Spurling, in "Recent Advances in Composites in the United States and Japan," J.R. Vinson and M. Taya, eds., ASTM, Philadelphia (1983).

14. S. Umekawa, C.H. Lee, J. Yamamoto and K. Wakashima, in "Recent Advances in Composites in the United States and Japan," J.R. Vinson and M. Taya, eds., ASTM, Philadelphia (1983).

15. S.K. Datta and H.M. Ledbetter, in "Mechanics of Material Interfaces," A.P.S. Selvadurai and G. Voyiadjis, eds., Elsevier, Amsterdam (1986).

16. J.A. Stratton, "Electromagnetic Theory," McGraw-Hill Book Company, Inc., New York (1941).

17. S.K. Datta, H.M. Ledbetter and V.K. Kinra, in "Composite Materials", K. Kawata and T. Akasaka, eds., Japan Society for Composite Materials, Tokyo (1981).

INFLUENCE OF THICKNESS SHEAR DEFORMATION ON FREE VIBRATIONS OF SIMPLY SUPPORTED ANTISYMMETRIC ANGLE-PLY LAMINATED CIRCULAR CYLINDRICAL PANELS

KOSTAS P.SOLDATOS

Department of Mathematics
University of Ioannina
Greece 45332.

ABSTRACT

This paper is concerned with the influence of thickness shear deformation and rotatory inertia on the free vibrations of antisymmetric angle-ply laminated circular cylindrical panels. The analysis is based on Love's approximations. A refined shell theory is employed which assumes a parabolic variation of thickness shear strains and stresses. For a simply supported panel, its equations of motion are solved by using Galerkin's method. For a family of graphite-epoxy panels, numerical results are obtained, discussed and compared with corresponding results based on a classical Love-type theory.

INTRODUCTION

Most of the linear or linearized analyses concerned with antisymmetric angle-ply laminates [1] are based on classical theories, either for flat plates [2-11] or for shells [12,13] and panels [14,15] with circular cylindrical configuration. However, as is wellknown, laminated composite plates and shells exhibit much larger thickness shear effects than corresponding structures made of a homogeneous isotropic material do. Hence, analyses of antisymmetric angle-ply laminates which take into account thickness shear deformation are already available in the literature but they are concerned with the flat plate configuration only [16-18]. The current article deals with the influence of thickness shear deformation and rotatory inertia on the free vibrations of antisymmetric angle-ply laminated circular cylindrical panels.

The problems of buckling and/or vibration of antisymmetric angle-ply laminated circular cylindrical shells and panels have been studied on the basis of the classical Donnell-type theory [12-15]. However, the accuracy of Donnell-type approximations is restricted to either short shells or shallow panels. Therefore, in the present paper, the analysis is based on a first-order shell theory which makes use of Love's approximations. The theory proposes a realistic parabolic variation of thickness shear strains which allows zero thickness shear stresses at the extreme fibers of the panel. Its equations of motion include, entirely, the equations of motion of the classical Love-type theory which, therefore, is entirely obtained as a special case.

For the solution of the free vibration problem, it is assumed that the cylindrical panel considered is subjected to a certain type of simply supported edge boundary conditions. Based on the already known exact solution of the corresponding flat plate problem [18], a Fourier series displacement model is considered and a solution of the differential equations of the present theory is then obtained by using Galerkin's method. Finally, for a family of graphite-epoxy panels, numerical results based on both classical and shear deformable theories are obtained, compared and discussed.

FORMULATION OF GOVERNING EQUATIONS

Figure 1 shows the nomenclature of the middle-surface of a circular cylindrical panel with radius R, axial length L_x and circumferential length L_s; the constant thickness of the panel is denoted by h. The parameters x,s and z denote length in the axial, circumferential and normal to the middle-surface direction, respectively.

The panel is an antisymmetric angle-ply laminated one [1]. As such, it is composed of an even number of layers perfectly bonded together. Each individual layer (say the kth one) is considered to behave macroscopically as a homogeneous, generally orthotropic (monoclinic) linearly elastic material whose state of stress is governed by the Hooke's law

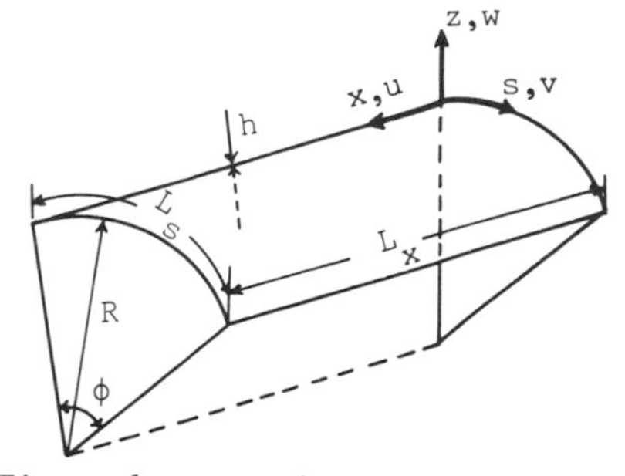

Figure 1. Nomenclature of a circular cylindrical panel.

$$\begin{bmatrix} \sigma_x^{(k)} \\ \sigma_s^{(k)} \\ \tau_{sz}^{(k)} \\ \tau_{xz}^{(k)} \\ \tau_{xs}^{(k)} \end{bmatrix} = \begin{bmatrix} Q_{11}^{(k)} & Q_{12}^{(k)} & 0 & 0 & Q_{16}^{(k)} \\ Q_{12}^{(k)} & Q_{22}^{(k)} & 0 & 0 & Q_{26}^{(k)} \\ 0 & 0 & Q_{44}^{(k)} & Q_{45}^{(k)} & 0 \\ 0 & 0 & Q_{45}^{(k)} & Q_{55}^{(k)} & 0 \\ Q_{16}^{(k)} & Q_{26}^{(k)} & 0 & 0 & Q_{66}^{(k)} \end{bmatrix} \begin{bmatrix} \varepsilon_x \\ \varepsilon_s \\ \varepsilon_{sz} \\ \varepsilon_{xz} \\ \varepsilon_{xs} \end{bmatrix}, \tag{1}$$

where Qs denote reduced stiffnesses [1,16]. However, due to the properties of an antisymmetric angle-ply laminate, the arrangement of these layers through the panel thickness is such that symmetry in thicknesses but antisymmetry in material properties exist, with respect to the panel middle surface. In more detail, every two layers that are symmetrically arranged with respect to the middle surface, have the same thicknesses and are constructed of the same orthotropic material, but their principal material directions form opposite angles to the panel axis (see for instance reference [1], Table 4-5).

The following displacement field is considered:

$$\begin{aligned} U(x,s,z;t) &= u(x,s;t) - z\left[w_{,x} - (1-4z^2/3h^2)u_1(x,s;t)\right], \\ V(x,s,z;t) &= (1+z/R)v(x,s;t) - z\left[w_{,s} - (1-4z^2/3h^2)v_1(x,s;t)\right], \\ W(x,s,z;t) &= w(x,s;t), \end{aligned} \tag{2}$$

where t denotes time. The functions u,v and w respresent the middle surface displacement components. The terms $zw_{,x}$ and $z(w_{,s}-v/R)$ are the standard terms which quarantee the validity of the Kirchhoff-Love assumptions in the classical Love-type theory; apparently, upon neglecting the remaining terms ($u_1=v_1=0$), the displacement field (2) reduces to the corresponding field used in the classical theory. These terms, involving u_1 and v_1 have been employed to disturb the assumption that normals to the undeformed middle surface still remain normal to it after deformation; they also remove the assumption of Timoshenko [19] and Mindlin [20] that these normals remain straight after deformation. As it can be seen from expressions (3) below, the functions u_1

and v_1 represent the action of the thickness shear strains on the shell middle surface.

Introducing expressions (2) into the linear version of the strain-displacement relations of the three-dimensional elasticity and restricting approximations in the limits of a first-order shell theory ($h/R<<1$), the strain components appearing in Hooke's law (1) are obtained in a power series form of the transverse co-ordinate z as follows:

$$(\varepsilon_x,\varepsilon_s,\varepsilon_{xs})=(e_x,e_s,e_{xs})+z(k_x,k_s,k_{xs})+\frac{4z^3}{3h^2}(\ell_x,\ell_s,\ell_{xs}) \quad ,$$

$$(\varepsilon_{xz},\varepsilon_{sz})=(1-\frac{4z^2}{h^2})(u_1,v_1) \quad , \tag{3}$$

where

$$e_x=u_{,x},\quad e_s=v_{,s}+w/R,\quad e_{xs}=v_{,x}+u_{,s},\quad k_x=u_{1,x}-w_{,xx}$$

$$k_s=v_{1,s}-w_{,ss}+v_{,s}/R \;,\quad k_{xs}=v_{1,x}+u_{1,s}-2w_{,xs}+v_{,x}/R \quad ,$$

$$\ell_x=-u_{1,x} \;,\quad \ell_s=-v_{1,s} \;,\quad \ell_{xs}=-(v_{1,x}+u_{1,s}) \quad . \tag{4}$$

Apparently, both thickness shear strains, ε_{xz} and ε_{sz}, vary parabolically through the shell thickness and take a zero value at the inner and outer surfaces of the shell. Accordingly, Hooke's law (1) suggests that although thickness shear stresses may present discontinuities at the interfaces between successive layers, they have also zero values at the extreme fibers of the shell. Hence, the boundary conditions of zero thickness shear stresses at the inner and outer shell surfaces are satisfied.

The following force and moment resultants are defined:

$$(N_x,N_s,N_{xs})=\int_{-h/2}^{h/2}(\sigma_x,\sigma_s,\tau_{xs})dz \quad , \quad (Q_x,Q_s)=\int_{-h/2}^{h/2}(\tau_{xz},\tau_{sz})(1-4z^2/h^2)dz \quad ,$$

$$(M_x,M_s,M_{xs})=\int_{-h/2}^{h/2}(\sigma_x,\sigma_s,\tau_{xs})zdz \;,\; (S_x,S_s,S_{xs})=\frac{4}{3h^2}\int_{-h/2}^{h/2}(\sigma_x,\sigma_s,\tau_{xs})z^3dz \;. \tag{5}$$

Introducing equations (1) into equations (5) and taking into account the geometric and material symmetries of an antisymmetric angle-ply laminate , the following constitutive equations are obtained:

$$\begin{bmatrix} N_x \\ N_s \\ N_{xs} \\ \hline M_x \\ M_s \\ M_{xs} \\ S_x \\ S_s \\ S_{xs} \end{bmatrix} = \left[\begin{array}{ccc|cccccc} A_{11} & A_{12} & 0 & 0 & 0 & B_{16} & 0 & 0 & B'_{16} \\ A_{12} & A_{22} & 0 & 0 & 0 & B_{26} & 0 & 0 & B'_{26} \\ 0 & 0 & A_{66} & B_{16} & B_{26} & 0 & B'_{16} & B'_{26} & 0 \\ \hline 0 & 0 & B_{16} & D_{11} & D_{12} & 0 & D'_{11} & D'_{12} & 0 \\ 0 & 0 & B_{26} & D_{12} & D_{22} & 0 & D'_{12} & D'_{22} & 0 \\ B_{16} & B_{26} & 0 & 0 & 0 & D_{66} & 0 & 0 & D'_{66} \\ 0 & 0 & B'_{16} & D'_{11} & D'_{12} & 0 & D''_{11} & D''_{12} & 0 \\ 0 & 0 & B'_{26} & D'_{12} & D'_{22} & 0 & D''_{12} & D''_{22} & 0 \\ B'_{16} & B'_{26} & 0 & 0 & 0 & D'_{66} & 0 & 0 & D''_{66} \end{array}\right] \begin{bmatrix} e_x \\ e_s \\ e_{xs} \\ \hline k_x \\ k_s \\ k_{xs} \\ \ell_x \\ \ell_s \\ \ell_{xs} \end{bmatrix} \quad , \tag{6}$$

$$(Q_s,Q_x)=(A_{44}e_{sz},A_{55}e_{xz}) \quad , \tag{7}$$

where the extensional, A_{ij}, coupling, B_{ij} and B'_{ij}, and bending, D_{ij}, D'_{ij} and D''_{ij}, stiffnesses (i,j= 1,2,6) are defined as follows :

$$(A_{ij},B_{ij},B'_{ij},D_{ij},D'_{ij},D''_{ij})=\int_{-h/2}^{h/2} Q_{ij}^{(k)}(1,z,\frac{4z^3}{3h^2}, z^2, \frac{4z^4}{3h^2}, \frac{16z^6}{9h^4})dz \; ; \tag{8}$$

the thickness shear stiffnesses are defined according to :

$$(A_{44},A_{55})=\int_{-h/2}^{h/2} (Q_{44}^{(k)}, Q_{55}^{(k)})(1-4z^2/h^2)^2 dz \; . \tag{9}$$

The govering differential equations of motion, derived by using Hamilton's principle, are given in terms of the afore-mentioned force and moment resultants as follows:

$$N_{x,x}+N_{xs,s}=[\rho_0 u+\rho_1(u_1-w,_x)-\frac{4}{3h^2}\rho_3 u_1],_{tt} \quad ,$$

$$N_{s,s}+N_{xs,x}+(M_{s,s}+M_{xs,x})/R=[\rho_0 v+\rho_1(v_1-w,_s+2v/R)$$
$$+\rho_2(v_1-w,_s+v/R)/R-4(\rho_3+\rho_4/R)/3h^2],_{tt} \quad ,$$

$$-N_s/R+M_{x,xx}+M_{s,ss}+2M_{xs,xs}=[\rho_0 w+\rho_1(u,_x+v,_s)$$
$$+\rho_2(u_{1,x}+v_{1,s}+v,_s/R-w,_{xx}-w,_{ss})+4\rho_4(u_{1,x}+v_{1,s})/3h^2],_{tt} \quad ,$$

$$(M_x-S_x),_x+(M_{xs}-S_{xs}),_s-Q_x=[\rho_1 u+\rho_2(u_1-w,_x)-4\rho_3 u/3h^2$$
$$+4\rho_4(w,_x-2u_1)/3h^2+16\rho_6 u_1/9h^4],_{tt} \quad ,$$

$$(M_s-S_s),_s+(M_{xs}-S_{xs}),_x-Q_s=[\rho_1 v_1+\rho_2(v_1-w,_s+v/R)-4\rho_3 v/3h^2$$
$$+4\rho_4(w,_s-2v_1-v/R)/3h^2+16\rho_6 v_1/9h^4],_{tt} \quad , \tag{10}$$

where the appearing inertia terms are defined according to :

$$\rho_i=\int_{-h/2}^{h/2}\rho z^i dz \; , \qquad i=0,1,\ldots,6 \; . \tag{11}$$

Introduction of expressions (4),(6) and (7) into (10) leads to a differential eigenvalue problem of the form :

$$[\bar{L}]\{\bar{\delta}\}=\{0\}, \qquad \{\bar{\delta}\}^T=\{u,v,w,L_x u_1,L_x v_1\} \quad , \tag{12}$$

where $[\bar{L}]$ is a 5×5 symmetric matrix of partial differential operators; a non-dimensional version of its components is given in a recent, and as yet unpublished, article [21].

Upon neglecting the contribution of all terms involving the unknown functions u_1 and v_1, expressions (3) and (4) reduce to the strain-displacement relations of the classical Love-type theory. Moreover, (6) and (10) reduce to the constitutive equations and equations of motion, respectively, of the classical theory, provided that only inertia terms containing ρ_0 are retained in (10). Hence, the differential eigenvalue problem corresponding to the classical theory is obtained in the form :

$$[\tilde{L}]\{\tilde{\delta}\} = \{0\}, \qquad \{\tilde{\delta}\}^T=\{u,v,w\} \; ; \tag{13}$$

the components of the 3×3 matrix $[\tilde{L}]$, can be obtained as special cases of the corresponding components of $[L]$, as is explained in [21] .

SOLUTION FOR SIMPLY SUPPORTED PANELS

It is assumed that the panel is subjected to the following simply supported edge boundary conditions :

$$u=w=v_1=N_{xs}+M_{xs}/R=M_x=S_x=0 \quad , \qquad \text{at} \quad x=0,L_x \quad ,$$

$$v=w=u_1=N_{xs}=M_s=S_s=0 \quad , \qquad \text{at} \quad s=0,L_s \quad . \tag{14}$$

For the solution of the free vibration problem, the following displacement model is considered :

$$(u,L_xv_1)=\cos(\omega t)\sum_{m=1}^{M}\sum_{n=1}^{N}(a_{mn},e_{mn})\sin(m\pi x/L_x)\cos(n\pi s/L_s) \quad ,$$

$$(v,L_xu_1)=\cos(\omega t)\sum_{m=1}^{M}\sum_{n=1}^{N}(b_{mn},d_{mn})\cos(m\pi x/L_x)\sin(n\pi s/L_s) \quad ,$$

$$w=\cos(\omega t)\sum_{m=1}^{M}\sum_{n=1}^{N}c_{mn}\sin(m\pi x/L_x)\sin(n\pi s/L_s) \quad , \tag{15}$$

where ω represents a certain unknown natural frequency, and $a_{mn},\ldots,e_{mn}$ are unknown constant coefficients. Although this displacement model is unable to satisfy exactly all of the boundary conditions (14), is at least satisfying the displacement ones. Accordingly, the Fourier-type expansions (15) form an appropriate displacement model which, through the application of Galerkin's method [14,15] on both (12) and (13), leads to an approximate solution obtained in the following algebraic eigenvalue form :

$$(\underset{\sim}{T}-\omega^2\underset{\sim}{H})\underset{\sim}{\Delta}=\{0\} \; . \tag{16}$$

Here, $\underset{\sim}{T}$ and $\underset{\sim}{H}$ represent proper square matrices while $\underset{\sim}{\Delta}$ represents a column matrix which contains the unknowns $a_{mn},\ldots,e_{mn}$. The components of all three matrices $\underset{\sim}{T},\underset{\sim}{H}$ and $\underset{\sim}{\Delta}$ are cited in [21], where the derivation of (16) is explained in detail.

NUMERICAL RESULTS

Since there is an infinite complexity of the class of antisymmetric angle-ply laminates, numerical examples concerning "regular" antisymmetric angle-ply laminates are only considered; that is, laminates composed of an even number equal-in-thickness layers constructed of the same orthotropic material and with principle material directions alternately oriented at angles $+\theta$ and $-\theta$ to the panel axis. For this kind of laminated composites it can be shown that the only non-vanishing coupling stiffnesses, B_{16},B_{26},B'_{16} and B'_{26}, are inversely proporsional to the shell number of layers. Thus, the coupling between bending and extension due to the panel lamination dies out rapidly as the number of layers increases and, therefore, a panel composed of an infinite number of layers behaves, macroscopically, like it was constructed of a homogeneous orthotropic material.

It must be noted here that the whole set of the results shown in this

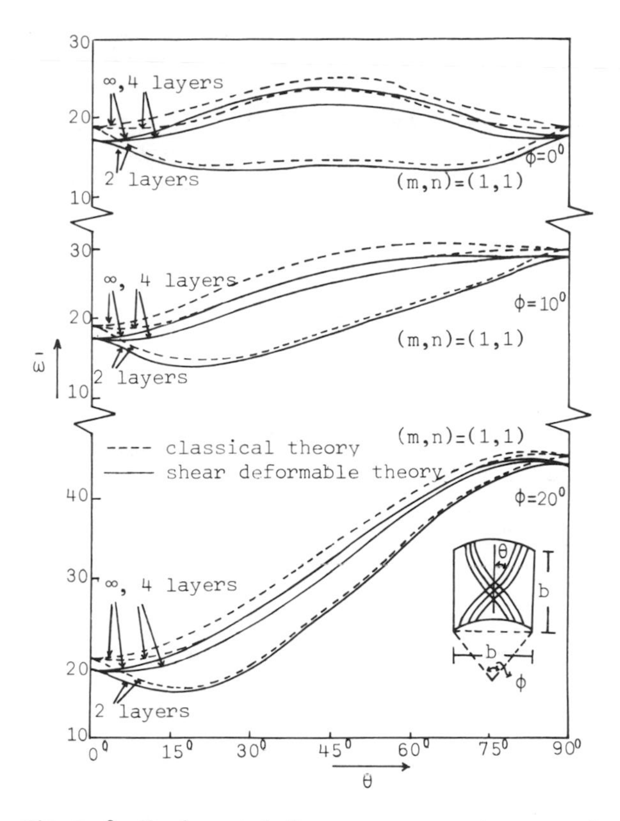

Figure 2. Fundamental frequency parameter as a function of the lamination angle ($\phi=0^0,10^0,20^0$).

section is a part of the numerical results presented in [21]. It is concerned with a family of graphite-epoxy panels with material properties

$$E_1/E_2=40, \quad G_{12}=G_{13}=G_{23}=E_2/2 \ , \quad \nu_{12}=0.25 \ . \tag{14}$$

The family consists of panels having the same axial length (L_x=b) and plan form (b×b), so that changing the value of the shallowness angle ϕ, the panel circumferential length changes according to :

$$L_s=b\phi/2\sin(\phi/2) \ . \tag{15}$$

The normalized fundamental frequency parameter

$$\bar{\omega}=\omega_{min}b^2(\rho_0/E_2h^3)^{\frac{1}{2}} \ , \tag{16}$$

as a function of the lamination angle θ, is shown in figures 2-4 for several values of the shallowness angle ϕ and for various numbers of layers. Results based on classical and shear deformable theory are drawn with dashed and solid lines, respectively.

Apparently, all fundamental frequencies predicted on the basis of the

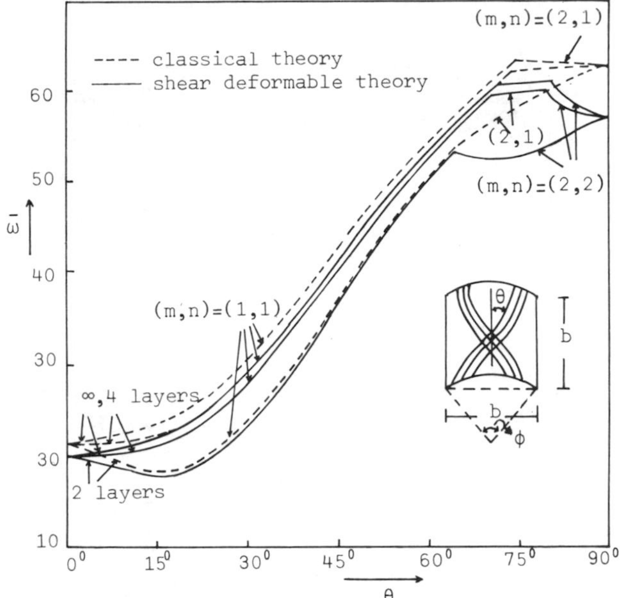

Figure 3. Fundamental frequency parameter as a function of the lamination angle ($\phi=30^0$).

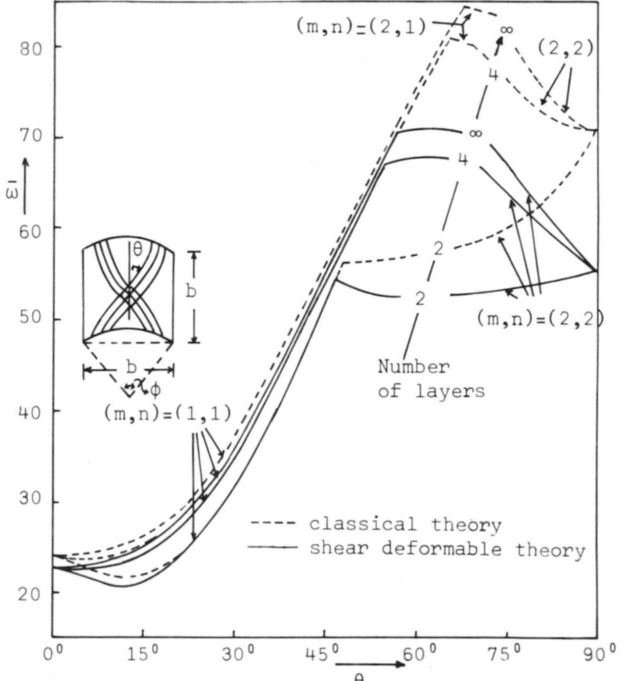

Figure 4. Fundamental frequency parameter as a function of the lamination angle ($\phi=45^0$).

thickness shear deformation theory are lower than the corresponding ones based on the classical theory. The fundamental frequency reduction due to the consideration of thickness shear deformation seems that depends on the value of several parameters as the shellowness angle, ϕ , the lamination angle, θ , and the number of layers. For instance, for small shallowness angles (figure 2), where the fundamental frequency has nominal half-waves (m,n)=(1,1), the afore-mentioned reduction is, in general, small and very rarely exceeds the value of the engineering admissible error (5%). On the contrary, upon increasing the value of the shallowness angle (figures 3 and 4) there are cases in which this reduction is increased considerably. In particular, this happens for large values of the lamination angle ($\theta>70^0$ for $\phi=30^0$ and $\theta>50^0$ for $\phi=45^0$) where the mode shape changes and the fundamental frequency occur for combinations of half-wave numbers different than (m,n)=(1,1).

CONCLUDING REMARKS

The free vibration problem of shear deformable angle-ply laminated circular cylindrical panels has been considered. The choise of the cylindrical panel geometry as a basic configuration was recommented from the fact that, dependent on the value of a shallowness angle, both the flat plate and the circular cylindrical shell configurations can be considered as special cases. However, most of the numerical results presented were concerned with open panels. Moreover, the present study was based on Love's first-order approximations while all relevant studies, already available in the literature, were based on a Donnell-type quasi-shallow classical shell theory. Under these considerations, it should be noted that the present article is a part of a recent, and as yet unpublished, paper [21]. There, classical and shear deformable theories based on Donnell's approximations are also used and their results are compared with corresponding results based on Love-type approximations. Moreover, Mindlin-type shear deformable theories, assuming a uniform thickness shear, have been considered. In [21], finally, further results concerned with the flat plate as well as the closed shell configurations are presented and compared with corresponding results available in the literature.

REFERENCES

1. Jones, R.M., Mechanics of Composite Materials. McGraw Hill, 1975, New York.

2. Reissner, E., Stavsky, Y., Bending and stretching of certain types of heterogeneous aeolotropic elastic plates. J.Appl.Mech. 28, 1961, 402.

3. Whitney, J.M., Bending-extensional coupling in laminated plates under transverse loading. J.Comp.Mater. 3, 1969, 20.

4. Whitney, J.M., Cylindrical bending of unsymmetrically laminated plates. J.Comp.Mater. 3, 1969, 715.

5. Whitney, J.M., Leissa, A.W., Analysis of heterogeneous anisotropic plates. J.Appl.Mech. 36, 1969, 261.

6. Whitney, J.M., Leissa,A.W., Analysis of simply supported laminated anisotropic rectangular plates. AIAA J. 8, 1970, 28.

7. Whitney, J.M., The effect of boundary conditions on the response of laminated composites. J.Comp.Mater. 4, 1970, 192.

8. Jones, R.M., Morgan, H.S., Whitney, J.M., Buckling and vibration of antisymmetrically angle-ply laminated rectangular plates. J. Appl. Mech. 40, 1973, 1143.

9. Fortier, R.C., Rossetos, J.N., Effects of inplane constrains and curvature on composite plate behavior. Int.J.Solids Struct. 10, 1974, 1417.

10. Sharma, S., Iyengar, N.G.R., Murthy, P.N., Buckling of antisymmetric cross- and angle-ply laminated plates. Int.J.Mech.Sci. 22, 1980, 607.

11. Sharma, S., Iyengar, N.G.R., Murthy, P.N., Buckling of antisymmetrically cross-ply and angle-ply rectangular plates. J.Eng.Mech.Div. ASCE 106, 1980, 161.

12. Hirano, Y., Buckling of angle-ply laminated circular cylindrical shells. J.Appl.Mech. 46, 1979, 233.

13. Soldatos, K.P., On the buckling and vibration of antisymmetric angle-ply laminated circular cylindrical shells. Int.J.Eng.Sci. 21, 1983, 217.

14. Soldatos, K.P., Free vibrations of antisymmetric angle-ply laminated circular cylindrical panels. Q.J.Mech. Appl.Math. 36, 1983, 207.

15. Whitney, J.M., Buckling of anisotropic laminated cylindrical plates, AIAA J. 22, 1984, 1641.

16. Whitney, J.M., Pagano, N.J.,Shear deformation in heterogeneous anisotropic plates. J.Appl.Mech. 37, 1970, 1031.

17. Bert, C.W., Chen, T.L.C., Effect of shear deformation on vibration of antisymmetric angle-ply laminated rectangular plates. Int.J.Solids Struct. 14, 1978, 465.

18. Reddy, J.N., Phan, N.D., Stability and vibration of isotropic, orthotropic and laminated plates according to a higher-order shear deformation theory. J.Sound.Vib. 98, 1985, 157.

19. Timoshenko, S., On the correction for shear of the differential equation for transverse vibrations of prismatic bars. Phil.Mag.,series 6, 41, 1921, 744.

20. Mindlin, R.D., Influence of rotatory inertia and shear on flexural motions of isotropic elastic plates. J.Appl.Mech. 18, 1951, 31.

21. Soldatos, K.P., Influence of thickness shear deformation on free vibration of simply supported antisymmetric angle-ply laminates. Tech. Rept. 128, Dept. of Math.,Univ. of Ioannina, 1986.

MICROLOCAL EFFECTS IN THE NONSTANDARD MODELLING OF PERIODIC COMPOSITES

S.J. MATYSIAK, Cz. WOŹNIAK

Department of Mechanics,
University of Warsaw,
Poland.

Abstract. The concepts of nonstandard analysis are applied to the homogenization of periodic linear-elastic composites. On this way not only mean but also local strain and stresses in every periodicity cell (microlocal effects) can be evaluated. Some applications to the wave propagation problems in laminated bodies are presented.

Introduction. Generally speaking, two lines of approach to the homogenized modelling of periodic composites can be distinguished in the recent investigations. One of them is based on rather heuristic assumptions, cf. [1,2] , and the other takes into account asymptotic theorems, cf. [3,4] . The approach we are going to outline here comprises characteristic features of both lines of approach and was developed as one from many examples of applications of the nonstandard analysis in the mathematical modelling of various problems within discrete and solid mechanics, [5] . The general idea of the proposed approach is based on the assumption that many real composite bodies with periodic material structures can be represented by continuous models in which all length dimensions of any periodicity cell are "infinitely small". The concept of actual "infinitely small" entities is well defined within the framework of the nonstandard analysis,[6]. That is why we shall use the nonstandard analysis as our mathematical tool. However, the resulting relations obtained via nonstandard method involve exclusively "standard" mathematical entities. Moreover, the method proposed will describe what are called "microlocal effects", such as strain and stress distributions in every periodicity cell.

The nonstandard method of modelling of periodic composites is very general and has been formulated both for thermo-elastic and thermo-unelastic materials within the scope of the theory of large strains and large temperature gradients, [7,8] . However, for the sake of simplicity, we confine ourselves here only to the case of linear elastic bodies. The note is intended to readers who do not know much about nonstandard analysis and we start with some basic "nonstandard" mathematical ideas which will be used in the sequel.

What is nonstandard analysis ? Roughly speaking, the nonstandard analysis can be treated as the analysis in which we deal not with the field R of real numbers but with a certain nonarchimedean field $^{*}R$ of reals, i.e., a field which apart from "standard" real numbers also contains infinitely small and infinitely large positive and negative numbers. To be more exact, the foundations of nonstandard analysis are based on the mathematical fact that for every full mathematical structure $\mathfrak{M}$ ([1]) there exists another mathematical structure $^*\mathfrak{M}$ which is an enlargement of $\mathfrak{M}$. All what we need in applications to mechanics is that such enlargement exists and has the following relevant properties, [6] :

1. Extension property: every mathematical notion which is meaningfull for $\mathfrak{M}$ is also meaningfull for $^*\mathfrak{M}$. It means that every entity e of $\mathfrak{M}$ extends uniquely and naturally to the entity $^*e$ of $^*\mathfrak{M}$; entity $^*e$ is called standard. If e is an individual of $\mathfrak{M}$ (for example a real number) then we shall identify $^*e$ with e, setting $^*e \equiv e$.

2. Enlargement property: every standard entity $^*\varrho$ of $^*\mathfrak{M}$, which is an infinite set (and only in this case) contains also nonstandard elements.

3. Externity property: every infinite set of standard entities is not an element of $^*\mathfrak{M}$ (i.e., $^*\mathfrak{M}$ is not a full structure).

4. Permanence property: every mathematical statement which holds (is meaningfull and true) in $\mathfrak{M}$ also holds in $^*\mathfrak{M}$, provided that it is interpreted exclusively in terms of entities belonging to $^*\mathfrak{M}$.

Assuming that $\mathfrak{M}$ contains the field R of real numbers, from the extension property we obtain in $^*\mathfrak{M}$ a set $^*R$ which, by virtue of the permanence property, also constitutes an ordered field. Let $^{\circ *}R$ stands for a set of all standard real numbers in $^*\mathfrak{M}$ (from the externity property it follows that $^{\circ *}R$ is not an element of $^*\mathfrak{M}$); then the enlargement property implies that $^*R \setminus {^{\circ *}R} \neq \emptyset$, i.e., that in $^*R$ we have also nonstandard real numbers. It can be shown that to $^*R$ belong infinite numbers, absolute value of which are greater than any standard number. Hence by means of the permanence property, to $^*R$ also belong inverses of infinite numbers which are infinitely small (their absolute values are smaller than any positive standard number). For any two finite numbers $a,b \in {^*R}$ (i.e., numbers which are not unfinite) we write $a \simeq b$ (a is infinitely close to b) if and only if $a-b$ is an infinitely small number.

By the nonstandard analysis we mean the mathematical analysis related to a certain enlargement $^*\mathfrak{M}$ of an arbitrary full structure $\mathfrak{M}$. Nonstandard analysis as a mathematical tool of mechanics enable us to describe simultaneously (avoiding the limit passages) physical entities which can be qualified as being of the different order (macro and micro

---

([1]) By a full mathematical structure we mean a pair $\mathfrak{M} = (X, T(X))$ where X is an infinite set of entities called individuals (they are nor sets neither relations) and $T(X)$ is a collection of all sets and all relations. In the sequel we shall assume that X coincides with a set R of real numbers.

entities). For the applications of nonstandard analysis in modelling of various material systems within mechanics the reader is referred to [5] where also the list of references can be found.

Nonstandard modelling. Let at the time instant $t = t_o$ a linear elastic body occupies the region Ω in R^3. Under the known denotations (2) the governing relations for the displacement fields $u_i(\cdot,t)$, and stress fields $\sigma_{ij}(\cdot,t)$, $t \in [t_o,t_1]$, can be given by

$$
(1)\quad
\begin{aligned}
&\int_{\Omega} \sigma_{ij}(x,t)v_{(i,j)}(x)dV = \oint_{\partial\Omega} t_i(x,t)v_i(x)dA + \\
&\qquad + \int_{\Omega} \left[f_i(x,t) - \rho(x)\ddot{u}_i(x,t)\right] v_i(x)dV \\
&\sigma_{ij}(x,t) = A_{ijkl}(x)\, u_{(k,l)}(x,t);\; x \in \Omega\,,\; t \in [t_o,t_1]\,,
\end{aligned}
$$

where $(1)_1$ has to hold for every pertinent test function $v_i(\cdot)$ defined on $\bar{\Omega}$ (the initial conditions are assumed to be known). Define $Y \equiv [0,Y_1] \times [0,Y_2] \times [0,Y_3]$ and let the material structure of the body under consideration be Y-periodic. Then $\rho(\cdot)$, $A_{ijkl}(\cdot)$ can be treated as Y-periodic functions defined on R^3. Let us also assume that $\max Y_i$ is very small as compared with the smallest characteristic length dimension of $\bar{\Omega}$. The problem of finding $u_i(\cdot)$ and $\sigma_{ij}(\cdot)$ formulated here will be denoted by $(\mathcal{P})$.

Let $(\mathcal{P}_n)$ for every $n \geqslant 1$, be a problem obtained from the problem $(\mathcal{P})$ via replacing Y-periodic material structure of the body by Y/n-periodic structure, where $Y/n \equiv [0,Y_1/n] \times [0,Y_2/n] \times [0,Y_3/n]$. It is easy to see that the governing relations of $(\mathcal{P}_n)$ can be obtained from Eqs. (1) by substituting $\rho(nx)$, $A_{ijkl}(nx)$ in the place of $\rho(x)$, $A_{ijkl}(x)$, respectively. Obviously, problems $(\mathcal{P})$ and $(\mathcal{P}_1)$ coincide.

The nonstandard method of homogenized modelling of the problem $(\mathcal{P})$ will be based on two following heuristic assumptions:

---

(2) Subscripts i,j,k,l run over $1,2,3$ and subscripts a,b run over $1,\dots,n$; summation convention holds. A comme denotes partial differentiation and $a_{(ij)} = 0.5(a_{ij} + a_{ji})$ for every second order tensor a_{ij}.

Assumptions 1. The solution of $(\mathcal{P})$ can be approximated by means of the formulas

$$
(2)\quad
\begin{aligned}
u_i(x,t) &= w_i(x,t) + \underline{q_{ai}(x,t)l_a(x)} \\
\sigma_{ij}(x,t) &= A_{ijkl}(x)\big[w_{(k,l)}(x,t) + \\
&\quad + q_{a(k}(x,t)l_{a,l)}(x) + \\
&\quad + \underline{q_{a(k,l)}(x,t)l_a(x)}\big] ,
\end{aligned}
$$

where $l_a(\cdot)$ are the postulated a priori Y-periodic oscillating function defined on R^3 and $w_i(\cdot)$, $q_{ai}(\cdot)$ are certain arbitrary sufficiently regular functions which are approximatively constant within every periodicity cell. It means that

$$
(3)\quad
\begin{aligned}
&(\forall y,z \in \Omega)\,\big[\,[|y_i - z_i| < Y_i\,] \Rightarrow \\
&\Rightarrow \big[w_i(y,t) \simeq w_i(z,t),\ q_{ai}(y,t) \simeq q_{ai}(z,t)\big]\big], \\
&t \in [t_0, t_1] ,
\end{aligned}
$$

where $\simeq$ stands for the intuitive relation "is nearly equal". Functions $w_i(\cdot,t)$ and $q_{ai}(\cdot,t)$ are called macrodisplacements and microlocal parameters, respectively. Postulated in every problem under consideration functions $l_a(\cdot)$ are called the shape functions; their choice depends of the material structure of the periodicity cell and has to be based on certain heuristic premises.

Assumption 2. The functions $w_i(\cdot)$, $q_{ai}(\cdot)$ in the problem $(\mathcal{P})$ determine the approximate solutions $u_i(\cdot)$, $\sigma_{ij}(\cdot)$ also for every problem $(\mathcal{P}_n)$, $n \geqslant 1$, provided that the shape functions $l_a(x)$ in Eqs. (2) are replaced by the shape functions $l_a^{(n)}(x) \equiv l_a(nx)/n$. It means that macrodisplacements $w_i(\cdot,t)$ and microlocal parameters $q_{ai}(\cdot,t)$ have a similar form in every problem $(\mathcal{P}_n)$.

The Y-periodic bodies which satisfy both aforementioned assumptions will be called microperiodic. The microperiodic character of a body depends not only on the values of parameters Y_i but also on the form of body loadings as well as on that of boundary and initial conditions.

It has to be emphasized that formulas (2) do not constitute the analytical basis for the calculating of approximate solution of the problem $(\mathcal{P})$, since the conditions (3) have only intuitive character and do not determine any well defined class of functions $w_i(\cdot)$, $q_{ai}(\cdot)$. However, treating the governing relations of problems $(\mathcal{P}_n)$, $n \geq 1$, as the unfinite sequence of relations of a certain full structure $\mathfrak{M}$, we shall pass with the sequence of problems $(\mathcal{P}_n)$, $n \geq 1$, from $\mathfrak{M}$ to ${}^*\mathfrak{M}$. Hence Ω, $f_i(\cdot)$, $t_i(\cdot)$ extend in ${}^*\mathfrak{M}$ to the known standard entities. Now setting $n = \breve{\omega}$, $\breve{\omega}$ being an arbitrary but fixed infinite positive integer, we obtain in ${}^*\mathfrak{M}$ the problem $(\mathcal{P}_{\breve{\omega}})$ governed by

$$\int_{{}^*\Omega} \sigma_{ij}(x,t) v_{(i,j)}(x,t) dV = \oint_{\partial^*\Omega} {}^*t_i(x,t) v_i(x) dA + $$

$$+ \int_{{}^*\Omega} \left[{}^*f_i(x,t) - \varrho(\breve{\omega} x) \ddot{u}_i(x,t) \right] v_i(x) dV \tag{4}$$

$$\sigma_{ij}(x,t) = A_{ijkl}(\breve{\omega} x) u_{(k,l)}(x,t), \quad x \in {}^*\Omega ,$$

$$t \in {}^*[t_o, t_1] ,$$

where $(4)_1$ has to hold for every pertinent test function $v_i(\cdot)$ defined on ${}^*\bar{\Omega}$.

Let us look for a solution of $(\mathcal{P}_{\breve{\omega}})$ in the form

$$u_i(x,t) = {}^*w_i(x,t) + {}^*q_{ai}(x,t) l_a^{(\breve{\omega})}(x),$$

$$l_a^{(\breve{\omega})}(x) \equiv \frac{1}{\breve{\omega}} l_a(\breve{\omega} x) , \tag{5}$$

$$\sigma_{ij}(x,t) = A_{ijkl}(\breve{\omega} x) \left[{}^*w_{(k,l)}(x,t) + \right.$$

$$+ {}^*q_{a(k}(x,t) l_{a,l)}^{(\breve{\omega})}(x) + {}^*q_{a(k,l)}(x,t) \cdot$$

$$\left. \cdot \, l^{(\breve{\omega})}(x) \right] ,$$

with standard macrodisplacements and microlocal parameters.

The crucial point of the approach is based on the fact that the problem $(\mathcal{P}_{\omega})$ given in $^*\mathfrak{M}$ by Eqs. (4) is well defined under conditions (5) and leads to the system of equations in $\mathfrak{M}$ for unknown functions $w_i(\cdot)$, $q_{ai}(\cdot)$. Setting

$$\langle\varphi\rangle = \frac{1}{Y_1Y_2Y_3}\int_Y \varphi(x)dV$$

for any integrable Y-periodic function $\varphi(\cdot)$, we shall formulate the following:

Homogenization statement. Functions $w_i(\cdot)$, $q_{ai}(\cdot)$ in Eqs. (4), (5) have to satisfy the system of equations in $\Omega\times[t_o,t_1]$ with constant coefficients, given by

$$\begin{aligned}
&\tau_{ij,j}(x,t) + f_i(x,t) = \langle\varrho\rangle\, \ddot{w}_i(x,t)\,,\\
&\zeta_{ai}(x,t) = 0\,,\\
(6)\quad &\tau_{ij}(x,t) = \langle A_{ijkl}\rangle w_{(k,l)}(x,t) +\\
&\qquad + \langle A_{ijkl}l_{a,k}\rangle\, q_{al}(x,t)\,,\\
&\zeta_{ai}(x,t) = \langle A_{ijkl}l_{a,j}\rangle\, w_{(k,l)}(x,t) +\\
&\qquad + \langle A_{ijkl}l_{a,j}l_{b,k}\rangle\, q_{bl}(x,t)\,.
\end{aligned}$$

For the proof of the statement cf. [7]. It is easy to see that Eqs. (6) lead to the system of 3n linear algebraic equations for microlocal parameters $q_{ai}(x,t)$ and 3 partial differential equations for macrodisplacements $w_i(x,t)$. Thus we have arrived to a certain class of homogenized models of the problem $(\mathcal{P})$, given by Eqs. (6). The form of every model depends on the choice of the shape functions $l_a(\cdot)$, $a = 1,\ldots,n$. It has been proved, [10], that a special choice of these functions leads to the classical homogenization theory obtained via asymptotic approach, [3,4].

After obtaining solution $w_i(\cdot)$, $q_{ai}(\cdot)$ of the pertinent boundary value problem for Eqs. (6) (which is implied by the problem $(\mathcal{P}_{\omega})$ and hence approximates problem $(\mathcal{P})$), we can use again formulae (2) and to verify conditions (3). In the microperiodic composites the underlined terms in Eqs. (2), i.e., the terms involving $l_a(x)$, are small and can be neglected. On the other hand terms involving the derivatives $l_{a,i}(x)$ are not small and describe the microlocal effects in the body under consideration. Hence we see that the microlocal effects do not affect displacements but they play an important role in the evaluation of stresses.

Wave propagation in laminated composites. The homogenized equations (6) may be used to study the propagation of waves in periodic composites. Here we confine attention to the periodic laminated body in which every lamina is composed of two homogeneous isotropic linear-elastic layers. Let λ_1, μ_1 and λ_2, μ_2 are Lamé constants, ϱ_1, ϱ_2 are the (constant) mass densities, h_1, h_2 are the thicknesses of the subsequent layers. Let $x \equiv (x_1, x_2, x_3)$ be the rectangular Cartesian coordinates such that the axis x_1 is normal to the layering.

Consider the shape function l_1 as the sectionaly linear $(0,\delta)$-periodic function defined by:

$$l_1(x_1) = \begin{cases} x_1 - 0.5h_1 & \text{for } 0 \leq x_1 \leq h_1 \\ -\frac{\eta}{1-\eta}x_1 - 0.5h_1 + \frac{h_1}{1-\eta} & \text{for } h_1 < x_1 \leq \delta \end{cases}, \tag{7}$$

where $\eta \equiv h_1/\delta$, $\delta \equiv h_1 + h_2$.

Using Eqs. (6) and (7), after some calculations, we obtain the following system of equations, cf. [9] :

$$\begin{aligned} &(\tilde{\lambda}+\tilde{\mu})(\operatorname{div} w)_{,1} + \tilde{\mu}\nabla^2 w_1 + ([\lambda] + 2[\mu])q_{11,1} + \\ &[\mu]\, q_{1k,k} + f_1 = \tilde{\varrho}\ddot{w}_1, \qquad k=2,3, \\ &(\tilde{\lambda}+\tilde{\mu})(\operatorname{div} w)_{,k} + \tilde{\mu}\nabla^2 w_k + [\lambda]q_{11,k} + [\mu]q_{1k,1} \\ &+ f_k = \tilde{\varrho}\ddot{w}_k, \\ &(\hat{\lambda}+2\hat{\mu})q_{11} + [\lambda]\operatorname{div} w + 2[\mu]w_{1,1} = 0, \\ &\hat{\mu}\, q_{1k} + [\mu](w_{k,1} + w_{1,k}) = 0, \end{aligned} \tag{8}$$

where

$$\begin{aligned} &\tilde{\lambda} \equiv \eta\lambda_1 + (1-\eta)\lambda_2, \quad [\lambda] \equiv \eta(\lambda_1 - \lambda_2), \\ &\tilde{\mu} \equiv \eta\mu_1 + (1-\eta)\mu_2, \quad [\mu] \equiv \eta(\mu_1 - \mu_2), \\ &\hat{\lambda} \equiv \eta(\lambda_1 + \eta(1-\eta)^{-1}\lambda_2), \quad \tilde{\varrho} \equiv \eta\varrho_1 + (1-\eta)\varrho_2 \end{aligned} \tag{9}$$

$$\hat{\mu} \equiv \eta(\mu_1 + \eta(1-\eta)^{-1}\mu_2).$$

The microlocal parameters $q_{1i}(\cdot)$ can be easily eliminated from Eqs. $(8)_{1,2}$ and hence the system of 3 partial differential equations for unknown macrodisplacements w_i may be obtained.
Consider now the two-dimensional problem of (shock or acceleration) wave front propagation in the composites under consideration. We assume that the singular surface S (i.e. the surface of a discontinuity in the first or the second order of derivative of the displacement vector) in space-time is given in the form:

$$\Psi(x_1,x_2) - t = 0 . \tag{10}$$

A straightforward calculation, using the results of the references, cf. [14] and Eqs. (8) reveals that this surface must satisfy the nonlinear partial differential equation:

$$\begin{aligned}(\Psi_{,1})^2(A_1+C) + (\Psi_{,2})^2(A_2+C) \pm \{[(\Psi_{,1})^2(A_1-C) -\\ - (\Psi_{,2})^2(A_2-C)]^2 + 4B^2(\Psi_{,1})^2(\Psi_{,2})^2\}^{1/2} - 2\tilde{\varrho} = 0\end{aligned} \tag{11}$$

where sign "+" in Eqs. (11) is for a longitudinal wave front and sign "-" in Eqs. (11) is for a transverse wave front and where

$$\begin{aligned}A_1 &\equiv \tilde{\lambda} + 2\tilde{\mu} - ([\lambda] + 2[\mu])^2(\hat{\lambda}+2\hat{\mu})^{-1} ,\\ A_2 &\equiv \tilde{\lambda} + 2\tilde{\mu} - [\lambda]^2(\hat{\lambda}+2\hat{\mu})^{-1} ,\\ B &\equiv \tilde{\lambda} + \tilde{\mu} - [\lambda]([\lambda] + 2[\mu])(\hat{\lambda}+2\hat{\mu})^{-1} - [\mu]^2\hat{\mu}^{-1} ,\\ C &\equiv \tilde{\mu} - [\mu]^2\hat{\mu}^{-1} .\end{aligned} \tag{12}$$

In the case $\mu_1 = \mu_2$, Eqs. (11) reduce to the following:

a) for a longitudinal wave front:

$$(\Psi_{,1})^2 + (\Psi_{,2})^2 - \tilde{\varrho}\,\bar{A}^{-1} = 0 , \quad \bar{A} = \tilde{\lambda} + 2\mu_1 - [\lambda]^2(\hat{\lambda}+2\mu_1)^{-1} ,$$

b) for a tranverse wave front:

$$(\Psi_{,1})^2 + (\Psi_{,2})^2 - \tilde{\varrho}\,\mu_1^{-1} = 0 .$$

The detailed analysis and the further results concerning of these problems can be found in [9,15-16].

<u>Final remarks.</u> Following [7], we shall mention some of the general features of the proposed homogenized models of periodic composite materials:

1° In every problem they can be based on certain physical premises, which make it possible to choose the proper form of the shape functions $l_a(\cdot)$.

2° They can be formulated in various forms involving more or less microlocal parameters $q_{ai}(\cdot)$.

3° They describe not only mean but also local stresses in every periodicity cell.

4° Their analytical structure is not more complicated then that of the governing relations of the linear elastic homogeneous bodies.

5° They can be also applied to linear as well as non-linear problems, including the cases of large deformation and large temperature gradients, as well as to elastic and unelastic materials and to rectilinear and curvilinear composite structures.

For the detailed analysis of the foundations and for the examples of applications of the nonstandard method of modelling of periodic composites the reader is referred to [7-13,15]

References

1. C.T.Sun, J.D.Achenbach, G.Herrmann, Continuum theory for a laminated medium, J. Appl. Mech., 35(1968), 467-475.
2. R.M.Christensen, Mechanics of composite materials, J.Wiley and Sons, New York, 1980.
3. A.Bensoussan, J.L.Lions, G.Papanicoleau, Asymptotic analysis for periodic structures, North Holl.Publ.Comp., Amsterdam 1978.
4. E.Sanchez-Palencia, Non-homogeneous media and vibration theory, Springer Verlag, Berlin 1980.
5. Cz.Woźniak, Nonstandard analysis in mechanics, Adv. in Mech., 9(1986), 3-35.
6. A.Robinson, Non-standard analysis, North Holl.Publ.Comp., Amsterdam 1976.
7. Cz.Woźniak, A nonstandard method of modelling of thermoelastic periodic composites, Int. J. of Engng Sci., in press.
8. Cz.Woźniak, On the nonstandard modelling of thermo--unelastic periodic composites, in press.
9. S.J.Matysiak, Cz.Woźniak, Micromorphic effects in a modelling of periodic multilayered elastic composites, Int. J. of Engng Sci., in press.
10. T.Lewiński, On the interrelation between microlocal parameter and homogenization approaches to the periodic elastic solids, Bull. Acad. Polon. Sci., Sér. Sci.Techn., in press.
11. S.J.Matysiak, Cz.Woźniak, On the modelling of heat conduction problem in laminated bodies, Acta Mechanica, in press.
12. S.J.Matysiak, M.Wągrowska, Certain axially-symmetric

problems of heat conduction in laminated bodies, Bull. Acad. Polon. Sci., Sér. Sci. Techn., in press.

13. M.Wągrowska, Certain solutions of axially-symmetric problems in linear elasticity with microlocal parameters, Bull. Acad. Polon. Sci., Sér. Sci. Techn., in press.

14. T.Y.Thomas, Plastic flow and fracture of solids, Academic Press, New York, London, 1961.

15. S.J.Matysiak, P.Pusz, Harmonic vibrations of microperiodic multilayered composites, to appear in Bull. Polon. Acad. Sci,, Sér. Sci. Techn.

16. E.Jakubowska, S.J.Matysiak, Propagation of plane harmonic waves in a microperiodic multilayered composites, to appear in Mech. Teor. Stos.

HOMOGENIZATION FOR A QUASI-STATIC PROBLEM OF TORSION OF A PRISMATIC PERIODIC FIBROUS BAR WITH COULOMB FRICTION ON THE FIBRE-MATRIX INTERFACE

W.S. Barański
School of Engineering and Engineering Technology, Federal University of Technology, Minna, P.O. Box 656, Nigeria

Z. Więckowski
Institute of Construction Engineering, Technical University of Łódź, Al. Politechniki 6, 93-590 Łódź, Poland

The macroscopic constitutive equation corresponding to the considered problem of torsion is derived using the method of homogenization theory. It is shown that the macroscopic behaviour of the composite is close to plasticity with strain hardening. Strain-stress paths illustrating the result are presented.

INTRODUCTION

In this paper we investigate the macroscopic constitutive equation corresponding to the problem of quasi-static torsion of a prismatic, periodic fibrous bar consisting of isotropic linearly elastic components. As in [1] it is assumed that the fibre-matrix interface behaves according to Outwater's model, i.e. that there is no cohesion between matrix and fibres and that the possible matrix-fibre slip is governed by Coulomb friction law with given, caused by shrinkage of the matrix, normal tractions on the interface.

The analysis is based on the homogenization theory for which we refer to [2]-[5]. Under appropriate hypotheses, concerning the mechanical and geometric properties of the bar, the homogenization theorem is proved. It is shown that the obtained macroscopic constitutive equation is close to plastic behaviour with strain hardening [6], [7]. The resulting strain-stress paths are similar to the experimental ones of Grimes and Francis [8].

The problem considered in the paper is somehow similar to that one investigated in [1]. However, it should be mentioned that previously we have considered only the static case which does not involve time into considerations.

NOTATION

Let R^2 denote the 2-dimensional Euclidean space with the inner product denoted by the dot $\cdot$, and the norm denoted by $| \; |$. We shall denote vectors in R^2 by x, y, z.

Let $Y = (y_1, y_2)$ be a given pair of linearly independent vectors in R^2. We call a function f Y-periodic if

$$f(x + y) = f(x) \qquad \forall x \in R^2 \qquad \forall y \in Y.$$

For a scalar α, a vector $x \in R^2$, subsets A, B of R^2 and a function f on R^2 we define

$$\alpha A = \{\alpha y : y \in A\}, \qquad x + A = \{x + y : y \in A\},$$

$$A + B = \{y + z : y \in A,\ z \in B\}, \qquad f^{\alpha}(x) = f(x/\alpha).$$

For each subset A of R^2 we denote its Lebesgue measure by meas A.

In the paper we shall deal with several Banach spaces [9], first of all with Sobolev spaces - for which we refer to [10]. We shall use also the space $M^1(\Omega)$ of bounded measures on Ω, and the space $BV(\Omega)$ of functions with bounded variation [11]. We shall denote the norm on a Banach space X by $\| \ \|_X$.

Assuming that $X(\Omega)$ is $L^2(\Omega)$ or $H^1(\Omega)$ we denote

$$X_{loc}(R^2) = \{f : f|_{\omega} \in X(\omega) \ \text{for all open bounded} \ \omega \subset R^2\},$$

$$L^2(\Omega; X_{loc}(R^2)) = \{f : f|_{\Omega\times\omega} \in L^2(\Omega; X(\omega)) \ \text{for all bounded} \ \omega \subset R^2\}.$$

BASIC ASSUMPTIONS AND VARIATIONAL FORMULATION OF THE QUASI-STATIC TORSION PROBLEM

Let us consider an infinitely long prismatic bar with a cross section Ω which fulfils the requirements of boundedness and appropriate regularity. Let ε, being a real number from the interval $]0,1]$, characterize the size of microstructure of the considered bar (see Fig. 1.). We assume that for

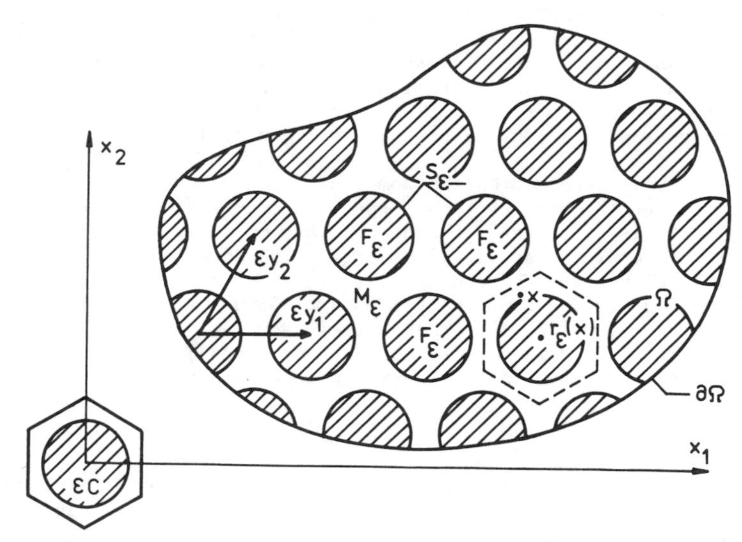

Fig. 1.

each such ε the microstructure is such that the cross section of the fibres, of the matrix and of the interface occupy the sets

$$F_{\varepsilon} = \Omega \cap \varepsilon F, \qquad M_{\varepsilon} = \Omega \cap \varepsilon M, \qquad S_{\varepsilon} = \Omega \cap \varepsilon S$$

with F,M,S being Y-periodic and sufficiently regular. We assume that for each $\varepsilon \in]0,1]$ the mechanical properties of material of corresponding composite are specified by the Kirchhoff modulus function $G^{\varepsilon}(x) = G(x/\varepsilon)$ and by the function of allowable friction stress $g^{\varepsilon}(x) = g(x/\varepsilon)$ with G and g being Y-periodic, positive and piecewise constant.

Let $\theta(t)$, being a given real valued function of time, denote the angle of twist per unite length of the bar at the moment $t \in I =]0,T[$. For each $\theta \in H^1(I)$ we shall look for the corresponding deplanation function $w_{\varepsilon}(t,x)$ and for the shear stress function $\sigma_{\varepsilon}(t,x)$ fulfilling the appropriate principles of mechanics [12], i.e. Hooke's law

$$\sigma_{\varepsilon} = G^{\varepsilon}(\text{grad}_{\varepsilon} w_{\varepsilon} + \theta a) \qquad \text{on} \quad I \times \Omega_{\varepsilon}, \tag{1}$$

the virtual work principle

$$\int_{\Omega_{\varepsilon}} \sigma_{\varepsilon} \cdot \text{grad}_{\varepsilon} v dx + \int_{S_{\varepsilon}} g^{\varepsilon} |[\![v]\!]_{\varepsilon}| ds \geq 0 \quad \forall v \in H^1(\Omega_{\varepsilon}) \quad \text{on} \quad I, \tag{2}$$

and the power balance condition

$$\int_{\Omega_{\varepsilon}} \sigma_{\varepsilon} \cdot \text{grad}_{\varepsilon} \dot{w}_{\varepsilon} dx + \int_{S_{\varepsilon}} g^{\varepsilon} |[\![\dot{w}]\!]_{\varepsilon}| ds = 0 \qquad \text{on} \quad I \tag{3}$$

with

$$a_{\alpha}(x) = -\in_{\alpha\beta} x_{\beta}, \qquad \Omega_{\varepsilon} = \Omega \setminus S_{\varepsilon}$$

and the brackets $[\![\ ]\!]_{\varepsilon}$ denoting the jump on S_{ε}.

To avoid the nonuniqueness of solutions we assume additionally that for each time $t \in I$

$$\int_{\Omega} w_{\varepsilon} dx = 0. \tag{4}$$

For the sake of completeness we should prescribe corresponding initial condition. In the considered quasi-static case it is sufficient to prescribe only the initial deformation state of the bar, for example

$$w_{\varepsilon}|_{t=0} = 0, \tag{5}$$

$$\theta|_{t=0} = 0. \tag{6}$$

We set the considered quasi-static problem as follows:

$P1_{\varepsilon}$: For given $\varepsilon \in]0,1]$ and $\theta \in H^1(I)$ fulfilling the condition (6) find $w_{\varepsilon} \in H^1(I; H^1(\Omega_{\varepsilon}))$ and $\sigma_{\varepsilon} \in H^1(I; L^2(\Omega; R^2))$ which satisfy (1)-(5).

The problems of type $P1_{\varepsilon}$ are known to have unique solutions [12], [13]. Hence, we can define

$$\gamma_{\varepsilon} = \text{grad}_{\varepsilon} w_{\varepsilon} \qquad \text{on} \quad I \times \Omega_{\varepsilon}, \tag{7}$$

$$\overline{\gamma}_{\varepsilon} = \text{grad}\, w_{\varepsilon} \qquad \text{on} \quad I \times \Omega. \tag{8}$$

Note that for each fixed time t the function $\gamma_{\varepsilon}(t,\cdot)$ can be considered as element of $L^2(\Omega; R^2)$. However, the possible slip on the interface makes $\gamma_{\varepsilon}(t,\cdot)$ an element of $M^1(\Omega; R^2)$.

In further analysis we shall consider the functions $w_\varepsilon, \sigma_\varepsilon, \gamma_\varepsilon, \overline{\gamma}_\varepsilon$ as defined a.e. on the whole $I \times R^2$ with zero value on $I \times (R^2 \setminus \Omega)$.

HOMOGENIZATION PROCEDURE

In this section we shall briefly present the adopted method of homogenization analysis, which is a generalization of that one established in [5]. The method involves the so-called averaged fields and local fields which are defined as follows:

$$\sigma_\varepsilon^{av}(t,x) = \varepsilon^{-2}(\text{meas } C)^{-1} \int_{r_\varepsilon(x)+\varepsilon C} \sigma_\varepsilon(t,z)dz$$

and

$$\overline{\gamma}_\varepsilon^{av}(t,x) = \varepsilon^{-2}(\text{meas } C)^{-1}\{ \int_{r_\varepsilon(x)+\varepsilon C} \gamma_\varepsilon(t,z)dz + \int_{r_\varepsilon(x)+\varepsilon(C\cap S)} [\![w_\varepsilon]\!]_\varepsilon(t,z)n_\varepsilon(z)ds_z\} \quad \text{a.e.} \quad (t,x) \in I \times R^2,$$

$$w_\varepsilon^{loc}(t,x,y) = w_\varepsilon(t,r_\varepsilon(x) + \varepsilon y),$$

$$\gamma_\varepsilon^{loc}(t,x,y) = \gamma_\varepsilon(t,r_\varepsilon(x) + \varepsilon y)$$

and

$$\sigma_\varepsilon^{loc}(t,x,y) = \sigma_\varepsilon(t,r_\varepsilon(x) + \varepsilon y) \qquad \text{a.e.} \quad (t,x,y) \in I \times R^2 \times R^2,$$

$$\tilde{w}_\varepsilon^{loc} = w_\varepsilon^{loc} + H^1(I; L^2(R^2; X))$$

with

$$N = \{\lambda_1 y_1 + \lambda_2 y_2 : \lambda_1, \lambda_2 = 0, \pm 1, \pm 2, \ldots\},$$

$$C = \{x \in R^2 : |x| < |x - y| \quad \forall y \in N\},$$

$$r_\varepsilon(x) = y \quad \text{if} \quad \exists y \in \varepsilon N \text{ such that } (x - y) \in \varepsilon C$$

and X being the space of constant real valued functions on R^2.

As in [1], [5] one can prove that the just defined fields fulfil the conditions

$$\text{meas } C\sigma_\varepsilon^{av}(t,x) = \int_C \sigma_\varepsilon^{loc}(t,x,y)dy \tag{9}$$

and

$$\text{meas } C\overline{\gamma}_\varepsilon^{av}(t,x) = \int_C \gamma_\varepsilon^{loc}(t,x,y)dy + \int_{C\cap S} [\![w_\varepsilon^{loc}(t,x,y)]\!]_1 n_1(y)ds_y \quad \text{a.e. } (t,x) \in I \times R^2, \tag{10}$$

$$\gamma_\varepsilon^{loc}(t,x,y) = \text{grad}_{1y}\tilde{w}_\varepsilon^{loc}(t,x,y) \tag{11}$$

and

$$\sigma_\varepsilon^{loc}(t,x,y) = G(y)[\gamma_\varepsilon^{loc}(t,x,y) + \theta(t)a(r_\varepsilon(x) + \varepsilon y)] \quad \text{if} \quad [r_\varepsilon(x) + \varepsilon y] \in \Omega, \tag{12}$$

$$\int_{R^2} dx \{\int_C \sigma_\varepsilon^{loc}(t,x,y)\cdot\dot{\gamma}_\varepsilon^{loc}(t,x,y)dy + \int_{C\cap S} g(y)|[\![\dot{\tilde{w}}_\varepsilon^{loc}(t,x,y)]\!]_1|ds_y = 0 \quad \text{a.e.} \quad t \in I, \tag{13}$$

$$\int_\Omega \varphi(x)dx \{\int_{R^2} \sigma_\varepsilon^{loc}(t,x,y)\text{grad}_{1y}v(y)dy + \int_S g(y)|[\![v(y)]\!]_1|ds_y\} \geq 0 \tag{14}$$

for each nonnegative $\varphi \in C_0^\infty(\Omega)$, for all $v \in H^1(R^2 \setminus S)$ with compact support and for ε so small that

$$(\operatorname{supp} \varphi + \varepsilon C + \varepsilon \operatorname{supp} v) \in \Omega,$$

$$\sigma_\varepsilon^{loc}\Big|_{t=0} = \gamma_\varepsilon^{loc}\Big|_{t=0} = \tilde{w}_\varepsilon^{loc}\Big|_{t=0} = 0, \tag{15}$$

$$\gamma_\varepsilon^{loc}(t,x,y+z) = \gamma_\varepsilon^{loc}(t,x+\varepsilon z,y) \tag{16}$$

and

$$\sigma_\varepsilon^{loc}(t,x,y+z) = \sigma_\varepsilon^{loc}(t,x+\varepsilon z,y) \quad \forall z \in N \quad \text{a.e. } (t,x,y) \in I \times R \times R^2. \tag{17}$$

Next we should find "a priori" estimates for the until now introduced fields. By virtue of the weak compactness principle [9], these estimates will enable us to perform the asymptotic analysis of the considered problem.

Using the known methods of estimates for problems with friction [12], [13] we get

$$\|\gamma_\varepsilon\|_{H^1(I;\, L^2(\Omega;\, R^2))} \leq c$$

with c depending on the data of the problem but not on ε. Next using identities of the type

$$\int_{R^2} dx \int_C |\gamma_\varepsilon^{loc}(t,x,y)| dy = \operatorname{meas} C \int_\Omega |\gamma_\varepsilon|^2 dx$$

we find that for each open bounded $\omega \subset R^2$

$$\|\gamma_\varepsilon^{loc}\|_{H^1(I;\, L^2(R^2;\, L^2(\omega;\, R^2)))} \leq c(\omega)$$

with $c(\omega)$ independent of ε. Consequently, we can deduce from (13) that

$$\|[\![\dot{\tilde{w}}_\varepsilon^{loc}]\!]_1\|_{L^1(I;\, L^1(R^2;\, L^1(\omega \cap S)))} \leq c(\omega)$$

and successively that

$$\|[\![\tilde{w}_\varepsilon^{loc}]\!]_1\|_{H^1(I;\, L^2(R^2;\, L^2(\omega \cap S)))} \leq c(\omega).$$

Thus, using identities of the type

$$\operatorname{meas} C \int_{S_\varepsilon} |[\![w_\varepsilon]\!]_\varepsilon|^2 ds = \varepsilon \int_{R^2} dy \int_{C \cap S} |[\![\tilde{w}_\varepsilon^{loc}]\!]_1|^2 ds_y$$

we obtain

$$\|w_\varepsilon\|_{H^1(I;\, BV(\Omega))} \leq c.$$

Consequently, we can perform the asymptotic analysis of the equations (2), (4), (5), (8) and (9)-(17). The result of such analysis is following:

THEOREM. There exist unique $w^m \in H^1(I;\, H^1(\Omega))$, $\bar{\gamma}^m \in H^1(I;\, L^2(\Omega;\, R^2))$,

$\sigma^m \in H^1(I;\ L^2(\Omega;\ R^2)$, $\quad w^\mu \in H^1(I;\ L^2(R^2;\ H^1_{loc}(R^2 \setminus S)/X))$,
$\gamma^\mu \in H^1(I;\ L^2(R^2;\ L^2_{loc}(R^2;\ R^2)))$ and $\sigma^\mu \in H^1(I;\ L^2(R^2;\ L^2_{loc}(R^2;\ R^2)))$
such that

$$
\begin{array}{llll}
w_\varepsilon \to w^m & \text{in} & H^1(I;\ BV(\Omega)) & \text{weak star and} \\
 & \text{in} & H^1(I;\ L^p(\Omega)) & \text{strongly for } p \in [1,2[, \\
\bar{\gamma}_\varepsilon \to \bar{\gamma}^m & \text{in} & H^1(I;\ M^1(\Omega;\ R^2)) & \text{weak star,} \\
\gamma_\varepsilon^{av} \to \gamma^m & \text{in} & H^1(I;\ L^2(\Omega;\ R^2)) & \text{strongly,} \\
\sigma_\varepsilon \to \sigma^m & \text{in} & H^1(I;\ L^2(\Omega;\ R^2)) & \text{weakly,} \\
\sigma_\varepsilon^{av} \to \sigma^m & \text{in} & H^1(I;\ L^2(\Omega;\ R^2)) & \text{strongly,} \\
\tilde{w}_\varepsilon^{loc} \to \tilde{w}^\mu & \text{in} & H^1(I;\ L^2(R^2;\ H^1(\omega \setminus S)/X)) & \text{strongly,} \\
\gamma_\varepsilon^{loc} \to \gamma^\mu & \text{in} & H^1(I;\ L^2(R^2;\ L^2(\omega;\ R^2))) & \text{strongly,} \\
\sigma_\varepsilon^{loc} \to \sigma^\mu & \text{in} & H^1(I;\ L^2(R^2;\ L^2(\omega;\ R^2))) & \text{strongly}
\end{array}
$$

as $\varepsilon \to 0$ provided that ω is an open bounded subset of R^2. Moreover, the limits fulfil the following conditions

(18) $\quad \bar{\gamma}^m = \operatorname{grad} w^m \qquad$ on $\quad I \times \Omega$,

(19) $\quad \int_\Omega \sigma^m \cdot \operatorname{grad} v\, dx = 0 \quad \forall v \in H^1(\Omega) \qquad$ on $\quad I$,

(20) $\quad w^m|_{t=0} = 0 \qquad$ on $\quad \Omega$,

(21) $\quad \sigma^m(t,x) = (\text{meas } C)^{-1} \int_C \sigma^\mu(t,x,y)dy \qquad$ and

(22) $\quad \bar{\gamma}^m(t,x) = (\text{meas } C)^{-1} \{ \int_C \gamma^\mu(t,x,y)dy + \int_{C\cap S} [\![\tilde{w}^\mu(t,x,y)]\!]_1 n(y) ds_y$

a.e. $(t,x) \in I \times \Omega$,

(23) $\quad \gamma^\mu(t,x,y) = \operatorname{grad}_{1y} \tilde{w}^\mu(t,x,y) \qquad$ and

(24) $\quad \sigma^\mu(t,x,y) = G(y)[\gamma^\mu(t,x,y) + \theta(t)a(x)] \qquad$ a.e. $(t,x,y) \in I \times \Omega \times R^2$,

(25) $\quad \int_\Omega dx \{ \int_C \sigma^\mu(t,x,y) \cdot \dot{\gamma}^\mu(t,x,y)dy + \int_{C\cap S} g(y) | [\dot{\tilde{w}}^\mu(t,x,y)]_1 | ds_y \} = 0$

a.e. $t \in I$,

(26) $\quad \int_{R^2} \sigma^\mu(t,x,y) \cdot \operatorname{grad}_1 v(y)dy + \int_S g(y) | [\![v(y)]\!]_1 | ds_y \geq 0$

a.e. $t \in I$ for all $v \in H^1(R^2 \setminus S)$ with compact support,

(27) $\quad \sigma^\mu|_{t=0} = \gamma^\mu|_{t=0} = w^\mu|_{t=0} = 0 \qquad$ on $\quad \Omega \times R^2$,

(28) $\gamma^{\mu}(t,x,y)$ is Y-periodic in y.

The system (18)-(28) can be considered as a set of laws of macroscopic mechanics of quasi-static torsion of considered bar with (18) as geometric equation, (19) as virtual work principle, (20) as initial condition and (21)-(28) as constitutive equation in implicit form. On the other hand the system (23)-(28) can be considered as defining the problem of quasi-static periodic shear of infinite periodic composite with the mean shear deformation given by (22).

MICROSCOPIC PROBLEM

To simplify the system (22)-(28) we introduce the function $\chi(t,x,y)$ which is such that $\chi \in H^1(I; L^2(\Omega; P))$ with

$$P = \{v \in H^1_{loc}(R^2 \setminus S): \int_C v dy = 0 \text{ and } v \text{ is Y-periodic}\}$$

and

$$\gamma^{\mu}(t,x,y) = \bar{\gamma}^{\mu}(t,x) + \text{grad}_{1y}\chi(t,x,y).$$

Then, dropping out the x variable as unessential parameter, one can bring the considered system in the form of following problem:

P2. For a given $\gamma^m = \bar{\gamma}^m + \theta a$ find $\sigma^{\mu} \in H^1(I; L^2_{loc}(R^2; R^2))$ and $\chi \in H^1(I; P)$ fulfilling the conditions

$$\sigma^{\mu}(t,y) = G(y)[\gamma^m(t) + \text{grad}_1\chi(t,y)] \quad \text{a.e.} \quad (t,y) \in I \times R^2,$$

$$\int_C \sigma^{\mu}(t,y) \cdot \text{grad}_1[v(y) - \dot{\chi}(t,y)]dy + \int_{C\cap S} g(y)\{|[\![v(y)]\!]_1| +$$

$$- |[\![\dot{\chi}(t,y)]\!]_1|\}ds \geq 0 \qquad \forall v \in P \quad \text{a.e.} \quad t \in I,$$

$$\chi|_{t=0} = 0 \quad \text{on } R^2.$$

As in the case of $P1_{\varepsilon}$, the problem P2 has the unique solution provided that $\gamma^m \in H^1(I)$ and $\gamma^m|_{t=0} = 0$.

For the aims of numerical analysis we set P2 in a dual form. To do this we define the convex of statically admissible microstresses

$$B = \{\tau \in L^2_{loc}(R^2; R^2): \tau \text{ is Y-periodic and}$$

$$\int_C \tau \cdot \text{grad}_1 v dy + \int_{C\cap S} g|[\![v]\!]_1|ds \geq 0 \quad \forall v \in P\},$$

the normal cone to B at $\sigma \in L^2(R^2; R^2)$

$$B_n(\sigma) = \{\rho \in L^2_{loc}(R^2; R^2): \rho \text{ is Y-periodic and}$$

$$[\rho|\tau - \sigma] \leq 0, \quad \forall\sigma \in B\}$$

and the tangent cone to B at σ

$$B_t(\sigma) = \{\tau \in L^2_{loc}(R^2;\ R^2):\ \tau \text{ is Y-periodic and}$$
$$[\tau|\rho] \leq 0, \quad \forall \rho \in B_n(\sigma)\}$$

with

$$[\rho|\tau] = \int_C G^{-1}\rho \cdot \tau \, dy.$$

Hence, using the methods of convex analysis [14] we can prove that P2 is equivalent to

P3. For a given $\gamma^m \in H^1(I)$ find $\sigma^\mu \in L^2(I;\ B)$ such that $\dot{\sigma}^\mu(t)$ minimizes the functional

$$f(\tau) = \frac{1}{2}[\tau|\tau] - [\tau|G\dot{\gamma}^m(t)]$$

on $B_t(\sigma)$ a.e. $t \in I$ and $\sigma^\mu|_{t=0} = 0$.

For the aims of mechanical interpretation we put P2 as

P4. For a given $\gamma^m \in H^1(I)$ find $\sigma^\mu \in L^2(I;\ B)$ such that

$$G\dot{\gamma}^m(t) - \dot{\sigma}^\mu(t) \in B_n(\sigma^\mu(t)) \quad \text{a.e. } t \in I \tag{29}$$

and $\sigma^\mu|_{t=0} = 0$.

Returning to the problem of macroscopic constitutive equation we see that now we can define it as the equation (21) with σ^μ being the solution for P4. Hence, we can conclude that the macroscopic behaviour of the considered rod is close to plasticity with hardening because the condition $\sigma^\mu(t) \in B$ can be considered as the condition of plastic admissibility and the condition (29) as the principle of maximum dissipation. However, it should be noticed that the plasticity convex B does not involve the macroscopic stress σ^m.

NUMERICAL EXPERIMENTS CONCERNING THE MACROSCOPIC CONSTITUTIVE EQUATION

It results from Section 5. that for a given macroscopic shear deformation $\gamma^m \in H^1(I)$ the macroscopic shear stress σ^m is given by [21] with σ^μ being the solution for P3. To obtain the corresponding numerical results we apply time-space discretization procedure with time discretization according to the finite difference method and space discretization according to the method of statically admissible finite element.

The applied time discretization leads to the problem

P4(B). For a given sequence $\gamma^m(t_1),\dots,\gamma^m(t_n)$ find a sequence $\sigma^\mu(t_1),\dots,\sigma^\mu(t_n)$ of elements of B such that for each $i = 1,\dots,n$, the microstress $\sigma^\mu(t_i)$ minimizes the functional

$$\Gamma(\tau) = [\tau|\tau] - [\tau|\sigma^\mu(t_{i-1}) + G\{\gamma^m(t_i) - \gamma^m(t_{i-1})\}]$$

on B.

On the second stage of discretization we approximate the convex B by a certain finite dimensional convex $B_h \subset B$ according to the method of statically admissible finite element [15], [16]. Such approximation leads to the problem $P4(B_h)$ which, from the point of view of nonlinear programming, is a problem of quadratic programming and can be effectively solved using the method of modified gradient [17].

The corresponding numerical calculations have been performed for the problem of cyclic shear with amplitude $6 \cdot 10^{-4}$ in the (y_1, y_2) plane for the composite with data $G_f = 80.38$ GPa, $G_m = 0.92$ GPa, $g = 3$ MPa and with the quarter of Brillonin like cell shown in Fig. 2. The resulting stress-deformation path is shown in Fig. 2b.

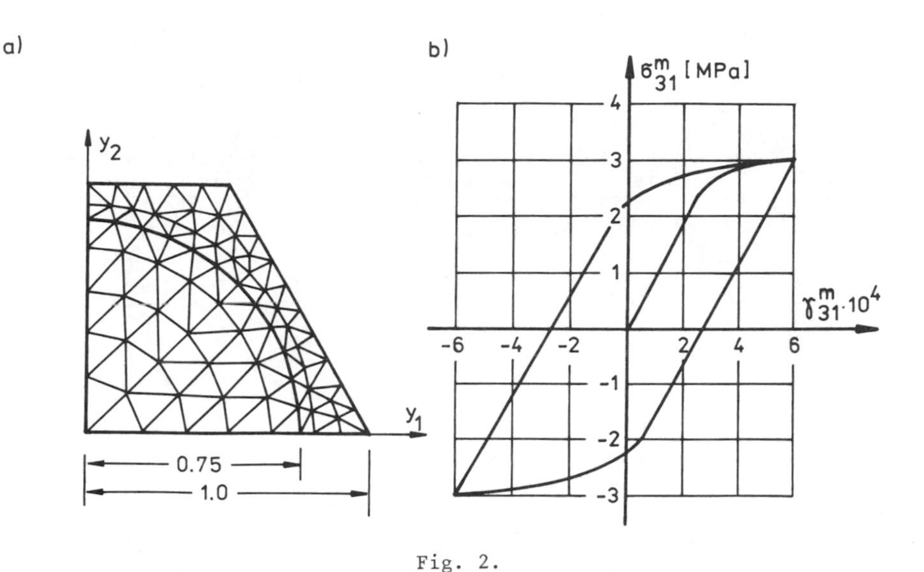

Fig. 2.

REMARK. The paper was partly suppoerted by the Polish Academy of Sciences within the Program MR I-23.

REFERENCES

1. BARAŇSKI W.S., On Homogenization for Fibrous Composite with Coulomb Friction on the Fibre-Matrix Interface, Third British-Polish Mechanics Symposium, Glasgow, 1985.
2. BENSOUSSAN A., LIONS J.L., PAPANICOLAOU G., Asymptotic Analysis for Periodic Structures, North-Holland Publ. Co., Amsterdam, 1978.
3. SANCHEZ-PALENCIA E., Non-Homogeneous Media and Vibration Theory, Springer V., Berlin, 1980.
4. SUQUET P., Plasticité et Homogénéization, Thèse de Doctorat d'Etat, Université Pierre et Marie Curie, Paris, 1982.
5. BARAŇSKI W.S., Microstresses in homogenization, Arch. Mech., (to appear).
6. HILL R., Mathematical Theory of Plasticity, Wiley and Sons, New York, 1950.
7. JOHNSON C., Mathematical and Numerical Analysis of Some Problems of Plasticity, in Functional Analysis Methods in Plasticity, (in Polish),

Ossolineum, Wrocław, 1981, pp.71-109.
8. WHITNEY J., DANIEL I.M., PIPER R.B., Experimental Mechanics of Fibre Reinforced Composite Materials, Prentice-Hall, London, 1982.
9. YOSIDA K., Functional Analysis, Springer V., Berlin, 1980.
10. ADAMS R.A., Sobolev Spaces, Academic Press, New York, 1975.
11. MIRANDA M., Comportamento delle successioni convergenti di frontiere minimali, Rend. Semin. Univ. Padova, 1967, pp.238-257.
12. DUVAUT G., LIONS J.L., Inequalities in Mechanics and Physics, Springer V., Berlin, 1976.
13. LICHT C., Un problème d'elasticité avec frottement visqueux nonlinéaire, J. Mec. Theor. Appl., vol. 4, 1, 1985, pp.15-26.
14. EKELAND I., TEMAM R., Convex Analysis and Variational Problems, North -Holland Publ. Co., Amsterdam, 1976.
15. GLOWINSKI R., LIONS J.L., TREMOLIERES R., Analyse Numérique des Inéquations Variationelles, Dunod, Paris, 1976.
16. WIĘCKOWSKI Z., Duality in Finite Element Method and Its Application for Some Linear and Non-Linear Problems of Mechanics of Composite Materials (in Polish), D. Thesis, Technical University of Łódź, 1986.
17. BAZARAA M.S., SHETTY C.M., Non-Linear Programming, Theory and Algorithms, John Wiley, New York, 1979.

A CONVERGENT MODEL OF NON-LINEARLY ELASTIC PERIODIC COMPOSITE

W.S. Barański

School of Engineering and Engineering Technology, Federal University of Technology, Minna, P.O. Box 656, Nigeria

O. Gajl

Institute of Construction Engineering Technical University of Łódź, Al. Politechniki 6, 93-590 Łódź, Poland

In the paper the previous ideas of the theory of convergent models of composite materials [2], [3] are developed for the case of nonlinearly elastic composite considered with assumption of linear geometric equation. A nonsimple homogeneous model is considered and conditions sufficient for its convergence are found.

INTRODUCTION

The paper concerns the problem of accurate approximations for solutions of problems of mechanics of periodic composites. It is possible to define many criteria of accuracy. In the paper we analyse the criterion of asymptotic convergence which can be understood as follows. If the structure of considered composite is periodic with ε characterizing its linear dimension then the corresponding exact solution can be denoted by s_ε, as depending on ε. A corresponding approximation s_ε^{app} is called *asymptotically convergent* if

$$\lim (s_\varepsilon - s_\varepsilon^{app}) \to 0 \qquad \text{as} \quad \varepsilon \to 0$$

in a certain topology. Obviously, we are interested in convergence in topology as strong as possible.

In fact, we are interested in evaluating the approximate solutions of problems corresponding to certain homogeneous materials, which are called model problems. The obtained in such a way model is called *convergent* if the corresponding approximation is asymptotically convergent.

In [3], [4] sufficient conditions for convergence of Cosserat and non simple models of certain linear problems have been found. The convergence of simple effective models of linear and non-linear problems have been analysed in numerous papers. We refer to [2], [5], [8], [9] for references. However, it occurs that the considered in the paper non-simple models provide approximations with stronger topology of convergence.

BASIC ASSUMPTIONS AND VARIATIONAL FORMULATION OF EQUILIBRIUM PROBLEM

Let Ω denote a domain occupied by considered composite. For each ε characterizing the dimension of mikrostructure of the composite we use the following notation

u_ε - displacement vector field,

E_ε - deformation tensor field,

T_ε - stress tensor field.

As in [2] we characterize the considered equilibrium problem by:

1° A manifold U_{ad} of admissible displacement fields which determines displacement boundary conditions.

2° Linear geometric equation

(1) $$E_\varepsilon = Du_\varepsilon \equiv \text{sym}\ (\text{grad}\ u_\varepsilon), \qquad \text{on}\ \ \Omega.$$

3° Constitutive equation

(2) $$T_\varepsilon(x) = \mathcal{T}^\varepsilon(x,E_\varepsilon(x)) \equiv \mathcal{T}(x/\varepsilon,E_\varepsilon(x)), \qquad \text{a.e.} \quad x \in \Omega.$$

4° Virtual work principle

(3) $$\int_\Omega Dv\cdot T_\varepsilon dx = \int_{\partial\Omega} v\,t\,ds + \int_\Omega v\cdot b_\varepsilon dx, \qquad \forall v \in V_{ad}$$

with given t,b_ε and V_{ad} defined by

$$V_{ad} = \{u - v : u,v \quad U_{ad}\}.$$

We assume that the following conditions hold:

A1. The domain Ω is open, bounded and Lipschitzean in R^n.

A2. The manifold $H^2(\Omega;R^n) \cap U_{ad}$ is dense in U_{ad} equipped with $H^1(\Omega;R^n)$ topology.

A3. The space V_{ad} constitutes a closed linear subspace of $H^1(\Omega;R^n)$ including $H^1_o(\Omega;R^n)$.

A4. The constitutive function $\mathcal{T}(y;E)$ is Y-periodic and measurable in y for all fixed E.

A5. There exist positive c_1,c_2 such that for all $E_1,E_2 \in S^n$, (the space of symmetric second order tensors on R^n)

$$(E_2 - E_1)\cdot[\mathcal{T}(y,E_2) - \mathcal{T}(y,E_1)] \geq c_1|E_2 - E_1|^2$$

and

$$|\mathcal{T}(y,E_2) - \mathcal{T}(y,E_1)| \leq c_1|E_2 - E_1|, \qquad \text{a.e.} \quad y \in R^n.$$

A6. The field $\mathcal{T}(\cdot,0)$ of initial stresses is locally square integrable.

A7. For each test function v from the space

$$N_{ad} = \{v \in V_{ad} : Dv = 0\}$$

the right hand side of (3.3) (i.e the functional of the work of external forces) vanishes.

To avoid nonuniqueness of displacement solutions we introduce

$$\tilde{U}_{ad} = U_{ad}/N_{ad}, \qquad \tilde{V}_{ad} = V_{ad}/N_{ad}.$$

We define the equilibrium problem as follows

P1$_\varepsilon$. For given $\varepsilon \in]0,1]$, $b_\varepsilon \in L^n(\Omega;R^n)$, $t \in L^2(\partial\Omega;R^n)$ find

$$\tilde{u}_\varepsilon \in \tilde{U}_{ad}, \quad E_\varepsilon \in L^n(\Omega;S^n), \quad T_\varepsilon \in L^2(\Omega;S^n)$$

satisfying the equations (3.1) - (3.3).

The problem $P1_\varepsilon$ is known to have unique solution [7]. Its asymptotic behaviour as $\varepsilon \to 0$ depends on asymptotic behaviour of volume forces b_ε which are assumed to fulfil the condition

A8. There exists $q > n$, $q \geq 2$ such that $b_\varepsilon \to b_o$ in $L^q(\Omega;R^n)$ weakly as $\varepsilon \to 0$.

LOCAL FIELDS

For every $\varepsilon \in]0,1]$ and each ε-cell we define a local coordinate system (y) by the coordinate transform: $x = \xi + \varepsilon y$ where $\xi \in \varepsilon N_r$ points at the center of considered ε-cell. From the practical point of view it is interesting to have the fields of deformations and stresses expressed as functions of local coordinates. Following [2] we shall call them local fields. Formally, we define them by

$$E_\varepsilon^{loc}(x,y) = \text{Ext } E_\varepsilon(r_\varepsilon(x) + \varepsilon y), \tag{4}$$

$$T_\varepsilon^{loc}(x,y) = \text{Ext } T_\varepsilon(r_\varepsilon(x) + \varepsilon y), \qquad (x,y) \in \Omega \times R^n \tag{5}$$

where Ext denotes the extension operator associating zero to points outside Ω.

By virtue of existence and uniqueness of solutions for problem $P1_\varepsilon$ we can consider the local fields to be well defined for all $\varepsilon \in]0,1]$.

RESULTS OF THE HOMOGENIZATION THEORY

We recall here basic results of the homogenization theory concerning the considered equilibrium problem [5], [9]. We shall start with the problem of periodic self-equilibrated state of stress of infinite composite defined by

P2. For a given $F \in S^n$ find

$$\chi_F \in H^1_{loc}(R^n;R^n), \quad E^P_F \in L^2_{loc}(R^n;S^n) \quad \text{and} \quad T^P_F \in L^2_{loc}(R^n;S^n)$$

such that

$$E^P_F = F + D\chi_F, \qquad \text{on } R^n, \tag{6}$$

$$T^P_F(y) = \mathcal{T}(y,E^P_F(y)), \qquad \text{a.e. } y \in R^n, \tag{7}$$

$$\int_{R^n} Dv \cdot T^P_F dy = 0, \qquad \forall v \in C^\infty_o(R^n;R^n), \tag{8}$$

$$\int_C \chi_F dy = 0, \tag{9}$$

$$\chi_F \text{ is Y-periodic.} \tag{10}$$

The problem P2 has unique solution, and the solution is Lipschitz continuous with respect to F, i.e. there exists a positive c such that for all $F_2, F_1 \in S^n$

$$\| \chi_{F_2} - \chi_{F_1} \|_{c;1} \leq c|F_2 - F_1|, \tag{11}$$

$$\| E^P_{F_2} - E^P_{F_1} \|_{c;o} \leq c|F_2 - F_1|, \tag{12}$$

$$\| T^P_{F_2} - T^P_{F_1} \|_{c;o} \leq c|F_2 - F_1|. \tag{13}$$

Existence and uniqueness of solutions for the problem P2 enable us to define the so-called effective constitutive function by

(14) $$\mathcal{T}^{eff}(E) = |C|^{-1} \int_C T_E^P dy,$$

which has been proved to fulfil condition A4.

Successively, we can introduce the homogenized equilibrium problem by

P3. For $t \in L^2(\partial\Omega;R^n)$ and $b_o \in L^2(\Omega;R^n)$ find $\tilde{u}_h \in \tilde{U}_{ad}$, $E_h \in L^2(\Omega;S^n)$, $T_h \in L^2(\Omega;S^n)$ such that

(15) $$E_h = D\tilde{u}_h, \qquad \text{on } \Omega,$$

(16) $$T_h = \mathcal{T}^{eff}(E_h), \qquad \text{on } \Omega,$$

(17) $$\int_\Omega Dv \cdot T_h dx = \int_\Omega v \cdot b_o dx + \int_{\partial\Omega} v \cdot t ds.$$

The problem P3 has unique solution which characterizes asymptotic behaviour of solutions for $P1_\varepsilon$ by conditions

(18) $$\tilde{u}_\varepsilon \to \tilde{u}_h \text{ in } H^1(\Omega;R^n)/N_{ad} \text{ weakly and in } L^2(\Omega;R^n)/N_{ad} \text{ strongly}$$

(19) $$E_\varepsilon \to E_h \text{ in } L^2(\Omega;S^n) \text{ weakly,}$$

(20) $$T_\varepsilon \to T_h \text{ in } L^2(\Omega;S^n) \text{ weakly,} \quad \text{as } \varepsilon \to 0.$$

Moreover, for each open bounded ω in R^n

(21) $$E_\varepsilon^{loc}|_{\Omega\times\omega} \to E_\mu|_{\Omega\times\omega} \text{ in } L^2(\Omega \times \omega;S^n) \text{ strongly,}$$

(22) $$T_\varepsilon^{loc}|_{\Omega\times\omega} \to T_\mu|_{\Omega\times\omega} \text{ in } L^2(\Omega \times \omega;S^n) \text{ strongly as } \varepsilon \to 0,$$

with

(23) $$E_\mu(x,y) \equiv E^P_{E_h(x)}(y)$$

and

(24) $$T_\mu(x,y) \equiv T^P_{E_h(x)}(y).$$

Thus, we can conclude that the homogenized body can be considered as convergent model with topology of convergence specified by (18) - (22).

A NON-SIMPLE ELASTIC MEDIA AS A CONVERGENT MODEL

To define the model we first introduce the manifold of admissible displacements

(25) $$U^m_{ad} = U_{ad} \cap H^2(\Omega;R^n)$$

and two constitutive functions

$$P : (S^n \times (S^n \times R^n)) \to R^n \times R^n$$

$$Q : (S^n \times (S^n \times R^n)) \to R^n \times R^n \times R^n$$

fulfilling the following assumptions

AM1. There exist positive constants c_1, c_2 such that for all $E_1, E_2 \in S^n$ and $G_1, G_2 \in S^n \times R^n$

$$(E_2 - E_1)\cdot[P(E_2,G_2) - P(E_1,G_1)] + \\ + (G_2 - G_1)\cdot[Q(E_2,G_2) - Q(E_1,G_1)] \geq \\ \geq c_1(|E_2 - E_1|^2 + |G_2 - G_1|^2),$$

$$|P(E_2,G_2) - P(E_1,G_1)| + |Q(E_2,G_2) - Q(E_1,G_1)| \leq \\ \leq c_2(|E_2 - E_1| + |G_2 - G_1|).$$

AM2. For all $E \in S^n$

$$P(E,0) = T^{eff}(E).$$

REMARK 1. Note that the constitutive functions

$$P(E,G) \equiv T^{eff}(E), \qquad Q(E,G) \equiv G$$

fulfil AM1 and AM2.

Next, we introduce the corresponding equilibrium problem $P4_\varepsilon$. For given $\varepsilon \in]0,1]$, $t \in L^2(\partial\Omega;R^n)$ and $b_o \in L^q(\Omega;R^n)$ find $\tilde{w}_\varepsilon \in \tilde{U}^m_{ad} = U^m_{ad}/N_{ad}$

(26) $$\int_\Omega [Dv\cdot P(Dw_\varepsilon, \varepsilon \text{ grad } Dw_\varepsilon) + \varepsilon \text{ grad } Dv\cdot Q(Dw_\varepsilon, \varepsilon \text{ grad } Dw_\varepsilon)]\, dx = \\ = \int_{\partial\Omega} t\cdot v dx + \int_\Omega b_o\cdot v dx, \qquad \forall v \in V^m_{ad},$$

with

$$V^m_{ad} = \{w - v : w,v \in U^m_{ad}\}.$$

Using the standard methods of analysis of quasilinear elliptic equations [5], [6] one can prove existence and uniqueness for problem $P4_\varepsilon$. Additionally, one can show estimates

$$\|Dw_\varepsilon\|_\Omega \leq c, \qquad \varepsilon\, \|\text{grad } Dw_\varepsilon\|_\Omega \leq c$$

which by virtue of Korn and Poincaré inequalities give

(27) $$\|\tilde{w}_\varepsilon\|_{\Omega;1} \leq c, \qquad \|\varepsilon w_\varepsilon\|_{\Omega;2} \leq c.$$

The estimates (27) provide an opportunity for analysis of asymptotic behaviour of w_ε as $\varepsilon \to 0$. But this is a subject of the theory of singular perturbations for quasilinear elliptic equations which in fact does not essentially differs from that one corresponding to linear elliptic equations. The result proved in Appendix 1 is following

(28) $$\tilde{w}_\varepsilon \to \tilde{u}_h \quad \text{in} \quad H^1(\Omega;R^n) \quad \text{strongly},$$

(29) $$\varepsilon\tilde{w}_\varepsilon \to 0 \quad \text{in} \quad H^2(\Omega;R^n) \quad \text{strongly}, \qquad \text{as} \quad \varepsilon \to 0.$$

So that, comparing (29) with (18)- (24) we see that the considered non-simple model can give approximate solutions for $P1_\varepsilon$ with the topology of convergence coinciding with that one of effective model.

Precisely, we have

(30) $$(\tilde{u}_\varepsilon - \tilde{w}_\varepsilon) \to 0 \quad \text{in} \quad H^1(\Omega;R^n)/N_{ad} \text{ weakly and in } L^2(\Omega;R^n)/N_{ad} \text{ strongly},$$

(31) $$(E_\varepsilon - D\tilde{w}_\varepsilon) \to 0 \quad \text{in} \quad L^2(\Omega;S^n) \quad \text{weakly}$$

(32) $\quad (T_\varepsilon - T^\varepsilon(D\tilde{w}_\varepsilon)) \to 0 \quad$ in $\quad L^2(\Omega;S^n) \quad$ weakly

and for each open bounded ω in R^n

(33) $\quad (E^{loc}_\varepsilon - E^{loc}_{\varepsilon,app}) \to 0 \quad$ in $\quad L^2(\Omega \times \omega;S^n) \quad$ strongly,

(34) $\quad (T^{loc}_\varepsilon - T^{loc}_{\varepsilon,app}) \to 0 \quad$ in $\quad L^2(\Omega \times \omega;S^n) \quad$ strongly, as $\varepsilon \to 0$

with

$$E^{loc}_{\varepsilon,app}(x,y) \equiv E^P_{Dw_\varepsilon(x)}(y).$$

$$T^{loc}_{\varepsilon,app}(x,y) \equiv T^P_{Dw_\varepsilon(x)}(y).$$

Moreover, the classical corrector type approximation

$$\bar{w} + \varepsilon\chi_{Dw_\varepsilon}$$

for $\tilde{u}_\varepsilon$ can be analysed. It is proved in Appendix 2 that if

A9. For all $E \in S^n$ the derivative $\partial\chi_E\, y/\partial E$ exists a.e. $y \in R^n$ and there exists a positive c such that for all $E_1,E_2 \in S^n$

$$|\text{grad}\ \chi_{E_2}(y) - \text{grad}\ \chi_{E_1}(y)| < c_3|E_2 - E_1|, \qquad \text{a.e. } y \in R^n,$$

then

(35) $\quad [\tilde{u}_\varepsilon - (\tilde{w}_\varepsilon + \chi^\varepsilon_{D\tilde{w}_\varepsilon})] \to 0 \quad$ in $\quad H^1(\Omega;R^n) \quad$ strongly as $\varepsilon \to 0$.

APPENDIX 1. SINGULAR PERTURBATIONS FOR A QUASILINEAR ELLIPTIC EQUATION

THEOREM 1. Suppose that assumptions A1 - A3, AM1, AM2 are fulfilled, the effective constitutive function T^{eff} satisfies A4 and the work done by (b_o,t_o) on rigid body displacements from N_{ad} vanishes. Then convergences (28), (29) hold.

P r o o f. By virtue of (27) and classical weak compactness principle [9], there exists a subset ξ of $]0,1]$ such that

$$\inf \xi = 0$$

(36) $\quad \tilde{w}_\varepsilon \to \tilde{w}_o \quad$ in $\quad H^1(\Omega;R^n)/N_{ad}$ weakly and in $L^2(\Omega;R^n)/N_{ad}$ strongly,

(37) $\quad \varepsilon w_\varepsilon \to d \quad$ in $\quad H^2(\Omega;R^n)/N_{ad}$ weakly and in $H^1(\Omega;R^n)/N_{ad}$ strongly as $\varepsilon \to 0$ in ξ.

Comparing (36) and (37) we get $d = 0$. Next, by the classical monotony arguments we get

(38) $\quad P(Dw_\varepsilon, \varepsilon\,\text{grad}\,Dw_\varepsilon) \to P(Dw_o,0) \quad$ in $\quad L^2(\Omega;R^n \times R^n)$ weakly as $\varepsilon \to 0$ in ξ.

Hence w_o is a solution for the problem

(39) $$\int_\Omega Dv\cdot P(D\tilde{w}_o,0)dx = \int_{\partial\Omega} t\cdot v ds + \int_\Omega b_o\cdot v dx, \qquad \forall v \in V^m_{ad}.$$

By A2 the limit $\tilde{w}_o$ is in $\tilde{U}_{ad}$, and V^m_{ad} is dense in V_{ad}.

Consequently, (36) is equivalent with P3 and $\tilde{w}_o = \tilde{u}_h$. Additionally, uniqueness of the limit implies that convergences (36), (37) hold as $\varepsilon \to 0$ in $]0,1]$.

It results from AM1 that it remains to show

$$\int_\Omega \{(Dw_\varepsilon - Du_h)\cdot[P(Dw_\varepsilon, \varepsilon \operatorname{grad} Dw_\varepsilon) - P(Du_h, 0)] + $$
$$+ \varepsilon \operatorname{grad} Dw_\varepsilon \cdot [Q(Dw_\varepsilon, \varepsilon \operatorname{grad} Dw_\varepsilon) - Q(Du_h, 0)]\} dx \to 0 \quad \text{as } \varepsilon \to 0.$$

But this can be easily proved using (26) and (38).

APPENDIX 2. CORRECTORS

In this section we shall prove that

$$(\tilde{u}_\varepsilon - \tilde{z}_\varepsilon) \to 0 \quad \text{in } H^1(\Omega;R^n) \text{ strongly} \quad \text{as } \varepsilon \to 0$$

with

$$\tilde{z}_\varepsilon(x) \equiv \tilde{w}_\varepsilon(x) + \varepsilon \chi^\varepsilon_{B_\varepsilon(x)}(x), \qquad B_\varepsilon \equiv D\tilde{w}_\varepsilon .$$

The function $\tilde{z}_\varepsilon$ is not an admissible displacement unless we consider the purely traction boundary value problem. To avoid this difficulty we introduce, following BENSOUSSAN et al. [5], the function

$$q_\varepsilon \equiv w_\varepsilon + \varepsilon m_\varepsilon \chi^\varepsilon_{B_\varepsilon}$$

with m_ε such that

$$(\tilde{z}_\varepsilon - \tilde{q}_\varepsilon) \to 0 \quad \text{in} \quad H^1(\Omega;R^n) \quad \text{strongly.} \tag{40}$$

So that it is sufficient to show that

$$(\tilde{u}_\varepsilon - \tilde{q}_\varepsilon) \to 0 \quad \text{in} \quad H^1(\Omega;R^n) \quad \text{strongly.}$$

But, by virtue of Korn inequality, this is equivalent to

$$(D\tilde{u}_\varepsilon - D\tilde{q}_\varepsilon) \to 0 \quad \text{in} \quad L^2(\Omega;S^n) \quad \text{strongly,}$$

and using again (40) we reduce the proof to the following convergence problem

$$(D\tilde{u}_\varepsilon - D\tilde{z}_\varepsilon) \to 0 \quad \text{in} \quad L^2(\Omega;S^n) \quad \text{strongly as} \quad \varepsilon \to 0.$$

Note that

$$D\tilde{z}_\varepsilon = E^{p,\varepsilon}_{B_\varepsilon} + \varepsilon \operatorname{grad} D\tilde{w}_\varepsilon \Psi_{B_\varepsilon} \qquad \text{with} \quad \Psi_E \equiv \frac{\partial \chi_E}{\partial E} .$$

So that by virtue of L^2-strong convergence to zero of $\varepsilon \operatorname{grad} Dw_\varepsilon$ it is sufficient to show that

$$I^1_\varepsilon \equiv \int_\Omega |Du_\varepsilon - E^{p,\varepsilon}_{B_\varepsilon}|^2 dx \to 0.$$

But introducing the local coordinates on each ε-cell we get

$$I^1_\varepsilon = \frac{1}{|C|} \int_{\Omega^*} \int_C |E^{loc}_\varepsilon(x,y) - E^p_{B_\varepsilon(r_\varepsilon(x) + \varepsilon y)}(y)|^2 dy dx.$$

The homogenization theorem states that

$$(E^{loc}_\varepsilon - E^p_{B_\varepsilon}) \to 0 \quad \text{in} \quad L^2(\Omega^* \times C;S^n) \quad \text{strongly.}$$

Thus, to finish the proof it is sufficient to show that

$$\int_{\Omega^*}\int_C |E^p_{B_\varepsilon(x)}(y) - E^p_{B_\varepsilon(r_\varepsilon(x)+\varepsilon y)}(y)|^2 dydx \to 0.$$

But, by virtue of A9, this is equivalent to

$$\int_{\Omega^*}\int_C |B_\varepsilon(x) - B_\varepsilon(r_\varepsilon(x)+\varepsilon y)|^2 dydx \to 0$$

or to

$$I^2_\varepsilon \equiv \int_{\Omega^*}\int_C |B_\varepsilon(x) - B^{ar}_\varepsilon(x)|^2 dydx \to 0$$

and

$$I^3_\varepsilon \equiv \int_{\Omega^*}\int_C |B^{ar}_\varepsilon(x) - B_\varepsilon(r_\varepsilon(x)+\varepsilon y)|^2 dydx \to 0 \qquad \text{as } \varepsilon \to 0$$

with

$$B^{ar}_\varepsilon(x) \equiv \frac{1}{|C|\varepsilon^n} \int_{r_\varepsilon(x)+\varepsilon C} B_\varepsilon(z)dz.$$

It holds

$$I^2_\varepsilon = I^3_\varepsilon = |C| \int_\Omega |B_\varepsilon(x) - B^{ar}_\varepsilon(x)|^2 dx$$

and the both functionals I^2_ε and I^3_ε converge to zero by virtue of the classical Finite Element Method theorem on approximation by piecewise constant functions.

REMARK. The paper was partly supported by the Institute of Fundamental Technological Research, Polish Academy of Sciences within Program I-23.

REFERENCES

1. ADAMS R.A., Sobolev Spaces, Academic Press, N.Y. 1975.
2. BARAŃSKI W.S., Microstresses in Homogenization, Arch. Mech., to appear.
3. BARAŃSKI W.S., GAJL O., Convergent models of steady heat flow in periodic composites, to appear.
4. BARAŃSKI W.S., WAGNER I., A Cosserat model of elastic periodic composite, to appear.
5. BENSOUSSAN A., LIONS J.L., PAPANICOLAOU G., Asymptotic Analysis for Periodic Structures, North-Holland, Amsterdam 1978.
6. KLUGE R., Nichtlineare Variationsungleichungen und Extremalaufgaben, Veb. Deutscher V. Berlin 1979.
7. LIONS J.L., Quelques méthodes de resolution des problèmes aux limites non linéaires, Dunod, Paris 1969.
8. SANCHEZ-PALENCIA E., Non Homogeneous Media and Vibration Theory, Springer V., Berlin 1980.
9. SUQUET P., Plasticité et homogénéisation, Thèse de Doctorat d'Etat, Université Pierre et Marie Curie, Paris 1982.
10. YOSIDA K., Functional Analysis, Springer V., Berlin 1980.

ANALYSIS AND CALCULUS OF VERY THIN SPACE FRAMES

Doina CIORANESCU - Jeannine SAINT JEAN PAULIN

The space frames studied here are used in many fields : civil engineering, aeronautics, electrotechnic. They are characterized by three aspects: periodic distribution (and period small compared with global dimensions), small thickness of the material (compared with the period) and high symmetry of the geometry. Fields of applications are: elasticity, thermic, electrotechnic, eigenvalue problems... The period can be cubic, triangular, hexagonal.

In the two-dimensional case we consider for instance the following structures :

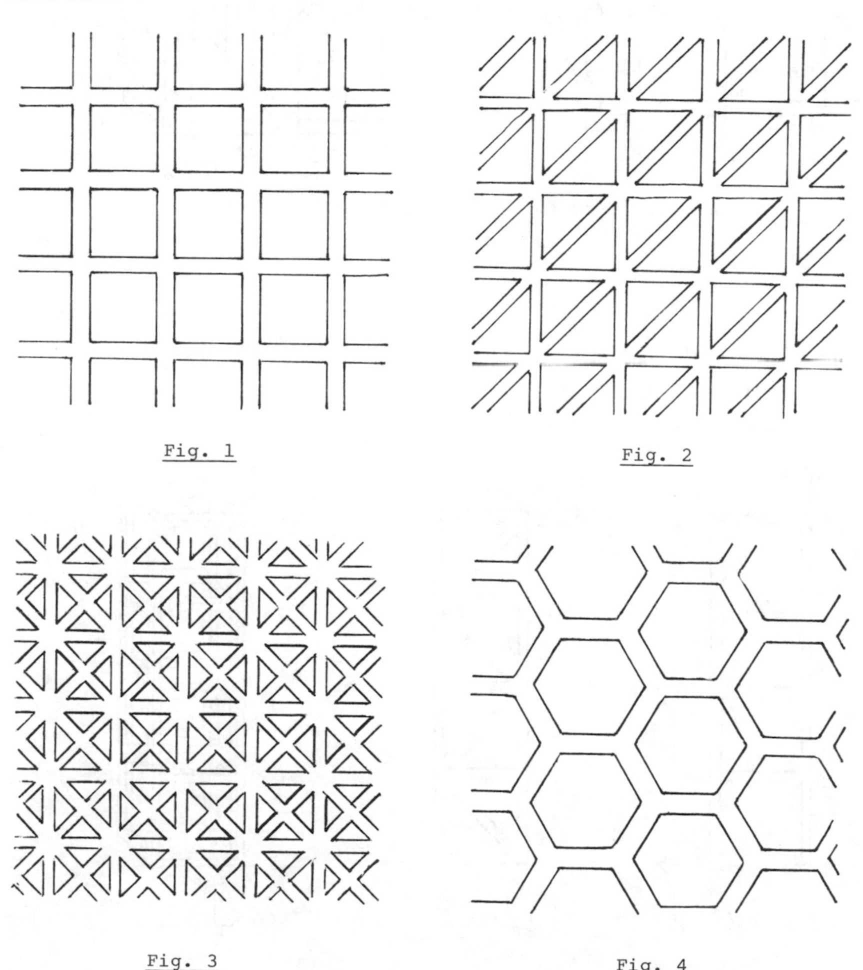

Fig. 1 Fig. 2

Fig. 3 Fig. 4

In the <u>three-dimensional</u> case, the structures we study are: honeycomb structures (material distributed along walls) and reinforced structures (material distributed along beams):

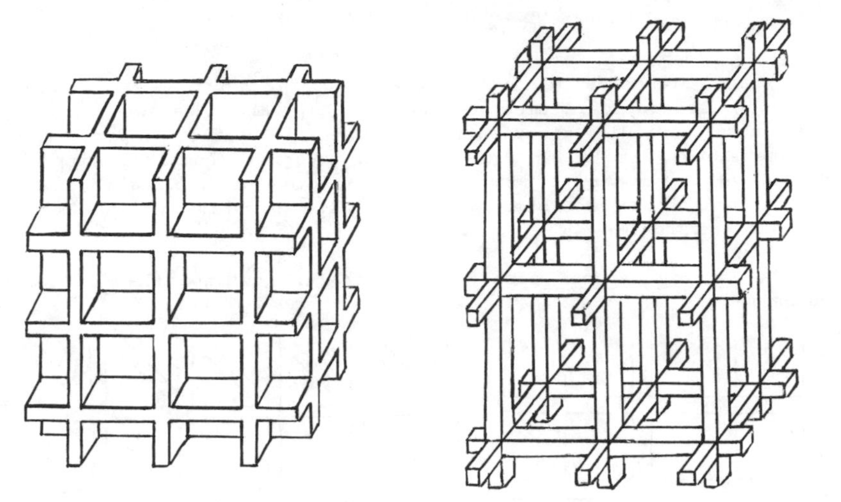

Figure 5 : Honeycomb structure   Figure 6 : reinforced structure

We also treat <u>tall structures</u> (where at least one of the dimensions is of the same size as the period):

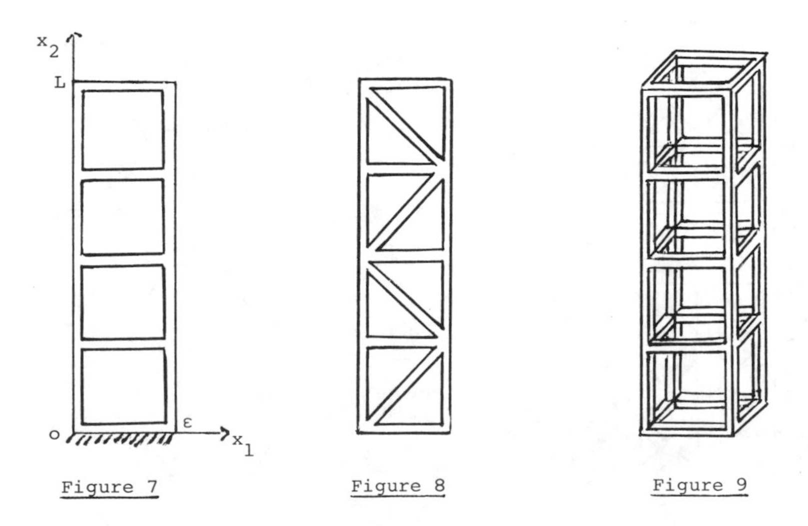

Figure 7   Figure 8   Figure 9

A direct computation to describe the behaviour of these frames would be very costly because of the rapidly oscillating character of the structures and of the very small thickness of the material. Our aim is to give the global behaviour without making long and expensive calculus.

Honeycomb and reinforced structures were studied by Panasenko [7][8][9], Bakhvalov-Panasenko [1]. They give explicit coefficients using asymptotic expansions.

Our methods are based on a variational approach. The periodic distribution is dealt with by a homogenization technique in domains with holes, see [2][6][10]. The result is a homogenized equation with constant coefficients which are calculated on a representative cell. Next we give a mathematical method to obtain the global coefficients of the frames, the small thickness of the material being the essential parameter, see [3][4][5].

We explicit the effective global coefficients. As we shall see, these coefficients are elementary expressions in terms of the characteristic constants of the material. We estimate the error made when replacing the physical problem by the one obtained through our mathematical method. Finally we show that the global coefficients are the sum of coefficients calculated successively in each direction of the frames.

STATEMENT OF THE PROBLEM

<u>As an example</u> we shall treat here a thermic problem for a three-dimensional honeycomb frame (see figure 5)

Let Ω be a bounded open subset of $\mathbb{R}^3$. In the cube $Y = [-1/2,1/2[\times [-1/2,1/2[\times [-1/2,1/2[$ we dig the cubic hole $T_\zeta = [-(1-\delta)/2,(1-\delta)/2] \times [-(1-\delta)/2,(1-\delta)/2] \times [-(1-\delta)/2,(1-\delta)/2]$. Denote by $Y_\delta = Y - T_\delta$ the part of Y occupied by the material. We cover $\mathbb{R}^3$ periodically by cells homothetic to Y, the ratio being ε:1. Let $\Omega^*_{\varepsilon\delta}$ denote the part of Ω occupied by the material. Thus we obtain the structure corresponding to figure 5, the period is ε and the material consists of $\varepsilon\delta$ -thick layers parallel to the planes of coordinates.

Let us consider the Neumann problem :

$$\begin{cases} -\operatorname{div} \sigma^{\varepsilon\delta} = f \text{ in } \Omega^*_{\varepsilon\delta} \\ \sigma^{\varepsilon\delta} \cdot n = 0 \quad \text{on the boundary of the holes} \\ \sigma^{\varepsilon\delta}_i = \sum_{j=1}^{3} a_{ij} \dfrac{\partial u^{\varepsilon\delta}}{\partial x_j} \\ u^{\varepsilon\delta} = 0 \quad \text{on } \partial\Omega \end{cases}$$

The normal n is directed towards the exterior of $\Omega^*_{\varepsilon\delta}$.

<u>Hypothesis H</u> : $f \in L^2(\Omega)$, a_{ij} are constant coefficients verifying the ellipticity condition, the holes do not intersect $\partial\Omega$.□

We begin by making $\varepsilon \to 0$. This corresponds to a classical homogenization problem. We recall the following result [2], [6], [10] :

Theorem 1 : Under the hypothesis H, there exist extensions

$P^{\varepsilon\delta} u^{\varepsilon\delta} \in H_0^1(\Omega)$, $Q^{\varepsilon\delta} \sigma^{\varepsilon\delta} \in L^2(\Omega)$ such that :

$P^{\varepsilon\delta} u^{\varepsilon\delta} \underset{\varepsilon\to 0}{\longrightarrow} u^\delta$ in $H_0^1(\Omega)$ weakly

$\sigma^{\varepsilon\delta} u^{\varepsilon\delta} \underset{\varepsilon\to 0}{\longrightarrow} \sigma^\delta$ in $L^2(\Omega)$ weakly

with

$$\begin{cases} -\operatorname{div} \sigma^\delta = (\text{meas } Y^*_\delta)\, f \text{ in } \Omega \\ \sigma_i^\delta = \displaystyle\sum_{j=1}^{3} q_{ij}^\delta \frac{\partial u^\delta}{\partial x_j} \\ u^\delta = 0 \text{ on } \partial\Omega \end{cases}$$

The "homogenized" coefficients q_{ij}^δ are defined by :

$$(1) \quad q_{ij}^\delta = \int_{Y^*_\delta} \left(a_{ij} - \sum_{k=1}^{3} a_{kj} \frac{\partial \chi_\delta^i}{\partial y_k}\right) dy$$

where the functions χ_δ^i are solutions of the adjoint system :

$$(2) \begin{cases} \displaystyle\sum_{k,m=1}^{3} -\frac{\partial}{\partial y_m}\left(a_{km} \frac{\partial(\chi_\delta^i - y_i)}{\partial y_k}\right) = 0 \text{ in } Y^*_\delta \\ \displaystyle\sum_{k,m=1}^{3} a_{km} \frac{\partial(\chi_\delta^i - y_i)}{\partial y_k} n_m = 0 \text{ on } \partial T_\delta \\ \chi_\delta^i \text{ periodic in } Y. \square \end{cases}$$

We shall study now the dependence of u^δ and σ^δ on the thickness δ and give their limit behaviour for $\delta \to 0$.

MAIN RESULTS - HONEYCOMB STRUCTURE

Theorem 2 Let $\delta \to 0$. The weak limits

$$(3) \quad u^* = w - \lim_{\delta\to 0} u^\delta \ (\text{in } H_0^1(\Omega)), \quad \sigma^* = w - \lim_{\delta\to 0} \delta^{-1} \sigma^\delta \ (\text{in } L^2(\Omega))$$

verify :

$$(4)\begin{cases} -\frac{1}{3} \ \text{div}\ \sigma^* = f & \text{in } \Omega \\ \sigma_i^* = \sum_{j=1}^{3} q_{ij}^* \frac{\partial u^*}{\partial x_j} & \\ u^* = o & \text{on } \partial\Omega \end{cases}$$

with :

$$(5) \quad q_{ij}^* = 3a_{ij} - \sum_{k=1}^{3} \frac{a_{ik}a_{kj}}{a_{kk}}$$

The matrix (q_{ij}^*) is coercive and in general non diagonal. □

Sketch of the proof

Since meas $Y_\delta^* = 3\delta - 3\delta^2 + \delta^3$, we obtain from system (2) the a priori estimate :

$$(6) \quad |\nabla\chi_\delta^i|_{[L^2(Y_\delta^*)]^3} \leq C\delta^{1/2}$$

which used in the definition (1) of q_{ij}^δ leads to :

$$\delta^{-1} \ q_{ij}^\delta \ \underset{\delta\to o}{\longrightarrow} \ q_{ij}^*$$

In formula (1) we decompose $Y_\delta^* = \bigcup_{\ell=1}^{3} \pi_\delta^\ell$, where π_δ^ℓ is the layer orthogonal to the axis y_ℓ (ℓ=1,2,3) :

$$(7) \quad \delta^{-1}q_{ij}^\delta = \delta^{-1} \ (\text{meas } Y_\delta^*) \ a_{ij} - \sum_{\ell=1}^{3} \delta^{-1} \int_{\pi_\delta^\ell} a_{kj} \frac{\partial\chi_\delta^i}{\partial y_k} \ dy + \delta \ O\ (1)$$

To get the limit of $\delta^{-1}q_{ij}^\delta$ as $\delta\to o$, we transform via affinities the δ - dependent domain π_δ^ℓ into the fixed domain Y. By these affinities the functions χ_δ^i are transformed into $\chi_{\delta,\ell}^i$ and the a priori estimate (6) shows that :

$$(8)\left.\begin{cases} \delta^{-1} \ \frac{\partial\chi_{\delta,\ell}^i}{\partial z_\ell} \ \to \ \xi_{(\ell,\ell)}^i & \text{(do not sum in } \ell \text{ !)} \\ \frac{\partial\chi_{\delta,\ell}^i}{\partial z_k} \ \to \ \xi_{(\ell,k)}^i & (k = \ell) \end{cases}\right\} \text{ in } L^2(Y) \text{ weakly}$$

$$(z_\ell = \delta^{-1} \ y_\ell \quad , \quad z_k = y_k \ (k \neq \ell)).$$

With these notations we have from (7) :

$$q^*_{ij} = (a_{ij}-a_{1j}\int_Y \xi^i_{(1,1)}dy)+(a_{ij}-a_{2j}\int_Y \xi^i_{(2,2)}dy)+(a_{ij}-a_{3j}\int_Y \xi^i_{(3,3)}dy) \qquad (9)$$

(remark that $\int_Y \xi^i_{(\ell,k)}\, dy = o$ for $\ell \neq k$, due to the Y-periodicity of X^i_δ).

To calculate explicitely $\int_Y \xi^i_{(\ell,\ell)} dy$, we multiply now the system (2) defining X^i_δ by test functions depending only on one variable. For instance if ϕ is a smooth function periodic in Y and depending only on y_ℓ, we obtain:

$$\delta^{-1} \int_{Y^*_\delta} \sum_{j=1}^{3} a_{j\ell} \frac{\partial X^i_\delta}{\partial y_j} \frac{\partial \phi}{\partial y_\ell}\, dy = \delta^{-1} \int_{Y^*_\delta} a_{i\ell} \frac{\partial \phi}{\partial y_\ell}\, dy$$

(no summation in ℓ !)

As $\delta \to o$ we get :

$$\sum_j a_{j\ell} \Big(\sum_{k\neq\ell} \int_Y (\xi^i_{(k,j)} \frac{\partial \phi}{\partial y_\ell}) dz\Big) + \frac{\partial \phi}{\partial y_\ell}(o)\ \Big[a_{\ell\ell} \int_Y \xi^i_{(\ell,\ell)}\, dz\Big] =$$

$$= a_{i\ell} \frac{\partial \phi}{\partial y_\ell}(o)$$

It is easy to check that this identity implies :

$$\int_Y \xi^i_{(\ell,\ell)}\, dz = \frac{a_{i\ell}}{a_{\ell\ell}}$$

which used in (9) gives the result : formula (5). It can be verified that the matrix (q^*_{ij}) is coercive so that (3) and (4) hold.

<u>Theorem 3</u> : (error estimate) Let f be sufficiently smooth. Then :

$$\frac{1}{(\text{meas }\ \Omega^*_{\varepsilon\delta})^{\frac{1}{2}}} \|u^{\varepsilon\delta} - u^*\|_{H^1(\Omega^*_{\varepsilon\delta})} \leq C(\delta^{\frac{1}{2}} + \varepsilon^{\frac{1}{2}}). \square$$

(for the proof see |5|)

GENERAL METHOD

Suppose that the layer $\pi^{\ell}_{\delta\ell}$ (ℓ=1,2,3) has the thickness δ_{ℓ} ($\delta_s \neq \delta_t$, s≠t). Using the same convergence method, we pass to the limit successively in the three directions. One proves that (no summation in ℓ) :

$$\lim_{\delta_\ell \to o} \left(\lim_{\delta_s \to o} \left(\lim_{\delta_t \to o} q_{ij}^{(\delta_\ell, \delta_s, \delta_t)} \right)\right) = q_{ij}^{\ell} = a_{ij} - a_{\ell j} \int_Y \xi^i_{(\ell,\ell)} \, dy$$

with (ℓ,s,t) a permutation of (1,2,3). The functions $\xi^i_{(\ell,\ell)}$ (ℓ=1,2,3) are given with by (8). It follows that :

$$q^*_{ij} = q^1_{ij} + q^2_{ij} + q^3_{ij}$$

Moreover we can show that it is sufficient to know q^{ℓ}_{ij} for <u>one</u> value of ℓ in (1,2,3), we get the other ones by rotations of angle $\frac{\pi}{2}$. We have the result :

<u>Theorem 4</u> : The limit coefficients q^*_{ij} of theorem 2 verify :

$$q^*_{ij} = q^1_{ij} + q^1_{\bar{i},\bar{j}} + q^1_{\overline{i+1},\overline{j+1}} \quad \text{with } \bar{\ell} = \ell + 1 \pmod 3). \square$$

<u>Remark</u> : It is this superposition method which enables us to deal with oblique layers by making additional rotations and summing the corresponding coefficients.

TWO-DIMENSIONAL CASE AND REINFORCED STRUCTURES

We give without proofs the results concerning the structures corresponding to figure 1 and figure 5 (for details see [4])

<u>Theorem 5 (Reinforced structures)</u> Denote by $\tilde{u}^{\delta}$, $\tilde{\sigma}^{\delta}$ the limits corresponding to the homogenized equation. Let $\delta \to o$. The weak limits:

$$u^* = w - \lim \tilde{u}^{\delta} \ (\text{in } H^1_o(\Omega)), \qquad \sigma^* = w - \lim \delta^{-2} \tilde{\sigma}^{\delta} \ (\text{in } L^2(\Omega))$$

verify equation (4) with q^*_{ij} defined by :

$$(10) \begin{cases} q^*_{ij} = 0 & \text{if } i \neq j \\ q^*_{ii} = \frac{1}{A_{ii}} & \text{(no summation in i)} \end{cases}$$

where $(A_{ij}) = (a_{ij})^{-1}$ denotes the inverse matrix of the matrix (a_{ij}).$\square$

The <u>two dimensional</u> case can be considered as a honeycomb structure or as a reinforced structure. Indeed we have the limit problem :

$$\begin{cases} -\frac{1}{2} \sum_{i,j} q^*_{ij} \dfrac{\partial^2 u^*}{\partial x_i \partial x_j} = f \text{ in } \Omega \\ u^* = o \quad \text{on} \quad \partial\Omega \end{cases}$$

with :

$$q^*_{ij} = 2a_{ij} - \sum_{k=1}^{2} \frac{a_{ik}a_{kj}}{a_{kk}} \quad \text{(see (5))}.$$

It can be verified that this matrix is diagonal and :

$$q^*_{ii} = \frac{1}{A_{ii}} \quad \text{(see (10))}. \ \square$$

The superposition method enables us to treat the other cases (fig.2-4).

<u>Example</u> : take $a_{ij} = \delta_{ij}$ (n=2) and the structure of figure 2. We get the limit problem :

$$-\frac{1}{2} \Delta u - \frac{\sqrt{2}}{2+\sqrt{2}} \frac{\partial^2 u}{\partial x_1 \partial x_2} = f \quad \text{in } \Omega \quad ; \quad u = o \quad \text{on } \partial\Omega.$$

TALL STRUCTURES

A method similar to the one used for honeycomb and reinforced structures gives convergence results for the structures of fig. 7 - 9. As an example we consider the Neumann problem for the tall structure of fig. 7 with $u^{\varepsilon\delta} = o$ on $x_2 = o$, $\sigma^{\varepsilon\delta}. n = o$ on the other boundaries. We have :

<u>Theorem 6</u> : There is an extension operator

$p^{\varepsilon\delta} \in \mathcal{L}(H^1(\Omega^*_{\varepsilon\delta}) , H^1(]o,\varepsilon[\times]o,L[)$

such that :

$$\lim_{\delta\to o} \left(\lim_{\varepsilon\to o} \frac{1}{\varepsilon} \int_o^{\varepsilon} p^{\varepsilon\delta} u^{\varepsilon\delta}(x_1,.)\, dx_1\right) = u^*(.) \text{ in } L^2(o,L) \text{ strongly}$$

where u^* verifies the equation :

$$\begin{cases} -\frac{1}{2}\, q^* \frac{\partial^2 u^*}{\partial x_2^2} = f\,(o,x_2) \text{ in } (o,L) \\ q^* \frac{\partial u^*}{\partial x_2}\,(L) = o \\ u^*(o) = o \end{cases}$$

with

$$q^* = a_{22} - \frac{a_{21}a_{12}}{a_{11}} \,. \square$$

We also have an error estimate analogous to that of theorem 3.

Remark : Our method is quite general and applies to Dirichlet and eigenvalue problems, elasticity systems...

REFERENCES

[1] N.S. BAKHVALOV, G.P. PANASENKO. Averaged processes in periodic media. Moscou, Nauka, 1984 (in russian).

[2] D. CIORANESCU, J. SAINT JEAN PAULIN. Homogenization in open sets with holes. Journal of Math. Anal. Appl., 71, 2 (1979), 590-607.

[3] D. CIORANESCU, J. SAINT JEAN PAULIN. Problèmes de Neumann et de Dirichlet dans des structures réticulées de faible épaisseur. Comptes-Rendus de l'Académie des Sciences I, 303 (1986) 7-10.

[4] D. CIORANESCU, J. SAINT JEAN PAULIN. Reinforced and honeycomb structures. To appear in Journal de Math. Pures et Appl. (Also in Publications du Laboratoire d'Analyse Numérique. Université Pierre et Marie Curie n° 85042, Paris, 1985).

[5] D. CIORANESCU, J. SAINT JEAN PAULIN. Calculus of tall structures. To appear.

[6] J.L. LIONS. Some methods in the mathematical analysis of systems and their control. Science Press, Beijing, China. Gordon Breach, New York, 1981.

[7] G.P. PANASENKO. The principle of averaged operator decomposition for a nonlinear equation set in a periodic and random skeletal construction. Dokl. Acad. Nauk USSR, 263 (1), 1982, 35 - 40 (in russian).

[8] G.P. PANASENKO. Averaged processes in skeletal constructions with random properties. J.V. Math. and Phys. Math. 23 (5), 1983, 1098 - 1109 (in russian).

[9] G.P. PANASENKO. Averaged processes in skeletal constructions. Mat. Sb. 122 (164), 2 (10), 1983, 220 - 231 (in russian).

[10] L. TARTAR. Problèmes d'homogénéisation dans les équations aux dérivées partielles. Cours Peccot, Collège de France, 1977.

Doina CIORANESCU
C.N.R.S.
Laboratoire d'Analyse Numérique LA 189
Université Pierre et Marie CURIE
4, place Jussieu
75252 PARIS CEDEX 05
FRANCE

Jeannine SAINT JEAN PAULIN
CNRS - L.E.M.T.A. - I.N.P.L.
2,rue de la Citadelle
B.P. 850
54011 NANCY CEDEX
FRANCE

THREE-DIMENSIONAL PROPERTIES OF A GENERALLY ORTHOTROPIC SYMMETRIC LAMINATE

GERALD R. KRESS

Dornier System GmbH
Postfach 1360
D-7990 Friedrichshafen

ABSTRACT

The classical lamination theory (CTL) views a laminated material as a thin plate and does not consider its through-the-thickness properties. However, thick-walled structural components exist for which the stiffness and expansion properties perpendicular to the layers are crucial for the overall performance of the structure.

This work presents exact equations for replacing symmetric laminates by a homogenous material the mechanical properties of which are defined in all three spatial directions.

The advantage of the method presented here lies in its potential to either greatly simplify or altogether eliminate costly numerical structural analyses where the through-the-thickness behaviour of thick laminates is involved.

Keywords: Laminate, Generally Orthotropic, Out-of-Plane Properties.

INTRODUCTION

It is the fibers which contribute the most to high-stiffness and high-strength properties of laminated fiber-matrix composites.

Since the fibers are oriented parallel to the laminate midplane these high-value properties can only be achieved in these directions. Consequently, laminates are mainly used as thin shells, plates or discs as reflected by the classical lamination theory (CLT). As Herakovich (1) states, little attention has been paid to laminate properties with respect to the through-the-thickness direction.

Yet laminated structural components exist, the out-of-plane properties of which, such as the stiffness E_z or the coefficients of thermal expansion α_z, play an essential role in the behaviour of the structure. Faced with the task of economocially including such components in a mathematical model the engineer may attempt to calculate the in-plane properties accurately, using CLT, whilst making assumptions for the out-of-plane properties.

Figure 1 Coordinates and Indices

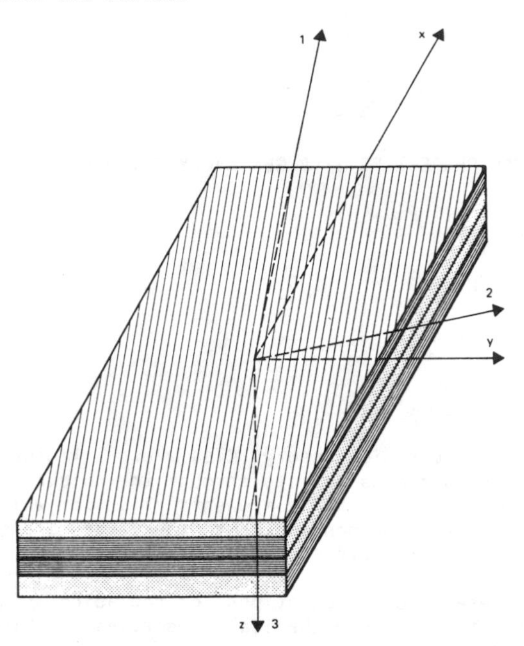

Simply adopting the uni-directional properties for the principal layer directions C_{33}, C_{44}, and C_{55} for the respective laminate stiffnesses would produce only small errors within a five-percent range. However, as already pointed out in the literature (1), the transverse-strain behaviour of a laminate as represented by poisson's ratios ν_{xz} and ν_{yz} can be extremely different from that of the uni-directional materials. Moreover, the coefficient of thermal expansion α_z of a quasi-isotropic laminate made from a typical carbon-fibre-prepreg material is almost 15 % higher than the uni-directional value α_3.

It should be recalled that it is not the mere thermal expansion of structural components which causes thermal stresses but rather the differences in thermal behaviour of different components within a structure. In view of this it becomes clear that an error of 15 % in the assumed value of α_z for one component can cause severe errors in the calculated temperature stresses within a structure.

This paper derives an exact presentation of the thirteen elements of the stiffness matrix of a generally orthotropic laminated plate. Also, the associated coefficients of thermal expansion α_x, α_y, α_z, and α_{xy} are calculated. The same equations hold for the coefficients of moisture expansion if the symbols α_{ij} are exchanged for β_{ij}. The procedure is similar to the CLT. The introductory book by Jones (2) provides the necessary background for this work.

The interested reader who misses a thorough comparison of results with those of other methods or a descriptive example of a useful application of the method presented here is referred to a previous paper [3].

One must, however, be careful in applying a smeared-out laminate model in terms of the orthotropic stiffness matrix. As pointed out by Herakovich the value of poissons ratios ν_{xz} and ν_{yz} as calculated under the typical assumptions of the CLT are usually invalid on free edges. The same applies for the other properties.

Even more important: the dependency of bending stiffnesses upon the stacking sequence must be considered. If bending occurs, only thick laminates having many evenly through-the-thickness distributed fiber-directions of the individual layers can be adequately described by the elastic moduli.

Errors, however small, must be taken into account if the laminate is subjected to pointloads.

THEORY

Problem Formulation and Assumptions

As in the classical lamination theory a perfect bond throughout the laminate is assumed; enforcing displacements being continuous through the thickness.

In contrast to the classical lamination theory it is not demanded that intersections normal to the midplane remain normal under load. Also, the thickness of the laminate is allowed to change.

These assumptions lead to constant distributions of the in-plane strains through the thickness:

$$\varepsilon_x = \varepsilon_x^\circ = \text{const} \tag{1}$$

$$\varepsilon_y = \varepsilon_y^\circ = \text{const} \tag{2}$$

$$\gamma_{xy} = \gamma_{xy}^\circ = \text{const} \tag{3}$$

and, relaxing the confinements of the classical lamination theory, the out-of-plane stresses are allowed to take on non-zero values, which are, as an equilibrium requirement, constant through the thickness:

$$\sigma_{zz} = \sigma_{zz}^\circ = \text{const} \tag{4}$$

$$\tau_{yz} = \tau_{yz}^\circ = \text{const} \tag{5}$$

$$\tau_{xz} = \tau_{xz}^\circ = \text{const} \tag{6}$$

In the following, the upper index ($^\circ$) serves as a reminder that the respective value is constant throughout the thickness while an upper (^{k}) indicates that the respective value may be different in different layers.

Development of Constitutive Equations on Laminate Level

The constitutive equations of an orthotropic lamina can be written in matrix notation:

$$[\bar{C}] \cdot \{\varepsilon\} = \{\sigma\} \tag{7}$$

with

$$\{\varepsilon\}^T = \{\varepsilon_x^{\circ} \quad \varepsilon_y^{\circ} \quad \varepsilon_z^k \quad \gamma_{yz}^k \quad \gamma_{xz}^k \quad \gamma_{xy}^{\circ}\} \tag{8}$$

$$\{\sigma\}^T = \{\sigma_x^k \quad \sigma_y^k \quad \sigma_z^{\circ} \quad \tau_{yz}^{\circ} \quad \tau_{xz}^{\circ} \quad \tau_{xy}^k\}. \tag{9}$$

The bar in $[\bar{C}]$ indicates that the principal material axes and the global x-y coordinates generally have different orientations.

The stiffness matrix $[\bar{C}]$ for arbitrary coordinates must be calculated from the matrix [C] which is valid for the principal material directions by a coordinate transformation:

$$[\bar{C}] = [T]^{-1} \; [C] \; [T]^{-T} . \tag{10}$$

The transformation matrix [T], the stiffness matrices [C] and $[\bar{C}]$, and the definition of the C_{ij} in terms of the engineering constants are taken from Jones (Ref. 2).

Inspecting Eqs. (8) and (9) one notices that both the strain vector $\{\varepsilon\}$ and the stress vector $\{\sigma\}$ contain components which are constant through the tickness, (upper index ($^{\circ}$)) and elements the magnitudes of which vary from layer to layer (upper index (^{k})).

In order to obtain the properties of multi-layered plates by an explicit integration through the thickness, Eq. (7) must be rearranged. Specifically, two new vectors $\{l\}^{\circ}$ and $\{r\}^k$ are formed. The left-hand-side vector $\{l\}^{\circ}$ contains the values which are taken to be constant through the thickness,

$$\{l\}^{\circ T} = \{\varepsilon_x^{\circ} \quad \varepsilon_y^{\circ} \quad \sigma_z^{\circ} \quad \tau_{xz}^{\circ} \quad \tau_{yz}^{\circ} \quad \gamma_{xy}^{\circ}\} , \tag{11}$$

while the right-hand-side vector $\{r\}^k$ contains the remaining values whose magnitudes vary from layer to layer:

$$\{r\}^{kT} = \{\sigma_x^k \quad \sigma_y^k \quad \varepsilon_z^k \quad \gamma_{xz}^k \quad \gamma_{yz}^k \quad \tau_{xy}^k\} . \tag{12}$$

The rearranged system of Eqs. (7) reads

$$[R]^k \cdot \{l\}^{\circ} = \{r\}^k , \tag{13}$$

where the elements of the [R] matrix are defined in Table A1. It is obvious that the [R] matrix can neither be correctly called stiffness matrix nor compliance matrix since some of its elements are stiffnesses, others are compliances and the rest are simply factors. However, it still presents the constitutive relations between the strains and stresses and, hence, may be called constitutive matrix [R].

The behaviour of a multi-layered plate can now be established by simply integrating Eqs. (13) through the thickness:

$$\frac{1}{t} \int_{-h}^{h} [R]^k \, dz \cdot \{1\}^\circ = \frac{1}{t} \int_{-h}^{h} \{r\}^k \, dz \,, \tag{14}$$

where t is the laminate thickness. The equation is divided by the plate thickness. Since all the values in Eqs. (14) are constant within one individual layer the integration rule results in a simple summation:

$$\frac{1}{t} \sum h^k \, [R]^k \, \{1\}^\circ = \frac{1}{t} \sum h^k \, \{r\}^k \,, \tag{15}$$

where the thickness of the k-th layer is denoted by h^k.

Table 1 Elements of the R-Matrix

$$R_{ij} = \overline{C}_{ij} - \overline{C}_{13} \, C_{3j}/C_{33} \qquad i,j = 1,2,6$$

$$R_{i3} = \overline{C}_{i3}/C_{33} \qquad i,j = 1,2,6$$

$$R_{3i} = -R_{i3}$$

$$R_{33} = 1/C_{33}$$

$$R_{44} = \overline{C}_{55}/\overline{C}_{44} \, \overline{C}_{55} - \overline{C}_{45} \, \overline{C}_{45})$$

$$R_{45} = -\overline{C}_{45}/(\overline{C}_{44} \, \overline{C}_{55} - \overline{C}_{45} \, \overline{C}_{45})$$

$$R_{55} = -\overline{C}_{44}/(\overline{C}_{44} \, \overline{C}_{55} - \overline{C}_{45} \, \overline{C}_{45})$$

An abbreviation of Eqs. (15) is

$$[P] \cdot \{1\}^\circ = [p] \tag{16}$$

with $[P] = \frac{1}{t} \sum h^k \, [R]^k$ (17)

and $\{p\} = \frac{1}{t} \sum h^k \, \{r\}^k$ (18)

The elements of vector {p}

$$\{p\}^T = \{\sigma_x \; \sigma_y \; \varepsilon_z \; \gamma_{xz} \; \gamma_{yz} \; \tau_{xy}\} \tag{19}$$

are averaged or smeared-out values.

Eqs. (16) can now be re-arranged using the same formalism used for transforming eqs. (7) into eqs. (13). The result is the desired threedimensional stiffness matrix for a generally orthotropic laminate:

$$[C]_{Laminate} \cdot \{\varepsilon\} = \{\sigma\} \tag{20}$$

Coefficients of Thermal Expansion

If a body is subjected to a temperature change the fixation of its surface will cause thermal stresses on the surface. Further, if the relationsship between stresses and strains with respect to the surface is known, the coefficients of thermal expansion are easily established even for nonhomogenous materials.

For a single lamina equ. (13) including thermal strains is valid:

$$\begin{bmatrix} R_{11} & R_{12} & R_{13} & R_{16} \\ R_{12} & R_{22} & R_{23} & R_{26} \\ R_{13} & R_{23} & R_{33} & R_{36} \\ R_{16} & R_{26} & R_{36} & R_{66} \end{bmatrix}^{k} \begin{bmatrix} ex^{\circ} - ax^{k}\,\Delta T \\ ey^{\circ} - ay^{k}\,\Delta T \\ \sigma z^{\circ} \\ \gamma xy^{\circ} - axy^{k}\,\Delta T \end{bmatrix} = \begin{bmatrix} \sigma x^{k} \\ \sigma y^{k} \\ \varepsilon z^{k} \\ \tau xy^{k} \end{bmatrix} \tag{21}$$

Considering the multi-layered laminate again, fixation of the boundaries is expressed by

$$\varepsilon_x^{\circ} = 0, \tag{22}$$

$$\varepsilon_y^{\circ} = 0, \tag{23}$$

$$\frac{1}{t} \sum \varepsilon_z^{k}\, t^{k} = 0, \tag{24}$$

and $\gamma^{\circ}xy = 0.$ (25)

It should be noted that the through-the-thickness strain ε_z^k within the individual layer k may deviate from zero though, especially for hybrid laminates.

Integrating eqs. (21) while incorporating eqs. (22) through (25) yields

$$\frac{1}{t} \sum \begin{bmatrix} R_{11} & R_{12} & R_{13} & R_{16} \\ R_{12} & R_{22} & R_{23} & R_{26} \\ R_{13} & R_{23} & R_{33} & R_{36} \\ R_{16} & R_{26} & R_{36} & R_{66} \end{bmatrix}^{k} \begin{bmatrix} -\alpha_x{}^{k}\,\Delta T \\ -\alpha_y{}^{k}\,\Delta T \\ \sigma_z{}^{\circ} \\ -\alpha_{xy}{}^{k}\,\Delta T \end{bmatrix} = \begin{bmatrix} \sigma_x{}^{T} \\ \sigma_y{}^{T} \\ 0 \\ \tau_{xy} \end{bmatrix}, \tag{26}$$

where $\sigma_z^{\circ T}$, $\sigma_x{}^T$, $\sigma_y{}^T$ and $\tau_{x_y}{}^T$ are unknown. From the third of eqs. (26) follows for $\sigma_z{}^T$:

$$\sigma_z^{\circ T} = [\Sigma t^k R_{33}{}^k]^{-1}\,[\Sigma t^k (R_{31}\alpha_x{}^k + R_{32}\alpha_y{}^k + R_{36}\alpha_{xy}{}^k]\Delta T \tag{27}$$

By insertion of the value of $\sigma_z^{\circ T}$ into the first, second, and fourth of eqs. (26) $\sigma_x{}^T$, $\sigma_y{}^T$, and $\tau_{xy}{}^T$ can be calculated.

Using the laminate stiffness matrix as derived above, the coefficients of thermal expansion are found upon solving the equations

$$\begin{bmatrix} C_{11} & C_{12} & C_{13} & C_{16} \\ C_{12} & C_{22} & C_{23} & C_{26} \\ C_{13} & C_{23} & C_{33} & C_{36} \\ C_{16} & C_{26} & C_{36} & C_{66} \end{bmatrix} \cdot \begin{bmatrix} \alpha_x \\ \alpha_y \\ \alpha_z \\ \alpha_{xy} \end{bmatrix} = - \begin{bmatrix} \sigma_x^T \\ \sigma_y^T \\ \sigma_z^T \\ \tau_{xy}^T \end{bmatrix} . \qquad (28)$$

The calculation of moisture-expansion coefficients is completely analogous.

CONCLUSIONS

- Thick, symmetric laminates consisting of many layers the fiber directions of which are distributed evenly through the thickness can be modeled as homogenous materials.

- The elements of the stiffness matrix of a generally orthotropic laminate as well as the temperature and moisture expansion coefficients in three dimensions can be calculated using a method similar to the well-known CLT-approach to the in-plane behaviour.

- In comparison to a full FEM-model the use of the homogenized model simplifies the analysis to a great extent.

- Great care must be exercised in applying the homogenized material model to laminates: thin laminates can be replaced by the model only if no bending is involved.

REFERENCES

1. C.T. Herakovich, Composite Laminates with Negative Through-the-Thickness Poisson's Ratios, Journal of Composite Materials, 18, 447-455 (1984)

2. R.M. Jones, Mechanics of Composite Materials, Scripta Book Company, Washington, D.C., 1975

3. G.R. Kress, Orthotropic Properties of Layered Fiber Composites, Proceedings of the ESTEC-Workshop 1984, Noordwijk, the Netherlands, 1985

ON THE HYGROTHERMOMECHANICAL BEHAVIOUR OF A COMPOSITE MATERIAL

HORIA I. ENE

Increst, Department of Mathematics,
Bd. Pacii 220,
79622 Bucharest,
Romania.

ABSTRACT

Starting from the equations for the hygrothermomechanical behaviour of each constituent in a periodic composite, the homogenization method is applied to derive macroscopic balance equations and constitutive equations. Explicit expressions are presented for the macoscopic coefficients. The equations are applicable to materials for which the stress deformation fields are strongly dependent on temperature and moisture content.

INTRODUCTION

For a periodic model of a composite material, the problem of the elastic behaviour under the effects of combined moisture and thermal environmments, may be considered from the point of view of the homogenization method. The periodic structure is associated with a small parameter ε, defined for practical problems as the ratio between a characteristic length of the period over a characteristic length of the macroscopic medium. In fact the asymptotic process, $\varepsilon \to 0$ implies that we are in the case of a great number of periods.

The hygrothermomechanical properties of the matrix and of the inclusions are different. We suppose that at the microscopic level the coupling effects are due to the presence of the strain-temperature, respectively strain-moisture terms in the energy equation, respectively in the equation for the moisture concentration. Consequently in each $\Omega_{\varepsilon i}$ $(i=1,2)$ we have:

$$\frac{\partial \sigma^{\varepsilon}_{ij}}{\partial x_j} - \rho^{\varepsilon} \frac{\partial^2 u_i}{\partial t^2} = -f_i \tag{1}$$

$$\frac{\partial}{\partial x_i}\left(k^{\varepsilon}_{ij} \frac{\partial T^{\varepsilon}}{\partial x_j}\right) - T_o \beta^{\varepsilon}_{ij} \frac{\partial e^{\varepsilon}_{ij}}{\partial t} - c^{\varepsilon} \frac{\partial T^{\varepsilon}}{\partial t} = -r \tag{2}$$

$$\frac{\partial}{\partial x_i}\left(d^{\varepsilon}_{ij} \frac{\partial H^{\varepsilon}}{\partial x_j}\right) - H_o \alpha^{\varepsilon}_{ij} \frac{\partial e^{\varepsilon}_{ij}}{\partial t} - b^{\varepsilon} \frac{\partial H^{\varepsilon}}{\partial t} = -h \tag{3}$$

$$\sigma^{\varepsilon}_{ij} = c^{\varepsilon}_{ijkh} e^{\varepsilon}_{kh} - \beta^{\varepsilon}_{ij} T^{\varepsilon} - \alpha^{\varepsilon}_{ij} H^{\varepsilon} \tag{4}$$

$$e^{\varepsilon}_{ij} = \frac{1}{2}\left(\frac{\partial u^{\varepsilon}_{i}}{\partial x_j} + \frac{\partial u^{\varepsilon}_{j}}{\partial x_i}\right) \tag{5}$$

$$\Omega_{\varepsilon i} = \{x;\ x \varepsilon \Omega,\ x \varepsilon \varepsilon Y_i\} \qquad (i=1,2) \tag{6}$$

where the period Y, in the space of the variables y_i (i=1,2,3), is formed by two parts Y_1 and Y_2 (the matrix and the inclusion), Ω is the domain of the composite material in the space of the variables x_i (macroscopic variables), $\sigma^{\varepsilon}_{ij}$ and e^{ε}_{ij} are, respectively, the linear stress and strain tensors, T^{ε} is the temperature, H^{ε} is the moisture concentration, u^{ε}_{i} are the components of the displacement vector, T_o and H_o are respectively, the absolute reference temperature and moisture content, f_i are the body force components, r and h are respectively, the heat and moisture supply, c^{ε}_{ijkh} is the stiffness tensor, β^{ε}_{ij} is the strain-temperature-tensor, $\alpha^{\varepsilon}_{ij}$ is the strain-moisture tensor, k^{ε}_{ij} is the thermal conductivity tensor, d^{ε}_{ij} is the hygroscopic conductivity tensor, c^{ε} is the specific heat at constant deformation, b^{ε} is the specific hygroscopic capacity and ρ^{ε} is the mass density.

All these tensors have the usual symmetry properties, and we look for Y-periodic functions in the variable $y=x/\varepsilon$:

$$c^{\varepsilon}_{ijkh} = c^{\varepsilon}_{khij} = c^{\varepsilon}_{jikh}\ ;\ \beta^{\varepsilon}_{ij} = \beta^{\varepsilon}_{ji}\ ,\ \alpha^{\varepsilon}_{ij} = \alpha^{\varepsilon}_{ji} \tag{7}$$

$$c^{\varepsilon}_{ijkh}(x) \equiv c_{ijkh}(x/\varepsilon),\ \beta^{\varepsilon}_{ij}(x) \equiv \beta_{ij}(x/\varepsilon),\ \alpha^{\varepsilon}_{ij}(x) \equiv \alpha_{ij}(x/\varepsilon)) \tag{8}$$

$$k^{\varepsilon}_{ij}(x) \equiv k_{ij}(x/\varepsilon),\ d^{\varepsilon}_{ij}(x) \equiv d_{ij}(x/\varepsilon),\ \rho^{\varepsilon}(x) \equiv \rho(x/\varepsilon)\ ,$$

$$c^{\varepsilon}(x) \equiv c(x/\varepsilon),\ b^{\varepsilon}(x) \equiv b(x/\varepsilon)\ .$$

Using the homogenization method, we obtain the macroscopic balance equations and the constitutive equation [1]:

$$\frac{\partial \sigma^{o}_{ij}}{\partial x_j} - \rho^{o} \frac{\partial^2 u^{o}_{i}}{\partial t^2} = -f_i \tag{9}$$

$$\frac{\partial}{\partial x_i}\left(k^{o}_{ij} \frac{\partial T^{o}}{\partial x_j}\right) - T_o \beta^{o}_{ij} \frac{\partial e^{o}_{ij}}{\partial t} - (c^{o}_{o} - T_o \gamma)\frac{\partial T^{o}}{\partial t} + T_o \delta \frac{\partial H^{o}}{\partial t} = -r \tag{10}$$

$$(11) \qquad \frac{\partial}{\partial x_i}(d^o_{ij}\frac{\partial H^o}{\partial x_j}) - H_o \alpha^o_{ij}\frac{\partial e^o_{ij}}{\partial t} - (b^o - H_o\lambda)\frac{\partial H^o}{\partial t} + H_o \delta \frac{\partial T^o}{\partial t} = -h$$

$$(12) \qquad \sigma^o_{ij} = c^o_{ijkh} e^o_{kh} - \beta^o_{ij} T^o - \alpha^o_{ij} H^o$$

where the mean value is defined by:

$$(13) \qquad a^o = \frac{1}{|Y|}\int_Y a(y)\,dy$$

The equation for the balance of momentum (9) and the constitutive equation (12) are of the same type as the microscopic ones (1), respectively (4). The macroscopic equation for temperature (10) contains the time derivative of the moisture and the macroscopic equation for moisture concentration (11) contains the time derivative of the temperature. Also, these equations contain new coefficients γ and λ. Then the macroscopic system of equations is coupled with new terms which do not appear at the microscopic level. This fact is a direct consequence of the homogenization technics. At the microscopic level we have a nonhomogeneous medium, with a periodic structure, but with coefficients depending effectively on x/ε. The homogenization technics for the study of composite materials is based on the study of periodic solutions of a system of partial differential equation and their asymptotic behaviour as $\varepsilon \to 0$. Following this way, at the macroscopic level we obtain a system of equation with constant coefficients which describe the behaviour of a "homogenized" material, and consequently at this level we have the coupled terms.

The convergence theorem for the homogenization process was proved in [1].

MACROSCOPIC COEFFICIENTS

The macroscopic coefficients which appear in the system (9)-(12), describe the microscopic structure of our medium, and they are defined by the following formula [1]:

$$(14) \qquad c^o_{ijkh} = [c_{ijkh} + c_{ijmn} e_{mn}(\underline{W}^{kh})]^o$$

(15) $\beta^o_{ij}=[\beta_{ij}+c_{ijmn}e_{mn}(\underline{\Theta})]^o=[\beta_{ij}+\beta_{mn}e_{mn}(\underline{W}^{ij})]^o$

(16) $\alpha^o_{ij}=[\alpha_{ij}+c_{ijmn}e_{mn}(\underline{\chi})]^o=[\alpha_{ij}+\alpha_{mn}e_{mn}(\underline{W}^{ij})]^o$

(17) $k^o_{ij}=[k_{ij}+k_{im}\frac{\partial w^j}{\partial y_m}]^o$

(18) $d^o_{ij}=[d_{ij}+d_{im}\frac{\partial h^j}{\partial y_m}]^o$

(19) $\gamma=[\beta_{ij}e_{ij}(\underline{\Theta})]^o$

(20) $\lambda=[\alpha_{ij}e_{ij}(\underline{\chi})]^o$

(21) $\delta=[\beta_{ij}e_{ij}(\underline{\chi})]^o=[\alpha_{ij}e_{ij}(\underline{\Theta})]^o$

where the vectors $\underline{W}^{kh}$, $\underline{\Theta}$, $\underline{\chi}$ and the functions w^j and h^j, are the solutions of $H^1_{per}(Y)$, with zero mean values, of the equations:

(22) $\int_Y c_{ijmn}e_{mn}(\underline{W}^{kh})e_{ij}(\underline{v})dy=\int_Y \frac{\partial c_{ijkh}}{\partial y_j}v_i dy \quad (\forall)\ \underline{v}\varepsilon H^1_{per}(Y)$

(23) $\int_Y c_{ijmn}e_{mn}(\underline{\Theta})e_{ij}(\underline{v})dy=\int_Y \frac{\partial \beta_{ij}}{\partial y_j}v_i dy \quad (\forall)\ \underline{v}\varepsilon H^1_{per}(Y)$

(24) $\int_Y c_{ijmn}e_{mn}(\underline{\chi})e_{ij}(\underline{v})dy=\int_Y \frac{\partial \alpha_{ij}}{\partial y_j}v_i dy \quad (\forall)\ \underline{v}\varepsilon H^1_{per}(Y)$

(25) $\int_Y k_{im}\frac{\partial w^j}{\partial y_m}\frac{\partial \varphi}{\partial y_i}dy=\int_Y \frac{\partial k_{jm}}{\partial y_m}\varphi dy \quad (\forall)\ \varphi\varepsilon H^1_{per}(Y)$

(26) $\int_Y d_{im}\frac{\partial h^j}{\partial y_m}\frac{\partial \varphi}{\partial y_i}dy=\int_Y \frac{\partial d_{jm}}{\partial y_m}\varphi dy \quad (\forall)\ \varphi\varepsilon H^1_{per}(Y)$

The macroscopic stiffness tensor (14) was first obtained in the classical linear elasticity [2]. The macroscopic strain-temperature tensor (15), the macroscopic thermal conductivity tensor (17) and γ (19) were first defined in the study of linear thermoelasticity of composite material [3], [4]. The other macroscopic coefficients are new and they take into consideration the hygroscopic properties.

The above formula give us explicit expressions for the computation of effective coefficients. For this it is necessary to solve some boundary values problems (22)-(26) of the elliptic type in each period, with the periodicity condition instead of a boundary condition, and after that to take the mean values on a perioad following the formula (14)-(21).

The macroscopic strain-temperature tensor (15) and the macroscopic strain-moisture tensor (16) depend also on the macroscopic stiffness, because they depend on $\underline{W}^{ij}$ from which we compute c^{o}_{ijkh}.

The macroscopic coefficients γ and λ are strictly negative definite. To see that it is sufficient to take as test function $\underline{\Theta}$ in (23), and respectively $\underline{\chi}$ in (24).

The macroscopic thermal conductivity tensor and the macroscopic hygroscopic conductivity tensor are different from the simply mean values of the microscopic tensors. The same remark is valid also for the macroscopic specific heat at constant deformation and the macroscopic specific hygroscopic capacity, because they are defined by:

$$c^{o}-T_{o}\gamma=[c-T_{o}\beta_{ij}e_{ij}(\underline{\Theta})]^{o} \tag{27}$$

$$b^{o}-H_{o}\lambda=[b-H_{o}\alpha_{ij}e_{ij}(\underline{\chi})]^{o} \tag{28}$$

In the formula (14)-(21) the mean values are calculated using (13), and then they are computed as mean values over the volume of the period.

A PROPERTY OF THE MACROSCOPIC COEFFICIENTS

It is known that the first term of the asymptotic expansion of the stress tensor verify the local equation [2], [5]:

$$\frac{\partial\sigma_{ij}}{\partial y_{j}} = 0$$

or in the equivalent form

$$\int_{Y}\frac{\partial\sigma_{ik}}{\partial y_{k}}y_{j}dy=0 \tag{29}$$

We consider the trivial equality:

$$(30)\quad \frac{1}{|Y|}\int_Y \frac{\partial}{\partial y_k}(\sigma_{ik}y_j)dy=\frac{1}{|Y|}\int_Y \frac{\partial\sigma_{ik}}{\partial y_k}y_j dy+\frac{1}{|Y|}\int_Y \delta_{jk}\sigma_{ik}dy$$

or in the equivalent form:

$$(31)\quad \sigma^o_{ij}=\frac{1}{|Y|}[\int_{\partial Y}\sigma_{ik}y_j h_k ds-\int_Y \frac{\partial\sigma_{ik}}{\partial y_k}y_j dy]$$

Then with (29) we have:

$$(32)\quad \sigma^o_{ij}=\frac{1}{|Y|}\int_{\partial Y}\sigma_{ik}y_j n_k ds$$

We denote by $\sum_k$ the face of the paralelipipedic cell Y othogonal to the direction y_k and with $\sum'_k$ the oposite face to $\sum_k$. Than, $\partial Y=\sum_1\cup\sum'_1\cup\sum_2\cup\sum'_2\cup\sum_3\cup\sum'_3$. If the dimensions of the cell are a_k (k=1,2,3), we have:

$$\frac{1}{|Y|}\int_{\partial Y}\sigma_{ik}y_1 n_k ds=\frac{1}{|Y|}[\int_{\sum_1}\sigma_{i1}a_1 dy_2dy_3-\int_{\sum'_1}\sigma_{i1}(-a_1)dy_2dy_3+$$

$$+\int_{\sum_2}\sigma_{i2}y_1dy_1dy_3-\int_{\sum'_2}\sigma_{i2}y_1dy_1dy_3+\int_{\sum_3}\sigma_{i3}y_1dy_1dy_2-$$

$$-\int_{\sum'_3}\sigma_{i3}y_1dy_1dy_2]=\frac{1}{|\sum_1|}\int_{\sum_1}\sigma_{i1}dy_2dy_3$$

the integrals on the other faces being zero by the periodicity condition. Then it is clear that:

$$(33)\quad \frac{1}{|Y|}\int_Y\sigma_{ik}y_j n_k ds = \frac{1}{|\sum_j|}\int_{\sum_j}\sigma_{ij}ds=\langle\sigma_{ij}\rangle$$

which represent the mean value on a face of the cell. From (32) we conclude:

$$(34)\quad \sigma^o_{ij}=\langle\sigma_{ij}\rangle$$

On the other hand we have:

$$(35)\quad \sigma_{ij}=[c_{ijkh}+c_{ijmn}e_{mn}(W^{kh})]e^o_{kh}-[\beta_{ij}+\beta_{mn}e_{mn}(\underline{W}^{ij})]T^o-$$

$$-[\alpha_{ij}+\alpha_{mn}e_{mn}(\underline{W}^{ij})]H^o$$

where the functions e^o_{kh} , T^o and H^o are independent of y. If we take the surface mean value of (35) we have:

$$<\sigma_{ij}>=<c_{ijkh}+c_{ijmn}e_{mn}(\underline{W}^{kh})>e^o_{kh}-<\beta_{ij}+\beta_{mn}e_{mn}(\underline{W}^{ij})>T^o-$$
$$-<\alpha_{ij}+\alpha_{mn}e_{mn}(\underline{W}^{ij})>H^o \quad (36)$$

which by virtue of (34) will be the same as (12). Then the macroscopic coefficients in the constitutive equation may be computed as mean values over the volume or as mean values over a surface of the cell.

CONCLUSION

The complet coupled system of equations which describe the hygrothermomechanical behaviour of a composite material was obtained. The macroscopic coefficients may be calculated directly by the explicite formula obtained here. It is known that the effect of temperature and moisutre is significant in the deformation and stress field, and consequently it is necessary to use the complet system of equations.

REFERENCES

[1] H.I. ENE, Int. J. Engng. Sci 24, 5, 841 (1986).
[2] E. SANCHEZ-PALENCIA, Non-Homogeneous Media and Vibration Theory, Lecture Notes in Physics, vol. 127, Springer, Berlin (1980).
[3] H.I. ENE, Int. J. Engng. Sci., 21, 5, 443 (1983).
[4] G.I. PASA, Int. J. Engng. Sci., 21, 11, 1313 (1983).
[5] H.I. ENE and G.I. PASA, Homogenization Method. Applications to the Theory of Composite Materials, Ed. Acad. Bucharest (1986) (in Romanian).

A THERMODYNAMIC VISCOPLASTICITY THEORY OF IDEAL FIBRE-REINFORCED MATERIALS

MIĆUNOVIĆ, MILAN

Faculty of Mechanical Engineering
Svetozar Markovic University
34000 Kragujevac, Yugoslavia

ABSTRACT

A metal-matrix fibre-reinforced viscoplastic body with elastically and plastically inextensible as well as incompressible but thermally extensible as well as compressible fibres is considered. For such an idealized composite body constitutive equations are restricted by means of Müller's entropy principle. For this sake quasikinematic constraints and constrains imposed by yield function are used.

The main results are:
- the proposed novel procedure gives rise to a rational theory,
- the continuity on the elastoplastic frontier and a generalization of Il'iushin and Drucker principles are accomplished and
- the heat flux and the entropy flux are not colinear so that the coldness function is not universal.

INTRODUCTION

In the paper [1] the thermodynamical approach of Coleman and Noll [2] has been used in order to formulate a rate independent thermoplasticity theory of metal matrix fibre-reinforced materials. Since time rates of temperature and plastic deformation are not considered the approach has been suitable. On the other hand in [3] a noncomposite viscoplastic body has been regarded with the above mentioned rates included into the list of constitutive variables. For such a case the approach [2] is not convenient any more and Müller's entropy principle [4] has to be used instead.

Here the last approach will be applied to a metal matrix fibre reinforced viscoplastic body. A relative motion between fibres and the matrix is neglected and the composite is considered as a strongly anisotropic continuum. The aim is to formulate a rational viscoplasticity theory for such materials.

1. GEOMETRIC AND QUASIKINEMATIC CONSTRAINTS

Let us consider a composite body, $\mathscr{B}$, composed of a metal matrix and one family of strong fibres continuously immersed into the matrix. In order to describe adequately a viscoplastic behaviour of $\mathscr{B}$ we need the following three configurations (cf. [5,6]): $\vec{\kappa}$ - a completely relaxed global reference configuration of $\mathcal{B}$ at some uniform reference temperature θ_o, $\vec{\chi}_t$ - the instant global observable configuration of $\mathcal{B}$ with stresses caused by external surface tractions, defects and inhomogeneous temperature field $\theta(\underline{X},t)$ and $\vec{\nu}_t$ - a completely relaxed local reference configuration of an infinitesimal part $\mathcal{N}$ of $\mathcal{B}$. Here X are material coordinates, t is time and the last configuration is usually referred to as natural state. Moreover, we assume that fibres are defined in $\vec{\chi}_t$, $\vec{\nu}_t$ and $\vec{\kappa}$ by means of unit tangent vectors $\vec{a}$, $\vec{A}_\nu$ and $\vec{A}$, respectively. Then thermoelastic and plastic distortion tensors as well as deformation gradient tensors are given, respectively, by mappings [1]

$$\mathbf{E}: \vec{\nu}_t \to \vec{\chi}_t, \quad d\vec{x} = \mathbf{E}\cdot d\vec{\xi}, \quad \vec{a} = \frac{1}{\lambda_E}\mathbf{E}\cdot\vec{A}_\nu, \quad \theta_\nu = \theta_o, \tag{1.1}$$

$$\mathbf{P}: \vec{\kappa} \to \vec{\nu}_t, \quad d\vec{\xi} = \mathbf{P}\cdot d\vec{X}, \quad \vec{A}_\nu = \frac{1}{\lambda_P}\mathbf{P}\cdot\vec{A}, \quad \theta_\kappa = \theta_o = \text{const}, \tag{1.2}$$

$$\mathbf{F}: \vec{\kappa} \to \vec{\chi}_t, \quad d\vec{x} = \mathbf{F}\cdot d\vec{X}, \quad \vec{a} = \frac{1}{\lambda}\mathbf{F}\cdot\vec{A}, \quad \theta_\chi = \theta(X,t), \tag{1.3}$$

where by λ_E, λ_P and λ (cf. [1,6]) elastic, plastic and total fibre extension ratios are designated.

Suppose, furthermore, a) that the composite behaves in such a way that during plastic deformation fibres do not change their lengths and that this deformation is isochoric, b) that material is elastically incompressible and inextensible in fibre direction and c) that material changes volume and fibres their lengths if a change of temperature takes place. These assumptions are represented by the following equations

$$\det\mathbf{P} - 1 = 0, \tag{1.4}$$

$$\lambda_P^2 - 1 = \vec{A}\cdot\mathbf{C}_P\cdot\vec{A} - 1 = 0, \tag{1.5}$$

$$\det\mathbf{C} - h_o(\theta-\theta_o) = 0, \tag{1.6}$$

$$\lambda^2 - h_1(\theta-\theta_o) = \vec{A}\cdot\mathbf{C}\cdot\vec{A} - h_1(\theta-\theta_o) = 0, \tag{1.7}$$

where $\mathbf{C}_P = \mathbf{P}^T\cdot\mathbf{P}$, $\mathbf{C} = \mathbf{F}^T\cdot\mathbf{F}$ are plastic and total right Cauchy-Green deformation tensors, while auxilliary conditions

$$h_o(0) = 1, \quad h_1(0) = 1$$

have to be introduced in order to meet assumptions b) and c).

In the above geometric constraints all quantities are functions of material point X and time t. If we write (1.4) - (1.7) for times t and t+dt and subtract the corresponding equations, we will obtain the following kinematic scalar constraints

$$D(\det \mathbf{P} - 1) = \mathrm{tr}\{\mathbf{P}^{-1} \cdot D\mathbf{P}\} \equiv \mathbf{P}^{-1} : D\mathbf{P} = 0, \tag{1.8}$$

$$D(\vec{A} \cdot \mathbf{C}_p \cdot \vec{A} - 1) = \vec{A} \cdot D\mathbf{C}_p \cdot \vec{A} \equiv \mathbf{A} : D\mathbf{C}_p = 0, \tag{1.9}$$

$$D[\det \mathbf{C} - h_o(\theta - \theta_o)] = h_o \mathbf{C}^{-1} : D\mathbf{C} - h_o' D\theta = 0, \tag{1.10}$$

$$D[\vec{A} \cdot \mathbf{C} \cdot \vec{A} - h_1(\theta - \theta_o)] = \vec{A} \cdot D\mathbf{C} \cdot \vec{A} - h_1' D\theta \equiv \mathbf{A} : D\mathbf{C} - h_1' D\theta = 0 \tag{1.11}$$

with the notation $\mathbf{A} \equiv \vec{A} \otimes \vec{A}$ introduced above.

In the same way, writing (1.4) - (1.7) for fixed t and infinitesimally adjacent particles X and X+dX by subtraction we arrive at vectorial constraints

$$\nabla(\det \mathbf{P} - 1) = \mathbf{P}^{-1} : \nabla \mathbf{P} = \vec{0}, \tag{1.12}$$

$$\nabla(\vec{A} \cdot \mathbf{C}_p \cdot \vec{A} - 1) = 2(\mathbf{A} \cdot \mathbf{P}) : \nabla \mathbf{P} = \vec{0}, \tag{1.13}$$

$$\nabla[\det \mathbf{C} - h_o(\theta - \theta_o)] = h_o \mathbf{C}^{-1} : \nabla \mathbf{C} - h_o' \nabla \theta = \vec{0}, \tag{1.14}$$

$$\nabla[\vec{A} \cdot \mathbf{C} \cdot \vec{A} - h_1(\theta - \theta_o)] = \mathbf{A} : \nabla \mathbf{C} - h_1' \nabla \theta = \vec{0}. \tag{1.15}$$

In (1.8) - (1.11) the letter D stands for time derivative with X kept fixed (the so-called material derivative), whereas in (1.12) - (1.15) ∇ denotes gradient with regard to $\vec{\kappa}$-configuration. In the sequel the last will be referred to as material gradient.

As we shall see soon the constraints (1.12) - (1.15) have the same purpose as kinematic constraints (1.8) - (1.11). For this sake (1.8) - (1.15) are named quasikinematic constraints.

2. CONSTITUTIVE EQUATIONS

Let the temperature-deformation state of the particle $\mathcal{N} \subset \mathcal{B}$ at time t be given by the set

$$\Sigma = \{\mathbf{C}, \theta, D\theta, \nabla\theta, \mathbf{P}, \mathbf{A}\} \in \mathcal{S}, \tag{2.1}$$

where $\mathcal{S}$ is intrinsic state space [3,7] and let the reaction of this particle at the same time be described by the set

$$\Gamma = \{\varepsilon, \eta, \mathbf{T}^*, \mathcal{T}, \vec{q}, \vec{\phi}\} \in \mathcal{G} \tag{2.2}$$

where $\mathcal{G}$ is the reaction space and

- ε - internal energy density,
- η - entropy density,
- $\mathbf{T}^*$ - the "elastic" second Piola-Kirchhoff stress tensor,
- $\mathcal{T}$ - the fourth order plastic viscosity tensor,
- $\vec{q}$ - the convective heat flux vector and
- $\vec{\phi}$ - the entropy flux vector (cf. [3-5]).

These tensors and vectors are defined by

$$\mathbf{T} = \mathbf{T}^{*} + \mathcal{T}{:}\mathrm{D}\mathbf{P} = \frac{\rho_o}{\rho}\,\mathbf{F}^{-1}\cdot\mathbf{T}_{\chi}\cdot\mathbf{F}^{-T}, \tag{2.3}$$

$$\vec{q} = \frac{\rho_o}{\rho}\,\mathbf{F}^{-1}\cdot\vec{q}_{\chi}, \quad \vec{\phi} = \frac{\rho_o}{\rho}\,\mathbf{F}^{-1}\cdot\vec{\phi}_{\chi} \tag{2.4}$$

with $\mathbf{T}$ - the total Piola-Kirchhoff stress tensor, $\mathbf{T}_{\chi}$ - the total Cauchy stress in $\vec{\chi}_t$-configuration of $\mathcal{B}$, $\mathcal{T}{:}\mathrm{D}\mathbf{P}$ - the viscoplastic overstress tensor, $\vec{q}_{\chi}$ - the heat flux vector in $\vec{\chi}_t$, $\vec{\phi}_{\chi}$ - the entropy flux vector in $\vec{\chi}_t$, ρ_o - the mass density in $\vec{\kappa}$ and ρ is the mass density in $\vec{\chi}_t$-configuration.

We now postulate constitutive equations by the following bijective mapping

$$\mathcal{R}:\mathcal{S}\to\mathcal{G} \qquad \text{or} \qquad \Gamma = \mathcal{R}(\Sigma) \equiv \hat{\Gamma}(\Sigma) \tag{2.5}$$

and it is easy to see that this mapping is objective (frame-indifferent) under the transformations of spatial coordinate frame

$$\vec{\mathbf{x}}' = \vec{c}(t) + \mathbf{Q}(t)\cdot\vec{\mathbf{x}}, \quad d\vec{\xi}' = d\vec{\xi}, \quad d\vec{X}' = d\vec{X}, \tag{2.6}$$

where $\mathbf{Q}$ is an arbitrary rotation tensor such that $\mathbf{Q}^{-1} = \mathbf{Q}^{T}$, $\det\mathbf{Q} = 1$.

These constitutive equations are too general and in what follows we shall define some restrictions on them.

It is worth noting that the presence of $\mathbf{A}$ in the list of constitutive variables is aimed to take into account fibre induced anisotropy.

3. DYNAMIC YIELD CONDITION

In most plasticity theories it is essential to assume that there exists a sharp boundary between elastic and plastic behaviour. Such a boundary is usually called yield limit (e.g. [8]). In others so-called endochronic plasticity theories (cf. [9]) there is no clear understanding of the physical process which takes place.

In this paper the first approach is promoted to new applications in the following way. Introduce [1,3,8] a scalar value function

$$f = \hat{f}(\mathbf{T},\theta,\mathbf{P},\mathbf{A}) \equiv \hat{f}(\mathbf{T}^{*}+\mathcal{T}{:}\mathrm{D}\mathbf{P},\theta,\mathbf{P},\mathbf{A}) \tag{3.1}$$

and name it the dynamic yield function. By means of it we can clearly classify the distinct behaviours such as:

$f > 0$, $\|\mathrm{D}\mathbf{P}\| \equiv (\mathrm{D}\mathbf{P}{:}\mathrm{D}\mathbf{P}^{T})^{1/2} > 0$ - plastic deformation beyond the yield limit,

$f = 0$, $\|\mathrm{D}\mathbf{P}\| = 0$ - the yield limit and

$f < 0$, $\|\mathrm{D}\mathbf{P}\| = 0$ - the elastic region interior.

Let the considered deformation be such that

$$\sup_{\tau \in (-\infty, t]} \|\mathbf{DP}(\tau)\| \ll 1. \tag{3.2}$$

Then $\hat{f}$ may be approximated by a linear function

$$f \approx \hat{f}(\mathbf{T}^*, \theta, \mathbf{P}, \mathbf{A}) + (\partial_{\mathbf{T}} \hat{f})_{\mathbf{DP}=\mathbf{O}} : \mathscr{T} : \mathbf{DP} \equiv f_o + (\partial_{\mathbf{T}^*} \hat{f}_o) : \mathscr{T} : \mathbf{DP}, \tag{3.3}$$

where

$$f_o \equiv \hat{f}[\hat{\mathbf{T}}^*(\Sigma), \theta, \mathbf{P}, \mathbf{A}) \equiv \hat{f}_o(\mathbf{T}^*, \theta, \mathbf{P}, \mathbf{A}) \equiv \hat{\psi}(\Sigma) \equiv \psi \tag{3.4}$$

is the corresponding statical yield function which does not take into account plastic deformation rate $\mathbf{DP}$. This function is nonpositive and obeys the following conditions according to the above three types of behaviour, i.e.

$$\begin{aligned} & f > 0, \quad \psi = 0, \\ & f = \psi = 0, \\ & f = \psi < 0. \end{aligned} \tag{3.5}$$

a) Consider, now, more specifically (3.3) and (3.5) in the case of plastic deformation. In such a case $f_o = \psi = 0$ and we arrive at the differential inequality

$$f = (\partial_{\mathbf{T}^*} \hat{f}_o) : \mathscr{T} : \mathbf{DP} > 0 \tag{3.5}$$

which restricts possible admissible plastic rate directions for $\mathbf{DP}$. The simplest solution of the above inequality is represented by

$$\mathbf{DP} = Da\, \mathscr{T}^{-1} : \partial_{\mathbf{T}^*} \hat{f}_o, \qquad Da \geq 0, \tag{3.7}$$

with a - nondecreasing scalar loading function which takes into account the plastic deformation history (cf. [3]). In this way, nine components of the rate tensor $\mathbf{DP}$ are replaced by only one scalar rate Da. It should be noted that in the special case when $\mathscr{T}$ is a fourth order unit tensor multiplied by an arbitrary scalar (3.7) is nothing but Drucker's principle which states that plastic strain rate is normal to yield surface in stress space (which is here generalized by the reaction space). This result has been already derived in [3]. It is noteworthy that for its derivation (3.1) and only (3.2) should be assumed.

b) Introduce the notions of elastic and plastic zones, respectively, by (cf. [3]) the following open sets:

$$\Pi_E(t) = \{X : f(X,t) < 0\}, \qquad \Pi_P(t) = \{X : f(X,t) > 0\} \tag{3.8}$$

as well as a discontinuous plastic strain indicator function

$$\langle f \rangle = \begin{cases} 1, & f > 0, \\ 0, & f \leqq 0. \end{cases} \tag{3.9}$$

Then, writing (3.5) for X and t as well as for X and t+dt we obtain the scalar kinematic constraint

$$<f>D\hat{\psi}(\Sigma) = 0. \quad (3.10)$$

Analogously, if (3.5) is written for X and t and then for X+dX and t subtracting the obtained equalities and inequalities the material gradient constraint

$$<f>\nabla\hat{\psi}(\Sigma) = \vec{0} \quad (3.11)$$

is met. The above two constrains are similar to the quasikinematic constraints and will be also used in the following section.

4. MÜLLER'S ENTROPY INEQUALITY AND CONSTITUTIVE RESTRICTIONS

In a series of papers (cf. e.g. [4]) Müller stated that in general entropy flux is not just heat flux divided by absolute temperature but should be considered as an additional variable as it has been taken in (2.2). Therefore, the corresponding second law of thermodynamics would read:

$$\sigma \equiv \rho_o D\eta + \nabla\cdot\vec{\phi} \geqq 0. \quad (4.1)$$

This inequality must hold for each thermodynamic process. According to Müller such a process is each solution of the equations of energy balance and momentum balance

$$\frac{1}{2}\mathbf{T}:D\mathbf{C} - \nabla\cdot\vec{q} - \rho_o D\varepsilon = 0, \quad (4.2)$$

$$\nabla\cdot(\mathbf{F}\cdot\mathbf{T}) - \rho_o D\vec{v} = \vec{0}, \quad (4.3)$$

where $\vec{v}$ is the velocity of the particle X at time t.

Solutions of these two equations must fulfil also constraint equations (1.8) - (1.15) as well as (3.10) - (3.11). Therefore, in the inequality (4.1) the elements of the sets $D\Sigma$ and $\nabla\Sigma$ are not independent because they have to obey (1.8) - (1.15), (3.10) - (3.11) and (4.2) - (4.3). However, extending (4.1) by introducing Lagrange multipliers Λ's and $\vec{\ell}$'s we arrive at the inequality

$$\begin{aligned}\sigma = {} & \rho_o D\eta + \nabla\cdot\vec{\phi} + \Lambda^{\varepsilon}(\frac{1}{2}\mathbf{T}:D\mathbf{C} - \nabla\cdot\vec{q} - \rho_o D\varepsilon) + <f>\Lambda^{\psi}D\psi + <f>\vec{\ell}^{\psi}\cdot\nabla\psi + \Lambda^{P}\mathbf{P}^{-1}:D\mathbf{P} + \\ & (\vec{\ell}^{P}\otimes\mathbf{P}^{-1})\vdots\nabla\mathbf{P} + \Lambda^{fP}(\mathbf{A}\cdot\mathbf{P}^{T}):D\mathbf{P} + 2(\vec{\ell}^{fP}\otimes\mathbf{A}\cdot\mathbf{P})\vdots\nabla\mathbf{P} + \\ & \Lambda^{E}(h_o\mathbf{C}^{-1}:D\mathbf{C} - h_o' D\theta) + h_o(\vec{\ell}^{E}\otimes\mathbf{C}^{-1})\vdots\nabla\mathbf{C} - h_o'\vec{\ell}^{E}\cdot\nabla\theta - \\ & \Lambda^{fE}(\mathbf{A}:D\mathbf{C} - h_1' D\theta) + (\vec{\ell}^{fE}\otimes\mathbf{A})\vdots\nabla\mathbf{C} - h_1'\vec{\ell}^{fE}\cdot\nabla\theta \geqq 0, \end{aligned} \quad (4.4)$$

in which all the elements of the sets $D\Sigma$ and $\nabla\Sigma$ except $D\theta$ and $\nabla\theta$ (included in Σ by themselves) and $D\mathbf{P}$ (given by the evolution equation (3.7)) are independent. Of course, we have made the tacit assumption that constitutive equations (2.5) are implemented in (4.4) and that composite differentiations have been already performed. In such a way the above inequality is linear in $D\mathbf{C}$, $DD\theta$, $D\nabla\theta$, $\nabla\mathbf{C}$, $\nabla\nabla\theta$ and $\nabla\mathbf{P}$ and the coefficients with these derivatives must vanish. Hence, we get the desired constitutive restrictions:

$$\rho_o(\partial_{\mathbf{C}}\hat{\eta}-\Lambda^{\varepsilon}\partial_{\mathbf{C}}\hat{\varepsilon}) + \frac{1}{2}\Lambda^{\varepsilon}(\mathbf{T}^{*}+\mathscr{T}:D\mathbf{P}) + <f>\Lambda^{\psi}\partial_{\mathbf{C}}\hat{\psi} + \Lambda^{E}h_o\mathbf{C}^{-1} + \Lambda^{fE}\mathbf{A} = \mathbf{O}, \tag{4.5}$$

$$\rho_o(\partial_{D\theta}\hat{\eta}-\Lambda^{\varepsilon}\partial_{D\theta}\hat{\varepsilon}) + <f>\Lambda^{\psi}\partial_{D\theta}\hat{\psi} = 0, \tag{4.6}$$

$$\rho_o(\partial_{\nabla\theta}\hat{\eta}-\Lambda^{\varepsilon}\partial_{\nabla\theta}\hat{\varepsilon}) + <f>\Lambda^{\psi}\partial_{\nabla\theta}\hat{\psi} + \partial_{D\theta}\hat{\vec{\phi}} - \Lambda^{\varepsilon}\partial_{D\theta}\hat{\vec{q}} + <f>\partial_{D\theta}\hat{\psi}\vec{\ell}^{\psi} = \vec{0}, \tag{4.7}$$

$$\partial_{\mathbf{C}}\hat{\vec{\phi}}-\Lambda^{\varepsilon}\partial_{\mathbf{C}}\hat{\vec{q}} + <f>\vec{\ell}^{\psi}\otimes\partial_{\mathbf{C}}\hat{\psi} + h_o\vec{\ell}^{E}\otimes\mathbf{C}^{-1} + \vec{\ell}^{fE}\otimes\mathbf{A} = 0, \tag{4.8}$$

$$(\partial_{\nabla\theta}\hat{\vec{\phi}}-\Lambda^{\varepsilon}\partial_{\nabla\theta}\hat{\vec{q}} + <f>\vec{\ell}^{\psi}\otimes\partial_{\nabla\theta}\hat{\psi})_{sym} = \mathbf{O}, \tag{4.9}$$

$$\partial_{\mathbf{P}}\hat{\vec{\phi}}-\Lambda^{\varepsilon}\partial_{\mathbf{P}}\hat{\vec{q}} + <f>\vec{\ell}^{\psi}\otimes\partial_{\mathbf{P}}\hat{\psi} + \vec{\ell}^{P}\otimes\mathbf{P}^{-1} + 2\vec{\ell}^{fP}\otimes\mathbf{A}\cdot\mathbf{P} = 0, \tag{4.10}$$

and the restricted dissipation inequality:

$$\begin{aligned}\sigma^{r} = {} & [\rho_o(\partial_{\theta}\hat{\eta}-\Lambda^{\varepsilon}\partial_{\theta}\hat{\varepsilon}) - \Lambda^{E}h_o' - \Lambda^{fE}h_1' + <f>\Lambda^{\psi}\partial_{\theta}\hat{\psi}]D\theta + \\ & + (\partial_{\theta}\hat{\vec{\phi}}-\Lambda^{\varepsilon}\partial_{\theta}\hat{\vec{q}} + <f>\partial_{\theta}\hat{\psi}\vec{\ell}^{\psi} - h_o'\vec{\ell}^{E} - h_1'\vec{\ell}^{fE})\cdot\nabla\theta + \\ & + Da[\rho_o(\partial_{\mathbf{P}}\hat{\eta}-\Lambda^{\varepsilon}\partial_{\mathbf{P}}\hat{\varepsilon}) + <f>\Lambda^{\psi}\partial_{\mathbf{P}}\hat{\psi} + \Lambda^{P}\mathbf{P}^{-1} + \Lambda^{fP}\mathbf{A}\cdot\mathbf{P}^{T}]:\mathscr{T}^{-1}:\partial_{\mathbf{T}}\hat{f}_o \geqq 0,\end{aligned} \tag{4.11}$$

where

$$2\mathbf{B}_{sym} \equiv \mathbf{B} + \mathbf{B}^{T}.$$

a) Due to restricted space let us concentrate on the first of the restrictions. Let $D\mathbf{P} = \mathbf{O}$ and $<f> = 0$ which means that we are still in elastic behaviour region. Then (4.5) reduces into

$$\mathbf{T}^{*} = 2\rho_o\left\{\partial_{\mathbf{C}}\hat{\varepsilon} - \frac{1}{\Lambda^{\varepsilon}}\partial_{\mathbf{C}}\hat{\eta}\right\} - 2\frac{\Lambda^{E}}{\Lambda^{\varepsilon}}h_o\mathbf{C}^{-1} - 2\frac{\Lambda^{fE}}{\Lambda^{\varepsilon}}\mathbf{A}. \tag{4.12}$$

Suppose that the considered elastic deformation process is near equilibrium, i.e. that $D\theta$ and $\nabla\theta$ are very small. For such a process Λ^{ε} called by Müller coldness function is approximately equal to inverse of absolute temperature T. Strictly speaking θ is not absolute but empirical temperature (cf. [4]). Then the free energy function $\varphi \equiv \varepsilon - T\eta$ may be introduced and (4.12) takes the familiar form (cf. e.g. [6,10]) with Λ^{E} and Λ^{fE} easily recognized to be connected to pressure and fibre reaction stress. Indeed, introducing the Cauchy extra stress $\mathbf{S}^{*}_{\chi}$, pressure p and fibre reaction stress t by

$$\mathbf{T}^{*}_{\chi} = \mathbf{S}^{*}_{\chi} - p\mathbf{1} + t\vec{a}\otimes\vec{a} = \frac{\rho}{\rho_o}\mathbf{F}\cdot\mathbf{T}^{*}\cdot\mathbf{F}^{T}, \tag{4.13}$$

with

$$\mathrm{tr}\mathbf{S}^{*}_{\chi} = 0, \quad \vec{a}\cdot\mathbf{S}^{*}_{\chi}\cdot\vec{a} = 0. \tag{4.14}$$

we see that

$$\mathbf{S}^*_\chi = \mathbf{S}_1 - \frac{1}{2}\,(\mathrm{tr}\mathbf{S}_1 - \vec{a}\cdot\mathbf{S}_1\cdot\vec{a})\mathbf{1} + \frac{1}{2}\,(\mathrm{tr}\mathbf{S}_1 - 3\vec{a}\cdot\mathbf{S}_1\cdot\vec{a})\vec{a}\otimes\vec{a}, \tag{4.15}$$

$$\Lambda^E h_o \equiv \Lambda^\varepsilon\,\left(\frac{1}{2}\,p + \frac{1}{4}\,\mathrm{tr}\mathbf{S}_1 - \frac{1}{4}\,\vec{a}\cdot\mathbf{S}_1\cdot\vec{a}\right), \tag{4.16}$$

$$\Lambda^{fE} h_1 \equiv -\Lambda^\varepsilon\,\left(\frac{1}{2}\,t + \frac{1}{4}\,\mathrm{tr}\mathbf{S}_1 - \frac{3}{4}\,\vec{a}\cdot\mathbf{S}_1\cdot\vec{a}\right) \tag{4.17}$$

where

$$\mathbf{S}_1 \equiv 2\rho\mathbf{F}\cdot\left(\partial_{\mathbf{C}}\hat{\varepsilon} - \frac{1}{\Lambda^\varepsilon}\,\partial_{\mathbf{C}}\hat{\eta}\right)\cdot\mathbf{F}^T.$$

It is natural to assume that stress should be a continuous function when we reach the yield limit. Accepting this assumption means that (4.12) is valid as well in the general case of viscoplastic behaviour. Hence, (4.5) and (4.12) allow the following evolution equation

$$\mathscr{T}:D\mathbf{P} = -2\langle f\rangle\,\frac{\dot{\Lambda}^\psi}{\Lambda^\varepsilon}\,\partial_{\mathbf{C}}\hat{\psi} \tag{4.18}$$

which has been already obtained in the identical form in [3]. The meaning of this relation is as follows: the viscoplastic overstress tensor would be normal to yield surface in intrinsic state space if all the other variables in $\hat{\psi}(\Sigma)$ are neglected except $\mathbf{C}$. In such a special case intrinsic state space reduces to deformation space and (4.18) becomes Il'iushin's principle (cf. [11]). It seems at first sight that fibres existence does not influence the viscoplastic overstress. This is not the case because the statical yield function depends also on $\mathbf{A} \equiv \vec{A}\otimes\vec{A}$ and fibres implicitly influence the overstress.

b) Let the heat flux and the entropy flux be given by the following special response functions

$$\vec{\phi} = \mathbf{L}\cdot\nabla\theta, \qquad \vec{q} = -\mathbf{K}\cdot\nabla\theta, \tag{4.19}$$

where $\partial_{\nabla\theta}\mathbf{L} = \partial_{\nabla\theta}\mathbf{K} = 0$. Müller called them [4] generalized Fourier materials and has shown that for such anisotropic thermoelastic materials $\vec{\phi} = \Lambda^\varepsilon\vec{q}$ holds and that Λ^ε is the so-called universal function of θ and $D\theta$ only. This means that it should not depend on the material considered but should have the same form for all materials.

In [3] it is shown that this does not hold for viscoplastic materials and that the following relation

$$\vec{\phi} = \Lambda^\varepsilon\vec{q} + (\mathbf{L}+\Lambda^\varepsilon\mathbf{K})_{asym}\cdot\nabla\theta - \langle f\rangle(\vec{\ell}^\psi\otimes\partial_{\nabla\theta}\hat{\psi})_{sym}\cdot\nabla\theta \tag{4.20}$$

(with $2\mathbf{L}_{asym} \equiv \mathbf{L}-\mathbf{L}^T$) should be used instead. Hence, the coldness function Λ^ε is not universal in general and vectors $\vec{\phi}$ and $\vec{q}$ do not have same direction.

CONCLUSIONS

a) The procedure given here coincides with the usual procedure of rational continuum mechanics applied up to now only to elastic and viscoelastic behaviour. Hence, the theory exposed here may be called a rational viscoplasticity theory.

b) Clear geometric distinction among plastic, elastic and thermal deformations can help in formulating simplified versions of constitutive equations like in the case of small elastic and finite plastic deformations, etc.

c) Drucker's postulate follows from the concept of dynamic yield function while Il'iushin's postulate originates from the second law of thermodynamics. The adjoint continuity of stress on yield limit is the expected result.

d) Eventual universality of the coldness function is still questioned and should be analysed especially in some simpler and special circumstances.

e) The approach may be quite well applied also to extensible fibres which could be elastic or elastoplastic. Of course, in such a case some of quasikinematic constraints (like that of elastic inextensibility) should be relaxed and an additional constitutive equation for fibre behaviour introduced.

REFERENCES

1. Mićunović, M., A thermodynamic theory of rate independent thermoplasticity of fibre-reinforced materials. International Symposium on Composite Materials and Structures, Beijing, 1986.

2. Coleman, B.D., Noll, W., The thermodynamics of elastic materials with heat conduction and viscosity. Arch. Rational Mech. Anal. 13, pp.167-179, 1963.

3. Mićunović, M., A thermodynamic approach to viscoplasticity. 1st National Congress on Mechanics, Athens, 1986.

4. Müller, I., The coldness, a universal function in thermoelastic bodies. Arch. Rational Mech. Anal. 41, pp.319-332, 1971.

5. Mićunović, M., A geometrical treatment of thermoelasticity of simple inhomogeneous bodies, I - Gemetrical and kinematical relations. Bull. Acad. Polon. Sci., Ser. Sci. Tech., 22, pp.579-588, 1982.

6. Mićunović, M., Thermoelasticity of ideal fibre-reinforced materials. J. Techn. Phys., 23, pp.59-67, 1982.

7. Perzyna, P., Thermodynamics of Dissipative Materials, CISM Lectures, No. 262, Springer, Wien 1980.

8. Mićunović, M., A yield condition for anisotropic materials. Transactions of 8th International Conference on Structural Mechanics in

Reactor Technology, Vol.L, pp.243-250, North-Holland, Amsterdam 1985.

9. Valanis, K.C., A theory of viscoplasticity without a yield surface. Arch. Mech. Stosow. 23, 4, pp.517-533, 1971.

10. Spencer, A.J.M., Deformations of Fibre-Reinforced Materials, Clarendon, Oxford 1972.

11. Il'iushin, A.A., On the postulate of plasticity. J. Appl. Math. Mech. (PMM), 25, pp.746-752, 1961.

PLASTIC BEHAVIOUR OF FIBER REINFORCED COMPOSITES IN LARGE DEFORMATION

DOGUI A. and SIDOROFF F.
GRECO Grandes Déformations et Endommagement
Ecole Centrale de Lyon
Ecully - France

ABSTRACT

The large strain plastic behaviour of a fiber reinforced composite is investigated under Voigt's hypothesis. The matrix is assumed rigid plastic while plastic and elastic fibers are considered. In both cases the initial yield surface and its evolution is described. The material response in an off axis tensile test is obtained in different cases.

INTRODUCTION

The work which is presented here is devoted to a simple mechanical analysis of the behaviour of a fiber reinforced composite material with a plastic matrix under large strain conditions. More precisely, the matrix will be assumed rigid plastic. Elastoplasticity could have been considered as well, but would require unnecessary complications. Furthermore, in most cases, elastic deformations are small enough to make this assumption reasonable. The reinforcement will be described by one or several unidirectional fiber systems, but their volume fractions will be assumed small enough to allow the Voigt homogenization technique to be used. In fact, it would be more appropriate to speak about string reinforcement.

From the point of view of composite materials therefore, the originality of this work neither lies in the homogenization which is rather crude, nor in the material, even if plastic composites are not so well understood [1,2]. It lies in the large strain analysis and the resulting geometric non linearities and evoluting anisotropy. Even if obtained from a simplified analysis, the results which are presented here will give a fair idea of what may happen when a metallic matrix composite undergoes large deformations.

From the continuum mechanics point of view, on the other hand, we are concerned here with finite strain anisotropic plasticity which is a very active and controversial field today [3,4,5]. Different kinds of models have been proposed and the basic issue is the choice of rotating frame in which the usual small strain constitutive equations are to be written [3]. The models which will be discussed in this paper may be considered as model anisotropic materials against which various phenomenological assumptions can be evaluated. We shall not emphasize this aspect here, but it also belongs to the motivation behind this work.

Standard finite strain notations will be used. We shall respectively denote by $\mathbf{F}$ the deformation gradient, $\mathbf{C}=\mathbf{F}^T\mathbf{F}$ the right Cauchy-Green tensor, $\mathbf{D}=(\dot{\mathbf{F}}\mathbf{F}^{-1})^S$ the rate of deformation tensor and $\mathbf{T}$ the (usual) Cauchy stress tensor. We shall also denote by $\mathbf{A}^D$ the deviatoric part of a second order tensor $\mathbf{A}$ and by $|\mathbf{A}|$ its euclidean norm ($|\mathbf{A}|^2=\mathbf{A}:\mathbf{A}=\mathrm{tr}\mathbf{A}\mathbf{A}^T$). In most applications attention will be focused on plane strain or stress situations [7] :

$$\mathbf{F} = \begin{bmatrix} f_1 & f_1\gamma & 0 \\ 0 & f_2 & 0 \\ 0 & 0 & f_3 \end{bmatrix} \qquad \mathbf{T} = \begin{bmatrix} N_1 & \tau & 0 \\ \tau & N_2 & 0 \\ 0 & 0 & 0 \end{bmatrix}$$

GENERAL BACKGROUND

The matrix is represented by an incompressible rigid plastic material obeying von Mises yield criterion:

$$f(\mathbf{T}) = |\mathbf{T}^D| - \sqrt{2/3}\ \sigma_o \leqq 0 \qquad \mathbf{D} = \lambda\ \partial f/\partial \mathbf{T} = \lambda\ \mathbf{T}^D/|\mathbf{T}^D| \tag{1}$$

This constitutive equation can also be written as:

$$\mathbf{T} = -p\ \mathbf{1} + \sqrt{2/3}\ \sigma_o\ \mathbf{D}/|\mathbf{D}| \qquad tr\mathbf{D} = 0 \tag{2}$$

where p is an arbitrary hydrostatic pressure. The tensile yield stress σ_o may be a constant (perfect plasticity) or a given function of the von mises equivalent strain $\sigma_o(\bar{\varepsilon})$ (isotropic hardening):

$$\dot{\bar{\varepsilon}} = \sqrt{2/3}\ |\mathbf{D}| \qquad \bar{\varepsilon} = \sqrt{2/3} \int_{-\infty}^{t} |\mathbf{D}|\ dt \tag{3}$$

A fiber system is defined by (Fig.1):

a)- Its initial orientation which is described by the corresponding unit vector $\vec{K}$ in the reference configuration C_o.

b)- Its density ξ which is the number of fibers per unit surface (normal to $\vec{K}$) in C_o.

c)- Its constitutive relation giving the force F in a fiber as a function of its strain history.

A material vector $\vec{K}\ ds_o$ in C_o becomes, in the actual configurations, $\vec{k}$ ds where $\vec{k}$ is a unit vector along the actual fiber direction (Fig.1):

$$\mathbf{F}\ \vec{K} = \lambda_k\ \vec{k} \quad ; \quad \lambda_k = ds/ds_o = |\mathbf{F}\ \vec{K}| = e^{\varepsilon_k} \tag{4}$$

where ε_k is the logarithmic strain of the fiber. The Cauchy stress tensor transmitted by the fiber system is a tensile stress along $\vec{k}$ and it can be written as:

$$\mathbf{T} = \xi\lambda_k\ F\ \vec{k}\otimes\vec{k} \tag{5}$$

where $\xi\lambda_k$ is the number of fibers per unit surface normal to $\vec{k}$ in C(t) (incompressibility).

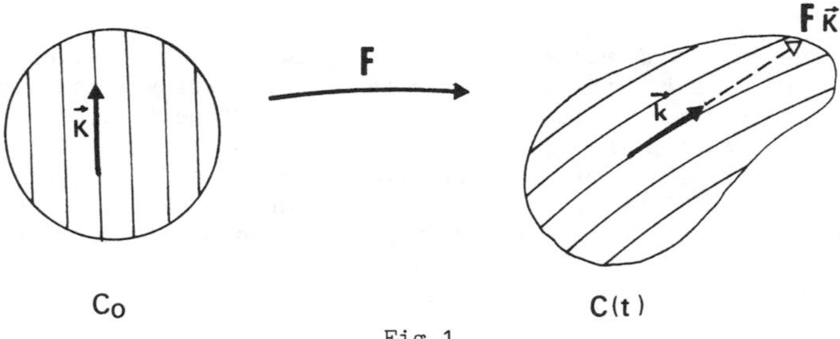

Fig.1

Homogenisation will be performed along Voigt's approximation which assumes that the matrix and the fibers are submitted to the same deformation while the total stress is the sum of the partial stresses transmitted by each component. The general constitutive equation can therefore be written as:

$$\mathbf{T} = -p\ \mathbf{1} + \sqrt{2/3}\ \sigma_o\ \mathbf{D}/|\mathbf{D}| + \sum_{\alpha=1}^{n} \xi_\alpha \lambda_{k\alpha} F_\alpha\ \vec{k}_\alpha \otimes \vec{k}_\alpha \tag{6}$$

where the sum in α is extended to all fiber systems.

The resulting behaviour will be analyzed according to the number, orientation and behaviour of the fiber systems (the behaviour being characterised by the dependance of the force F on the strain ε).

PLASTIC FIBERS

A rigid plastic fiber is defined by the following constitutive relation:

$$F = F_o(\varepsilon_k)\ \mathrm{sgn}(\varepsilon_k) \tag{7}$$

where F_o denotes the limit force and where sgn(a) is equal to +1 if a>0, -1 if a<0 and may take any value between -1 and +1 if a=0. It should be noted that this relation corresponds to a symmetric behaviour in tension and compression. This is certainly not reasonable but a non symmetric behaviour could be treated in the same way. The time derivative $\dot{\varepsilon}_k$ is easily obtained from (4):

$$\dot{\varepsilon}_k = \vec{k}\ \mathbf{D}\ \vec{k} = \mathbf{k:D} \qquad \text{with } \mathbf{k} = \vec{k}\otimes\vec{k} \tag{8}$$

The corresponding Cauchy stress tensor is then obtained from (5) as:

$$\mathbf{T} = \hat{\sigma}(\varepsilon_k)\ \mathrm{sgn}(\mathbf{D:k})\ \mathbf{k} \qquad \hat{\sigma}(\varepsilon_k) = \xi\ \lambda_k\ F_o(\varepsilon_k) \tag{9}$$

The general constitutive equation of the composite is then:

$$\mathbf{T} = -p\ \mathbf{1} + \sqrt{2/3}\ \sigma_o \frac{\mathbf{D}}{|\mathbf{D}|} + \sum_{\alpha=1}^{n} \hat{\sigma}_\alpha \mathrm{sgn}(\mathbf{D:k}_\alpha)\ \mathbf{k}_\alpha \tag{10}$$

At a given stage $\sigma_o(\bar{\varepsilon})$ and $\hat{\sigma}_\alpha(\varepsilon_{k\alpha})$ are fixed and $\mathbf{T}^D$ is a function of $\mathbf{D}$ which is positively homogeneous of degree 0 (rate independant plastic behaviour) and it follows that the Cauchy stress $\mathbf{T}$ satisfies one scalar relation which is the yield condition [5] defining the yield surface. There are many ways for obtaining this condition:
- direct elimination of $\mathbf{D}$ from (1) expressed in a properly chosen coordinate system;
- Legendre transformation of the dissipation potential

$$\mathbf{T} = \partial\Omega/\partial\mathbf{D} \quad ; \quad \Omega = \sqrt{2/3}\ \sigma_o\ |\mathbf{D}| + \sum_{\alpha=1}^{n} \hat{\sigma}_\alpha |\mathbf{k}_\alpha\mathbf{:D}| \quad ; \ \mathrm{tr}\mathbf{D}=0$$
$$\omega(\mathbf{T}) = \sup_{\mathbf{D}} \left[\mathbf{T:D} - \Omega(\mathbf{D})\right] \tag{11}$$

where ω is the indicator function of the plasticity convex C (ω=0 inside C,$+\infty$ outside)[8];
- evaluation of the scalar anisotropic invariants of $\mathbf{D}$ in terms of those of $\mathbf{T}$ and elimination of the latter [5].

In any case, an anisotropic plastic yield surface is obtained which only depends on the partial yield stress (σ_o,σ_α) and on the actual direction of the fiber system. In this case, large strain are taken into account through the evolution of the material fiber directions. It also follows from (11) that a standard plasticity law satisfying normality is obtained.

In case of one fiber system, the formalism of [3] is directly obtained with an usual small strain formalism written in the rotating frame following the material vector $\vec{k}$. Moreover if σ_o and $\hat{\sigma}_1$ are constant, perfect plasticity is obtained with a fixed plasticity convex in the rotating frame. After some calculations, the equation of this convex is obtained as

$$\frac{3}{2}\ \mathbf{T}^D\mathbf{:T}^D + \left\langle\frac{3}{2}|\mathbf{T}^D\mathbf{:k}| - \hat{\sigma}_1\right\rangle^2 - \frac{9}{4}(\mathbf{T}^D\mathbf{:k})^2 \leq \sigma_o^2 \tag{12}$$

with <a> denoting the positive part of a (<a>=a if a>0, 0 if a≦0). The derivation of this equation is given in the Appendix.

In case of several fiber systems, a more complex situation is obtained since the relative orientations of the fibers change. This implies some kind of induced anisotropy to be combined with the basic initial anisotropy. The rotating frame to be used is not so obvious.

An important special case is obtained for two identical perfectly plastic fiber systems. In this case, the material is and remains orthotropic with respect to the fiber bissectrix.

EVOLUTION OF THE YIELD SURFACE

We shall now present some examples describing the yield surface and its evolution in some simple cases. We restrict ourselves to plane stress situations and the plastic convex will be represented in the principal stress plane N_1,N_2 ($\tau=0$). Because of anisotropy, this repesentation will depend on the orientation of the principal stress directions with respect to the material.

In case of one fiber system, this orientation will be defined by the angle Θ between the fiber direction and the N_1 axis ($0\leqq\Theta\leqq\pi/4$). The corresponding convex is easily obtained from (12) and represented in Figure 2 for some values of Θ with $\sigma_o=\hat{\sigma}_1$. It should be noted however that these are different representations of the same plastic convex which remains the same with respect to the material. Its evolution with Θ simply follows from the rotation of the fiber direction, which occurs, for instance, in an off axis tensile test. Some details about the construction of this surface can be found in the Appendix.

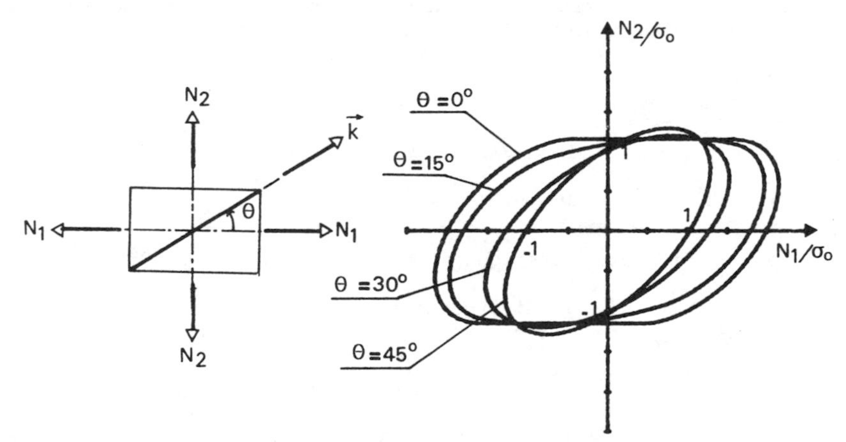

Figure 2: Evolution of the yield surface - 1 plastic fiber system.

Another simple case is obtained for two identical perfectly plastic fiber systems: in this case, the material is and remains orthotropic with respect to the fiber bissectrix. The orientation of the principal stress directions can then be defined by the angle Θ between the N_1 direction and this bissectrix. However the plastic convex now changes due to the variation of the angle 2α between the two fiber systems. This variation is represented in Figure 3b for $\Theta=0$ and some values of α. This represents for instance the evolution of the plastic convex during a tensile test along the bissectrix.

If the two fiber systems are perpendicular, then the material is also orthotropic with respect to the fiber directions. If the principal stress axis coincide with the fiber directions ($\Theta=\pi/4$), there will be no rotation of fibers, and consequently, no change in the plastic convex which is represented in Figure 3a.

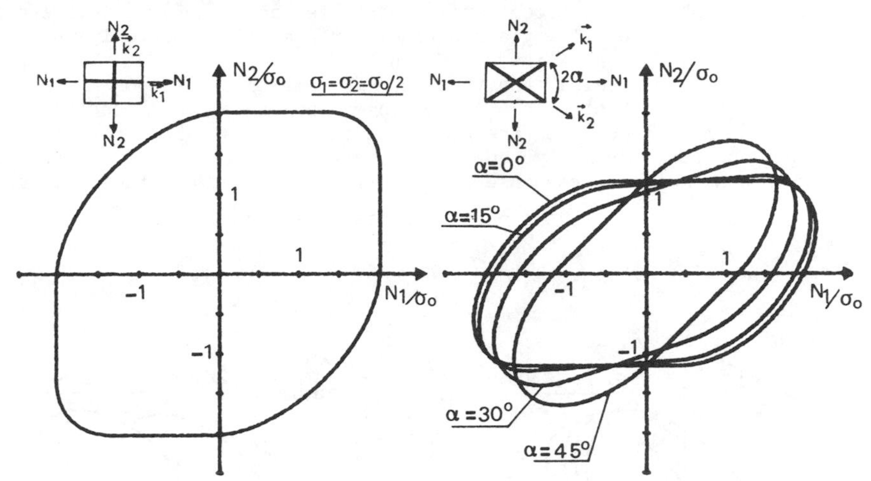

Figure 3: Evolution of the yield surface - 2 plastic fiber systems.
a: perpendicular fibers b: non perpendicular fibers

ELASTIC FIBERS

In an elastic fiber, the transmitted force F only depends on the strain ε_k:

$$F = F(\varepsilon_k) \;;\; \mathbf{T} = \sigma(\varepsilon_k)\,\mathbf{k} \;;\; \sigma = \xi e^{\varepsilon_k} F(\varepsilon_k) \qquad (13)$$

The constitutive equation (6) then becomes:

$$\mathbf{T} = -p\,\mathbf{1} + \sqrt{2/3}\,\sigma_o \frac{\mathbf{D}}{|\mathbf{D}|} + \sum_{\alpha=1}^{n} \sigma_\alpha(\varepsilon_{k\alpha})\,\mathbf{k}_\alpha \qquad (14)$$

The stress results from the superposition of a matrix transmitted plastic part and a fiber transmitted elastic part. This is a kinematic hardening model and the plasticity convex can be written as:

$$|\mathbf{T}^D - \mathbf{X}| \leq \sqrt{2/3}\,\sigma o \qquad (15)$$

with a back stress $\mathbf{X}$ given by:

$$\mathbf{X} = \sum_{\alpha=1}^{n} \sigma_\alpha(\varepsilon_{k\alpha})(\mathbf{k}_\alpha - \frac{1}{3}\,\mathbf{1}) \qquad (16)$$

The actual yield surface therefore depends on both the orientation and strain of each fiber systems, and it is entirely defined by the evolution of the back stress $\mathbf{X}$ which is the essential variable to be followed. This evolution however cannot be directly characterized, and its determination requires the individual analysis of each fiber system.

Some examples are given in Figure 4 and 5 which show the evolution of the yield surface in the N_1, N_2 plane ($\tau=0$) resulting from tensile tests.

Figure 4 corresponds to an off axis tensile test on a unidirectional composite (Θ_o denoting the initial orientation of the fibers with respect to the tensile axis N_1). Figure 5 corresponds to a tensile test along the bissectrix of the fiber directions in a two directional composite. These situations are similar to those depicted in Figures 2 and 3 for plastic fibers, but it should be noted that there is an important difference between these two cases. The yield surface for plastic fibers only depends on the orientation of the fibers so that Figures 2 and 3 have been obtained without a precise analysis of the material response in the investigated tensile tests. This is no longer possible here because the yield surface also depends on the strain in the fibers. The results are presented here for simplicity, but their derivation requires the precise analysis of the material response which will be presented in the following.

These results assume an identical linear elastic law for each fiber system

$$\sigma_\alpha = E\varepsilon_\alpha \tag{17}$$

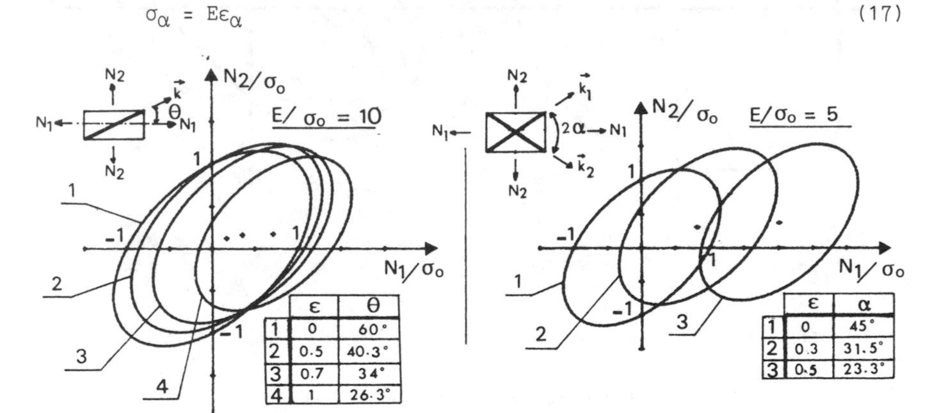

Evolution of the yield surface resulting from tensile tests
Figure 4: 1 fiber system   Figure 5: 2 fibers systems

TENSILE TEST ANALYSIS

We now proceed to the analysis of the material response in some homogeneous plane stress situations. We shall focus our attention on the tensile test which is commonly used, but other situations, like a torsion test for instance, can be treated in the same way [7]. Generally speaking a tensile test in an incompressible material can be described in the laboratory frame by

$$\mathbf{F} = \begin{bmatrix} e^{\varepsilon} & \gamma e^{\varepsilon} & 0 \\ 0 & e^{-\eta} & 0 \\ 0 & 0 & e^{-(\varepsilon-\eta)} \end{bmatrix} ; \quad \mathbf{T} = \begin{bmatrix} \sigma & 0 & 0 \\ 0 & 0 & 0 \\ 0 & 0 & 0 \end{bmatrix} \tag{18}$$

where ε is the longitudinal strain which usually is the control variable, η describes the transverse contraction and the shear strain γ describes the rotation ψ of the final sections which occurs in an off axis tensile test (Fig.6). The Langford ratio r is also often introduced:

$$\gamma e^{\varepsilon+\eta} = \tan\psi \quad ; \quad r = \eta/(\varepsilon-\eta) = \varepsilon_2/\varepsilon_3 \tag{19}$$

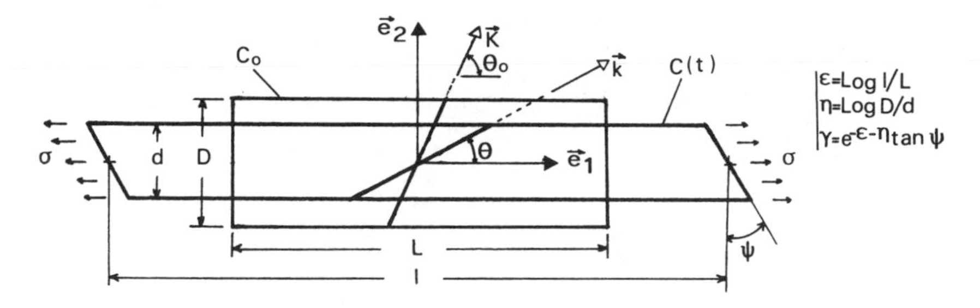

Figure 6: Off axis tensile test

Straightforward calculations give:

$$\mathbf{L} = \begin{bmatrix} \dot{\varepsilon} & \dot{\gamma} e^{\varepsilon+\eta} & 0 \\ 0 & -\dot{\eta} & 0 \\ 0 & 0 & \dot{\eta}-\dot{\varepsilon} \end{bmatrix} ; \quad \mathbf{F}\,\vec{K} = \begin{bmatrix} e^{\varepsilon}(\cos\theta_o+\gamma\sin\theta_o) \\ e^{-\eta}\sin\theta_o \\ 0 \end{bmatrix} \tag{20}$$

$$\tan\theta = e^{\varepsilon+\eta}\,\frac{\sin\theta_o}{\cos\theta_o+\gamma\sin\theta_o}$$

where, for a given fiber direction, θ_o and θ respectively denote the angle of $\vec{K}$ and $\vec{k}$ with respect to the axis x_1, i.e. the initial and actual orientation of the fiber system with respect to the tensile direction. The general constitutive equation (6) becomes

$$N_1 = \sigma = \sqrt{\tfrac{2}{3}}\,\sigma_o\,\frac{2\dot{\varepsilon}-\dot{\eta}}{|\mathbf{D}|} + \sum_\alpha \sigma_\alpha \cos^2\theta_\alpha \tag{21}$$

$$\left.\begin{aligned} N_2 &= 0 = -\sqrt{\tfrac{2}{3}}\,\sigma_o\,\frac{2\dot{\eta}-\dot{\varepsilon}}{|\mathbf{D}|} + \sum_\alpha \sigma_\alpha \sin^2\theta_\alpha \\ 2\tau &= 0 = \sqrt{\tfrac{2}{3}}\,\sigma_o\, e^{\varepsilon+\eta}\,\frac{\dot{\gamma}}{|\mathbf{D}|} + \sum_\alpha \sigma_\alpha \sin 2\theta_\alpha \end{aligned}\right] \tag{22}$$

where $|\mathbf{D}|$ is given by:

$$|\mathbf{D}|^2 = 2\dot{\varepsilon}^2 + 2\dot{\eta}^2 - 2\dot{\varepsilon}\dot{\eta} + \tfrac{1}{2}e^{(\varepsilon+\eta)}\,\dot{\gamma}^2$$

$$|\mathbf{D}| = \sqrt{2}\,|\dot{\varepsilon}|\sqrt{1+(d\eta/d\varepsilon)^2 - d\eta/d\varepsilon + (1/4)e^{\varepsilon+\eta}\,d\gamma/d\varepsilon} \tag{23}$$

Assuming the stress σ_α transmitted by each fiber system to be known, the substitution of (23) into (22) provides a system of two equations which allows the determination of $d\eta/d\varepsilon$ and $d\gamma/d\varepsilon$. After some calculations, these quantities are obtained as:

$$2\,d\eta/d\varepsilon = 1 + 3b/u \quad ; \quad d\gamma/d\varepsilon = -6e^{-(\varepsilon+\eta)}c/u \tag{24}$$

The stress σ is then given by (21):

$$\sigma = a + (u-b)/2$$

where:

$$a = \sum_\alpha \sigma_\alpha \cos^2\theta_\alpha \quad ; \quad b = \sum_\alpha \sigma_\alpha \sin^2\theta_\alpha \quad ; \quad c = \tfrac{1}{2}\sum_\alpha \sigma_\alpha \sin(2\theta_\alpha)$$

$$u = (4\sigma_o^2 - 3b^2 - 12c^2)^{\frac{1}{2}}$$

In the case of elastic fibers the stresses σ_α in each fiber are known at each step, and (24) provides a differential system for the determination of η and γ. Figures 7 a and b shows the behaviour of a unidirectional composite (E/σ_o=10) in an off axis tensile test with initial fiber orientation θ_o=60°.

Figure 7a gives the stress σ/σ_o, the langford coefficient r, the orientation of the fiber Θ and of the final section ψ as a function of the stress ε. Evolution of the anisotropy is depicted in Figure 7b which shows the evolution of the orientation dependence of the tensile yield stress σ_Θ/σ_o with increasing ε.

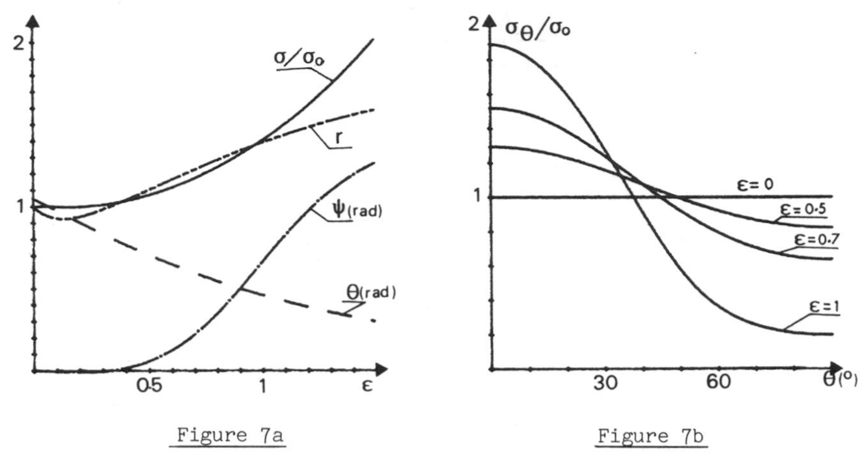

Figure 7a Figure 7b

In the case of plastic fibers, the situation is more complex, because σ_α also depends on $d\eta/d\varepsilon$ and $d\gamma/d\varepsilon$ through sgn(**k:D**) and a specific algorithm must be developed for following the plastic state of each fiber system (extending, contracting or undeforming).

REFERENCES

1 F. WEIL - O. DEBORDES, "Homogénéisation en plasticité, un exemple de calculs", Rapport GRECO Grandes Déformations et Endommagement, 107, 1983.

2 J. ABOUDI - Y. BENVENISTE, "The mechanical behaviour of elastic-plastic fiber reinforced plates", in Mechanical characterization of load bearing fibre composite laminates, p. 65, ed. A. H. CARDON - G. VERCHERY, Elsevier, London, 1985.

3 A. DOGUI - F. SIDOROFF, "Rhéologie anisotrope en grandes déformations", 19ème Colloque du Groupe Français de Rhéologie, Paris 1984, to be published.

4 Y. F. DAFALIAS, "Corotational rates for kinematic hardening at large plastic deformations", J. Applied Mech., 50, p. 561, 1983.

5 J. P. BOEHLER - J. RACLIN, "Ecrouissage anisotrope des matériaux orthotropes prédéformés", J. Méc. Thé. et Appl., numéro spécial, p. 23, 1982.

6 F. SIDOROFF, "Cours sur les grandes déformations", Rapport GRECO Grandes Déformations et Endommagement, 51, 1982.

7 A. DOGUI,"Traction hors axes et torsion en plasticité anisotrope et grandes déformations", 7 ème Congrés Français de Mécanique, Bordeaux, 1985.

8 P. GERMAIN, "Cours de mécanique des milieux continus", tome 1, Masson, Paris, 1973.

APPENDIX: DERIVATION OF THE YIELD SURFACE

There are many ways to derive the plastic convex (12) described in Figure 2 in case of one fiber system. We shall present here a geometrical construction in the plane stress space $(N_1, N_2, \sqrt{2}\,\tau)$. The constitutive equation (10) then reduces to

$$\left.\begin{aligned} N_1 &= \sqrt{2/3}\,\sigma_o \frac{2D_{11}+D_{22}}{|\mathbf{D}|} + \hat{\sigma}_1 \mathrm{sgn}\dot{\varepsilon}_1 \cos^2\Theta = N_{1m} + N_{1f} \\ N_2 &= \sqrt{2/3}\,\sigma_o \frac{2D_{22}+D_{11}}{|\mathbf{D}|} + \hat{\sigma}_1 \mathrm{sgn}\dot{\varepsilon}_1 \sin^2\Theta = N_{2m} + N_{2f} \\ \tau &= \sqrt{2/3}\,\sigma_o \frac{D_{12}}{|\mathbf{D}|} + \hat{\sigma}_1 \mathrm{sgn}\dot{\varepsilon}_1 \sin\Theta\cos\Theta = \tau_m + \tau_f \end{aligned}\right] \quad \text{(A-1)}$$

$$|\mathbf{D}|^2 = \sqrt{2(D^2_{11} + D^2_{22} + D_{11}D_{22} + D^2_{12})}$$

$$\dot{\varepsilon}_1 = D_{11}\cos^2\Theta + D_{22}\sin^2\Theta + 2D_{12}\sin\Theta\cos\Theta$$

Since these 3 quantities only depend on two variables, for instance $D_{11}/|\mathbf{D}|$ and $D_{22}/|\mathbf{D}|$, (A-1) provides the parametric équation of the yield surface. For the matrix ($\hat{\sigma}_1=0$) this results in the usual von Mises ellipsoid:

$$N^2_{1m} + N^2_{2m} - N_{1m}N_{2m} + 3\tau^2_m = \sigma^2_o \quad \text{(A-2)}$$

For the composite material ($\hat{\sigma}_1>0$) the points with $\dot{\varepsilon}_1>0$ are translated of $+\hat{\sigma}_1$ in the direction $(\cos^2\Theta, \sin^2\Theta, \sqrt{2}\sin\Theta\cos\Theta)$, while those with $\dot{\varepsilon}_1<0$ are translated of the same quantity in the opposite direction. The boundary between these two regions is given by $\dot{\varepsilon}_1=0$ which from (A-1) can also be written as:

$$(2\cos^2\Theta - 2/3)N_{1m} + (2\sin^2\Theta - 2/3)N_{2m} - 2\tau\sin\Theta\cos\Theta = 0 \quad \text{(A-3)}$$

The geometrical construction of the yield surface directly follows from this: von Mises ellipsoid is cut along the plane (A-3) and its two halves are translated of $+\hat{\sigma}_1$ and $-\hat{\sigma}_1$ in the direction $(\cos^2\Theta, \sin^2\Theta, \sqrt{2}\sin\Theta\cos\Theta)$ defining a yield surface consisting in two half-ellipsoids and a cylinder. The intersection with the $\tau=0$ plane which is represented in Fig. 2 therefore consists in 4 elliptic segments.

If the x_1 axis is chosen along the fibers ($\Theta=0$), then the cutting plane is defined by $N_{2m}=2N_{1m}$ and translation occurs in the direction N_1. The intersection of the cylindral part with the plane $\tau=0$ then reduces to the straight lines which are observed in Fig. 2 for $\Theta=0$. The analytical form (12) may also be inferred from this special case but a more precise derivation is obtained by using invariants: the constitutive equation (10) can be written as

$$\mathbf{T}^D = \sqrt{2/3}\,\sigma_o \frac{\mathbf{D}}{|\mathbf{D}|} + \sigma_1\,\mathbf{k}^D \quad ; \quad \sigma_1=\hat{\sigma}_1 \mathrm{sgn}(\mathbf{k}:\mathbf{D}) \quad \text{(A-4)}$$

The invariants $\mathbf{k}:\mathbf{T}^D$ and $\mathbf{T}^D:\mathbf{T}^D$ are then obtained as:

$$\mathbf{k}:\mathbf{T}^D = \sqrt{2/3}\,\sigma_o\,\mathbf{k}:\mathbf{D}/|\mathbf{D}| + (2/3)\sigma_1 \quad \text{(A-5)}$$

$$\begin{aligned} \mathbf{T}^D:\mathbf{T}^D &= 2(\sigma^2_o+\sigma^2_1)/3 + 2\sqrt{2/3}\,\sigma_o\sigma_1\,\mathbf{k}:\mathbf{D}/|\mathbf{D}| \\ &= 2(\sigma^2_o-\sigma^2_1)/3 + 2\sigma_1\,\mathbf{k}:\mathbf{T}^D \end{aligned} \quad \text{(A-6)}$$

If $\mathbf{k}:\mathbf{D}\neq 0$ then $|\sigma_1|=\hat{\sigma}_1$ and it follows from (A-5) that

$$\sigma_1\,\mathbf{k}:\mathbf{T}^D > 0 \quad ; \quad |\mathbf{k}:\mathbf{T}^D| > 2\hat{\sigma}_1/3 \quad \text{(A-7)}$$

so that the yield condition is obtained from (A-6) as

$$\mathbf{T}^D:\mathbf{T}^D = 2(\sigma^2_o-\hat{\sigma}^2_1)/3 + 2\hat{\sigma}_1|\mathbf{k}:\mathbf{T}^D| \quad \text{(A-8)}$$

If $\mathbf{k}:\mathbf{D}=0$ then $|\sigma_1|<|\hat{\sigma}_1|$ and (A-5) reduces to

$$\mathbf{k}:\mathbf{T}^D = 2\sigma_1/3 \quad ; \quad |\mathbf{k}:\mathbf{T}^D| < 2\hat{\sigma}_1/3 \tag{A-9}$$

The yield condition is then obtained from (A-6) as

$$\mathbf{T}^D:\mathbf{T}^D = 2\sigma_0^2/3 + 3(\mathbf{k}:\mathbf{T}^D)^2/2 \tag{A-10}$$

Finally the yield condition is given by (A-8) or (A-10) according to whether $|\mathbf{k}:\mathbf{T}^D|$ is greater than $2\hat{\sigma}_1/3$ (the fibers are deformed) or smaller than $2\hat{\sigma}_1/3$ (no deformation in the fibers). By introduction of the positive part $<|\mathbf{k}:\mathbf{T}^D|-2\hat{\sigma}_1/3>$ these two equations can be unified in one single relation which directly gives the analytical form (12).

A similar construction can be realized in the case of several fiber systems by cutting and translating the original matrix von Mises ellipsoid. The special cases which are represented in Figure 3 are such that the translations are all situated in the N_1, N_2 plane so that its intersection with the cylindrical parts of the yield surface are straight lines. The derivation of an explicit yield condition similar to (12) is much more difficult.

ON CERTAIN SOLUTIONS IN THE THEORY OF PLASTICITY WITH MICRO-LOCAL PARAMETERS

M. WAGROWSKA

Department of Mechanics,
University of Warsaw,
Poland.

Abstract. The nonstandard method of homogenization, [1,2,3] of periodic elastic-plastic composites is applied to the case of the Prandtl-Reuss elastic-plastic materials with kinematic work hardening. Problems are analysed within the assumptions of the small deformation theory. The cylindrical tube under pressure is taken as an example of application of the approach.

Introduction. This note is an example of application of non-standard methods of modeling of periodic composites, [1,2], to the Prandtl-Reuss theory of elastic-plastic materials with kinematic work hardening within the assumption of small deformation theory. The foundations and details concerning the nonstandard homogenization method used in the note can be found in [1,2,3].

Nonstandard modeling. Let at the time instant $t = t_0$ an elastic-plastic body in its natural state occupies the region B in the Euclidean 3-space of points $x = (x^i)$. We introduce in B curvilinear coordinates $X = (X^\alpha) \in \Omega$, introducing the invertible smooth mapping $\varkappa : \Omega \to R^3$, such that $\overline{\varkappa(\Omega)} = \bar{B}$. The body is made of N homogeneous materials and hence we introduce the decomposition $\bar{\Omega} = \cup \bar{\Omega}_A$, $A = 1, \ldots, N$, $\Omega_A \cap \Omega_B = \emptyset$ for every $A \neq B$. Under the known denotations [1] the equations of motion of the body will be assumed here (for the quasi-static case) in the weak form:

$$(1) \quad \int_\Omega \sigma^{\alpha\beta}(X,t) u_{(\alpha|\beta)}(X) dV = \oint_{\partial\Omega} t^\alpha(X,t) u_\alpha(X) dA + \int_\Omega b^\alpha(X,t) u_\alpha(X) dV , \quad t \in [t_0, t_1],$$

---

(1) Sub- and superscripts $\alpha, \beta, \gamma, \delta$ run over 1,2,3 and are related to the curvilinear coordinates $X = (X^\alpha)$ in B with a metric tensors $G(X) = \nabla\varkappa^T(X) \nabla\varkappa(X)$, $X \in \Omega$. Subscripts A, B run over $1, \ldots, N$ and subscripts a, b over $1, \ldots, n$.

Summation convention holds with respect to $\alpha, \beta, \gamma, \delta$ and a,b A comma denotes a partial and a vertical line stands for a covariant differentiation (in a metric $G(X)$, $X \in \Omega$).

which has to hold for every pertinent test function $u(\cdot)$ defined on $\bar{\Omega}$.
Let $v_\alpha(X,t)$, $X \in \Omega$ be the velocity fields defined on Ω .
Also define

$$s^{\alpha\beta}(X,t) \equiv \sigma^{\alpha\beta}(X,t) - \tfrac{1}{3} G^{\alpha\beta}(X)\, \sigma^{\gamma}_{\gamma}(X,t) \;,$$

$$\bar{s}^{\alpha\beta}(X,t) \equiv \bar{\sigma}^{\alpha\beta}(X,t) - \tfrac{1}{3} G^{\alpha\beta}(X)\, \bar{\sigma}^{\gamma}_{\gamma}(X,t) \;,$$

$$f(X,t) \equiv \bar{s}^{\alpha\beta}(X,t)\, \bar{s}_{\alpha\beta}(X,t) - k^2(X) \;,$$

$$\xi(f(X,t)) \equiv \begin{cases} 0 & \text{if} \quad f(X,t) < 0, \\ 1 & \text{if} \quad f(X,t) \geq 0, \end{cases} \tag{2}$$

$$\dot{\varepsilon}_{\alpha\beta}(X,t) \equiv v_{(\alpha|\beta)}(X,t) \;,$$

$$W(X,t) \equiv \begin{cases} \bar{s}^{\alpha\beta}(X,t)\, \dot{\varepsilon}_{\alpha\beta}(X,t) & \text{if} \quad \bar{s}^{\alpha\beta}(X,t)\, \dot{\varepsilon}_{\alpha\beta}(X,t) \geq 0, \\ 0 & \text{if} \quad \bar{s}^{\alpha\beta}(X,t)\, \dot{\varepsilon}_{\alpha\beta}(X,t) < 0. \end{cases}$$

The constitutive relations will be based on those proposed in [4] and have to hold for every $X \in \Omega_A$ and for $A = 1, \dots, N$.
Hence, the yield condition is $\bar{s}^{\alpha\beta}(X,t)\, \bar{s}_{\alpha\beta}(X,t) - k^2(X) = 0$, the material properties are determined by $\mu(X)$, $\lambda(X)$, $m(X)$, $k(X)$ and the relations are:

$$\sigma^{\alpha\beta}(X,t) = \bar{\sigma}^{\alpha\beta}(X,t) + \beta^{\alpha\beta}(X,t) \;,$$

$$\dot{\bar{\sigma}}^{\alpha\beta}(X,t) = 2\mu(X)\, \dot{\varepsilon}^{\alpha\beta}(X,t) + \lambda(X) G^{\alpha\beta}(X)\, \dot{\varepsilon}^{\gamma}_{\gamma}(X,t) - \frac{2\,\xi(f(X,t))\, W(X,t)\, \mu(X)}{k^2(X)}\, \bar{s}^{\alpha\beta}(X,t) \;, \tag{3}$$

$$\dot{\beta}^{\alpha\beta}(X,t) = \frac{2\,\xi(f(X,t))\, W(X,t)\, \mu(X)\, m(X)}{(\mu(X) + m(X))\, k^2(X)}\, \bar{s}^{\alpha\beta}(X,t) \;.$$

Functions $k(X)$, $\mu(X)$, $\lambda(X)$, $m(X)$, $X \in \Omega$, are assumed to be constant in every Ω_A . Define $Y \equiv [0, Y^1] \times [0, Y^2] \times [0, Y^3]$ and assume that max Y_i is very small as compared with the smallest characteristic length dimension of $\bar{\Omega}$. Let the material structure of the body under consideration be Y-periodic, i.e. $k(\cdot)$, $\mu(\cdot)$, $\lambda(\cdot)$, $m(\cdot)$ can be treated as

Y -periodic functions defined on R^3. The problem of finding $v_\alpha(\cdot)$, $\sigma^{\alpha\beta}(\cdot)$, $\beta^{\alpha\beta}(\cdot)$, govered by Eqs. (3.1), (3.3) will be denoted by $(\mathcal{P})$.

Using an approach outlined in [3] we shall "approximate" functions $v_\alpha(\cdot, t)$ by means of

$$(4) \qquad v_\alpha(X,t) = \dot{w}_\alpha(X,t) + \dot{q}^a{}_\alpha(X,t)\, l_a(X), \quad X \in \Omega,$$

where $l_a(\cdot)$ are a priori postulated Y -periodic oscillating functions defined on R^3 (shape functions) and linear in every $X + Y \cap \Omega_A$, [2]; functions $q^a{}_\alpha(\cdot)$ and $w_\alpha(\cdot)$ are unknown microlocal parameters and macro-displacements, respectively, [3], which are "nearly constant" in every periodicity cell. Using the procedure given in [3], from problem $(\mathcal{P})$ we obtain a certain nonstandard problem $(\mathcal{P}_{\breve{\omega}})$ (where $\breve{\omega}$ is an infinitive natural number) in which we deal with $Y/\breve{\omega}$ periodic body. The problem $(\mathcal{P}_{\breve{\omega}})$ is governed by the following weak form of the quasi-stationary equations of motion:

$$(5) \quad \int_{{}^*\Omega} \sigma^{\alpha\beta}(X,t)\, u_{(\alpha|\beta)}(X)\, dV = \oint_{\partial {}^*\Omega} t^\alpha(X,t)\, u_\alpha(X)\, dA + \int_{{}^*\Omega} b^\alpha(X,t)\, u_\alpha(X)\, dV, \quad t \in {}^*[t_0, t_1],$$

which has to hold for every pertinent test function $u(\cdot)$ defined on ${}^*\bar{\Omega}$, and by the constitutive relations:

$$\sigma^{\alpha\beta}(X,t) = \bar{\sigma}^{\alpha\beta}(X,t) + \beta^{\alpha\beta}(X,t),$$

$$(6) \quad \dot{\bar{\sigma}}^{\alpha\beta}(X,t) = 2\mu(\breve{\omega}X)\dot{\varepsilon}^{\alpha\beta}(X,t) + \lambda(\breve{\omega}X)\, G^{\alpha\beta}(X)\, \dot{\varepsilon}^\gamma_\gamma(X,t) - \frac{2\xi(f(X,t))\, W(X,t)\, \mu(\breve{\omega}X)}{k^2(\breve{\omega}X)}\, \bar{s}^{\alpha\beta}(X,t),$$

$$\dot{\beta}^{\alpha\beta}(X,t) = \frac{2\xi(f(X,t))\, W(X,t)\, \mu(\breve{\omega}X)\, m(\breve{\omega}X)}{(\mu(\breve{\omega}X) + m(\breve{\omega}X))\, k^2(\breve{\omega}X)}\, \bar{s}^{\alpha\beta}(X,t).$$

Let $\Omega^{\breve{\omega}}_A$ be a part of ${}^*\Omega_A$ occupied by A-th material component of $Y/\breve{\omega}$ -periodic body under consideration. Define

$$l_a^{(\breve{\omega})}(X) \equiv \frac{1}{\breve{\omega}}\, l_a(\omega X), \quad X \in {}^*R^3.$$

We shall look for the solution of $(\mathcal{P}_{\breve{\omega}})$ in the form

$$v_\alpha(X,t) = {}^*\dot{w}_\alpha(X,t) + {}^*q^a{}_\alpha(X,t)\, l_a^{(\breve{\omega})}(X), \quad X \in {}^*\Omega ,$$

(7) $$\sigma^{\alpha\beta}(X,t) = {}^*\sigma_A^{\alpha\beta}(X,t), \quad X \in \Omega_A^{\breve{\omega}}, \quad A=1,\ldots,N,$$

$$\beta^{\alpha\beta}(X,t) = {}^*\beta_A^{\alpha\beta}(X,t), \quad X \in \Omega_A^{\breve{\omega}}, \quad A=1,\ldots,N,$$

with standard macro-displacements ${}^*w_\alpha(\cdot)$, standard microlocal parameters ${}^*q^a{}_\alpha(\cdot)$, standard stresses ${}^*\sigma_A^{\alpha\beta}$ and standard parameters ${}^*\beta_A^{\alpha\beta}(\cdot)$. It can be shown by the detailed calculations that the problem $(\mathcal{P}_{\breve{\omega}})$ under condition (7) leads to the system of equations for the unknown functions $w_\alpha(X,t)$, $q^a{}_\alpha(X,t)$, $\sigma_A^{\alpha\beta}(X,t)$, $\beta_A^{\alpha\beta}(X,t)$, $A=1,\ldots,N$, $X \in \Omega$, $t \in [t_0,t_1]$.
To this aid we shall define by k_A, μ_A, λ_A, m_A the values of $k(X)$, $\mu(X)$, $\lambda(X)$, $m(X)$ for $X \in \Omega_A$, respectively, and we introduce the following denotations:

$$\eta_A \equiv \frac{1}{Y^1Y^2Y^3}\int_{Y\cap\Omega_A} dV, \qquad \Lambda^A_{a\alpha} \equiv l_{a,\alpha}(X) \text{ for } X \in \Omega_A$$

$$W_A(X,t) = \begin{cases} \bar{s}_A^{\alpha\beta}(X,t)\, v^A_{\alpha\beta}(X,t) & \text{if } \bar{s}_A^{\alpha\beta}(X,t)\, v^A_{\alpha\beta}(X,t) \geq 0, \\ 0 & \text{if } \bar{s}_A^{\alpha\beta}(X,t)\, v^A_{\alpha\beta}(X,t) < 0, \end{cases}$$

(8) $$v^A_{\alpha\beta}(X,t) \equiv \dot{w}_{(\alpha|\beta)}(X,t) + \dot{q}^a{}_{(\alpha}(X,t)\,\Lambda^A_{a\beta)},$$

$$\bar{s}_A^{\alpha\beta}(X,t) \equiv \bar{\sigma}_A^{\alpha\beta}(X,t) - \tfrac{1}{3}\, G^{\alpha\beta}(X)\, \bar{\sigma}_A{}^\gamma{}_\gamma(X,t),$$

$$f_A(X,t) \equiv \bar{s}_A^{\alpha\beta}(X,t)\, \bar{s}_{A\alpha\beta}(X,t) - k_A^2,$$

$$\langle \varphi \rangle \equiv \frac{1}{Y^1Y^2Y^3}\int_Y \varphi(X)\, dV,$$

where $\varphi(\cdot)$ is an arbitrary Y-periodic integrable function.
After that, we shall formulate the following homogenization statement :

Functions $w_\alpha(X,t)$, $q^a{}_\alpha(X,t)$, $\sigma_A^{\alpha\beta}(X,t)$ and $\beta_A^{\alpha\beta}(X,t)$, $X \in \Omega$, $t \in [t_0,t_1]$ for the

Prandtl–Reuss elastic-plastic materials with kinematic work hardening have to satisfy the system of following equations:

1° Equations of motion (for the quasi static case):

$$\tau^{\alpha\beta}|_{\beta}(X,t) + f^{\alpha}(X,t) = 0 ,$$

$$(9) \qquad \zeta^{a\alpha}(X,t) = 0,$$

where

$$\tau^{\alpha\beta}(X,t) \equiv \sum_{A=1}^{N} \eta_A \sigma_A^{\alpha\beta}(X,t) ,$$

$$(10) \qquad \zeta^{a\alpha}(X,t) \equiv \sum_{A=1}^{N} \eta_A \Lambda^{A}_{a\beta} \sigma_A^{\alpha\beta}(X,t) ;$$

2° Constitutive relations of the A-th component of a body, which have to hold for $A = 1,\ldots, N$:

$$\sigma_A^{\alpha\beta}(X,t) = \bar{\sigma}_A^{\alpha\beta}(X,t) + \beta_A^{\alpha\beta}(X,t) ,$$

$$\dot{\bar{\sigma}}_A^{\alpha\beta}(X,t) = 2\mu_A v^{A\alpha\beta}(X,t) + \lambda_A G^{\alpha\beta}(X) v^{A}{}_{\gamma}^{\gamma}(X,t) -$$

$$(11) \qquad - \frac{2\,\xi(f_A(X,t))\, W_A(X,t)\,\mu_A}{k_A^2}\, \bar{s}_A^{\alpha\beta}(X,t) ,$$

$$\dot{\beta}_A^{\alpha\beta}(X,t) = \frac{2\,\xi(f_A(X,t))\, W_A(X,t)\,\mu_A m_A}{(\mu_A + m_A)\, k_A^2}\, \bar{s}_A^{\alpha\beta}(X,t) .$$

The detailed discussion of the relations obtained will be given in [5] .

Example. As an example of application of Eqs. (9), (10), (11) we shall take the axially symmetric problem of the multi-layered periodic elastic-plastic incompressible composite in the shape of the cylindrical tube given on Fig.1, and loaded by the internal pressure

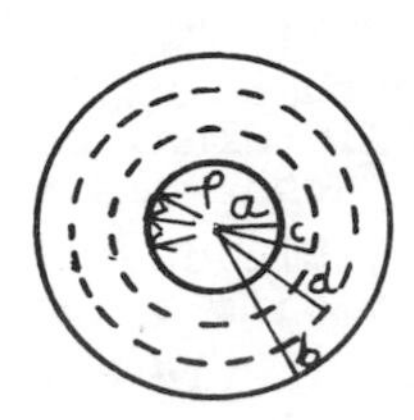

Every layer is assumed to have a thickness δ and is composed of two homogeneous sublayers with thickness δ_1 and δ_2 , $\delta = \delta_1 + \delta_2$. Also define $\eta_A \equiv \delta_A / \delta$.
In the problem under consideration the shape function can be assumed in the form:

$$
l(r) = \begin{cases} \dfrac{r}{\eta_1} + \dfrac{\delta}{2} & \text{for} \quad r \in (-\delta_1, 0) , \\[2ex] -\dfrac{r}{\eta_2} + \dfrac{\delta}{2} & \text{for} \quad r \in (0, \delta_2) . \end{cases}
$$

for $r \in (-\delta_1, \delta_2)$ and extended to R by means of $l(r) = l(r + n\delta)$ for every integer n .
The material properties of the layers are determined by functions $\mu(r)$, $\lambda(r)$, $k(r)$ which are constant in every sublayer taking the values μ_A , λ_A , k_A for $A = 1, 2$, respectively.
We shall assume that $k_1 > k_2$.
For a simple loading we obtain the three kinds of zones (cf. Fig.1)

1° for $r \in (a, c)$ both components of the composite are in a plastic state,

2° for $r \in (c, d)$ the first component is in an elastic state and the second one is in the plastic state,

3° for $r \in (d, b)$ both components are in the elastic state.

The problem under consideration can be solved under the known approximate condition that condition $2\sigma^{zz}(r) = \sigma^{rr}(r) + \sigma^{\vartheta\vartheta}(r)$ holds also in the plastic state. Then it can be shown that from Eqs. (3.7), (3.8) the stresses $\sigma_A^{\alpha\beta}(X, t)$ can be eliminated and we obtain a system of equations for radial macrovelocities $\dot{w}(r, t)$ and the microlocal parameters $q(r, t)$ in every zone.
The unknown radii c, d for the plastic zones can be found from the known continuity conditions. The detailed calculations concerning this problem can be found in [5].

References

1. Cz.Woźniak, A nonstandard method of modelling of thermo-elastic periodic composites, Int. J. Engng Sci., in press.

2. Cz.Woźniak, Microlocal parameters in the modelling of composite materials with internal constraints, Bull.Acad. Polon. Sci., Sér. Sci. Techn., in course of publication.

3. S.J.Matysiak, Cz.Woźniak, Microlocal effects in the non-standard modelling of periodic composites, this issue.

4. T.Tokuoko, Prandtl-Reuss plastic material with kinematic work hardening, Proc. of the XXI Japan Congr. on Mater. Res.

5. M.Wągrowska, Modelling with microlocal parameters of elastic-plastic periodic composites, Mech. Teor. Stos., in preparation.

ELASTIC-VISCO-PLASTIC COMPOSITES UNDER AXISYMMETRIC LOADING

C. LICHT, P. SUQUET

Laboratoire de Mécanique Générale des Milieux Continus
Université des Sciences et Techniques du Languedoc
Place Eugène Bataillon
34060 MONTPELLIER Cedex
and
GRECO "Comportement mécanique des composites à fibres"

ABSTRACT

This paper derives macroscopic constitutive equations governing the behaviour of unidirectional and laminated elastic/viscoplastic composites under axisymmetric loading. It is shown that the overall behaviour is equivalent to that of a generalized nonlinear Kelvin-Voigt solid. The macoscopic stresses are shown to be deriveable from two macroscopic thermodynamic potentials which are averages of the corresponding potentials for the individual phases.

PRELIMINARIES

The present paper is devoted to the study of the macroscopic constitutive law of a viscoplastic composite made of the assembly of a linearly elastic phase, hereafter called the reinforcement,and of a purely viscoplastic matrix. The governing equations of each phase are

<u>Reinforcement</u> $\sigma_{ij} = \lambda \mathrm{Tr}\ \varepsilon(u)\delta_{ij} + 2\mu\ \varepsilon_{ij}(u)$

<u>Matrix</u> $\mathrm{div}\ u = 0 \quad , \quad \sigma_{ij} = -p\ \delta_{ij} + \sigma^D_{ij}$ (1)

$$\sigma^D_{ij} = \eta|\dot{\varepsilon}(u)|^{s-2}\ \dot{\varepsilon}_{ij}(u) \quad \text{where} \quad |\dot{\varepsilon}| = (\dot{\varepsilon}_{ij}\ \dot{\varepsilon}_{ij})^{1/2}$$

The mean stress $-p$ is the unknown Lagrange multiplier arising from the incompressibility of the matrix. The constitutive laws of both the reinforcement and the matrix can be defined more generally by two thermodynamical potentials, the free energy associated with reversible processes, and the potential of dissipation for irreversible evolutions :

$$\sigma_{ij} = \frac{\partial w}{\partial \varepsilon_{ij}}(\varepsilon) + \frac{\partial d}{\partial \dot{\varepsilon}_{ij}}(\dot{\varepsilon}) \quad (2)$$

The choices of w and d for the reinforcement and the matrix are the following :

<u>Reinforcement</u> $w(\varepsilon) = \frac{\lambda}{2}(\mathrm{Tr}\ \varepsilon)^2 + \mu(\varepsilon_{ij}\ \varepsilon_{ij})$

$d(\dot{\varepsilon}) = 0$

<u>Matrix</u> $w(\varepsilon) = 0$ if $\mathrm{Tr}\ \varepsilon = 0$, $w(\varepsilon) = +\infty$ if $\mathrm{Tr}\ \varepsilon \neq 0$ (+)

$d(\dot{\varepsilon}) = \frac{\eta}{s}\ |\dot{\varepsilon}|^s$

(+) The differential of w with respect to ε is to be understood in the sense of subdifferentials.

The macroscopic or "homogenized" constitutive law relates the macroscopic stresses and strains defined as the averages over a representative volume element (r.v.e.) V of the microscopic stresses and strains :

$$\Sigma_{ij} = \langle\sigma_{ij}\rangle = \frac{1}{|V|}\int_V \sigma_{ij}\, dx = \frac{1}{|V|}\int_{\partial V} \sigma_{ip}\, n_p\, x_j\, ds$$
$$E_{ij} = \langle\varepsilon_{ij}\rangle = \frac{1}{|V|}\int_V \varepsilon_{ij}\, dx = \frac{1}{|V|}\int_{\partial V} \frac{1}{2}(u_i n_j + u_j n_i) ds \qquad (3)$$

The *local problem* which is to be solved in order to determine the homogenized constitutive law is (1) + (3) and additional boundary conditions describing the state of stress and strain on the boundary of the r.v.e. : homogeneous state of stress or strain on ∂V (HILL [1]) in the case of a random composite and a large r.v.e. , periodicity conditions on the unit cell for σ and ε in the case of a periodic composite (SUQUET [2]) .

Without further assumptions the problem of deriving the macroscopic constitutive law is nonlinear and extremely complex, and even approximate solutions are welcome (ABOUDI [3]). A first attempt to overcome the difficulty could be to compute (or at least to approximate) instead of the local fields σ and ε , the macroscopic thermodynamical potentials, free emergy W and potential of dissipation $\mathcal{D}$ for the composite, defined as

$$W = \langle w(\varepsilon)\rangle \qquad \mathcal{D} = \langle d(\dot{\varepsilon})\rangle \quad . \qquad (4)$$

In the case of voided viscoplastic materials the approximation of $\mathcal{D}$ has been carried out by DUVA & HUTCHINSON [4] . However the presence of an elastic phase further complicates the problem by coupling the microscopic internal energy and the microscopic dissipation. This coupling is generally achieved through internal variables describing the irreversible processes at the microscopic level : slips, microscopic inelastic strains (RICE [5] , MANDEL [6]) . However it can be shown that an infinite number of internal variables are required for a rigorous description of the homogenized law (SUQUET [2]) . Therefore the potentials W and $\mathcal{D}$ are not functions only of E and $\dot{E}$ respectively but also of these internal variables.

The problem greatly simplifies if we restrict the attention to specific geometries and specific loadings : the present paper considers the macroscopic behavior of a unidirectional fiber composite under axisymmetric loading, and the behavior of a laminate composed of elastic and viscoplastic layers. In both cases it is shown that the derivation of the homogenized constitutive law *does not require the introduction of internal variables*, and that the obtained rheological behavior is similar to a nonlinear Kelvin-Voigt model.

UNIDIRECTIONAL COMPOSITES

We consider in this section a viscoplastic matrix reinforced by unidirectional fibers aligned in the direction $\vec{e}_3$. The loading is assumed to be invariant by rotation around $\vec{e}_3$ and the macroscopic stress and strain states are

$$\Sigma_{11} = \Sigma_{22} = T \ , \quad \Sigma_{33} = S \qquad \text{other } \Sigma_{ij} = 0 \ ,$$
$$E_{11} = E_{22} = E_{TT} \ , \quad E_{33} = E_{zz} \ \text{other } E_{ij} = 0 \ .$$

We study here the most simple modelling of the elementary volume of the composite, namely a composite cylinder made from an elastic fiber surrounded by the viscoplastic matrix and subjected to the (S,T) loading as described on figure 1.

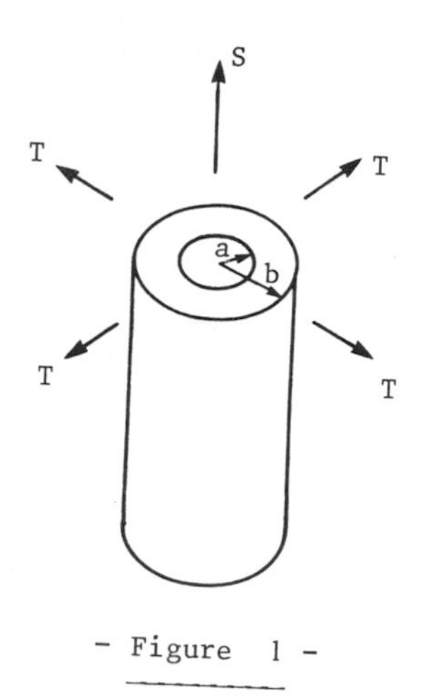

- Figure 1 -

The simplicity of the geometry allows to give the solution in a partially closed form. The displacement field in the composite has a simple structure and can be expressed in cylindrical coordinates (r,z) by

$$\vec{u}(r,z) = u_r(r)\vec{e}_r + E_{zz}\,\vec{e}_z$$

In the matrix the incompressibility condition yields

$u_r = -E_{zz}\dfrac{r}{2} + \dfrac{A}{r}$ and therefore

$$\dot{\varepsilon}_{rr} = -\frac{\dot{E}_{zz}}{2} - \frac{\dot{A}}{r^2} \qquad \dot{\varepsilon}_{\theta\theta} = -\frac{\dot{E}_{zz}}{2} + \frac{\dot{A}}{r^2} \qquad \dot{\varepsilon}_{zz} = \dot{E}_{zz}$$

where $\dot{A}$ is a function of t only. The stress field deduced from the constitutive law reads as

$$\sigma_{rr} = -p + \eta\,|\dot{\varepsilon}|^{s-2}\left(-\frac{\dot{E}_{zz}}{2} - \frac{\dot{A}}{r^2}\right) \quad \text{where} \quad |\dot{\varepsilon}| = \left(\frac{3}{2}\dot{E}_{zz}^2 + \frac{2\dot{A}^2}{r^4}\right)^{1/2}$$

$$\sigma_{\theta\theta} = -p + \eta\,|\dot{\varepsilon}|^{s-2}\left(-\frac{\dot{E}_{zz}}{2} + \frac{\dot{A}}{r^2}\right), \quad \sigma_{zz} = -p + \eta\,|\dot{\varepsilon}|^{s-2}\,\dot{E}_{zz}$$

where the mean stress $-p$ is an unknown function of r and t. The constant A can be expressed in terms of the transverse and axial strains

$$E_{11} = E_{22} = E_{TT} = \frac{u_r(b)}{b} = -\frac{E_{zz}}{2} + \frac{A}{b^2}$$

i.e. $A = b^2\left(E_{TT} + \dfrac{E_{zz}}{2}\right)$.

In the fiber the equilibrium equations together with Hoske's law yield the classical form of the radial displacement

$$u_r = Br + C/r$$

and, to avoid singularities on the fiber axis the constant C must vanish :

$$u_r = Br \qquad \varepsilon_{rr} = \varepsilon_{\theta\theta} = B \quad , \quad \varepsilon_{zz} = E_{zz} .$$

The stress field in the fiber reads as

$$\sigma_{rr} = \sigma_{\theta\theta} = \lambda(2B + E_{zz}) + 2\mu\,B \quad , \quad \sigma_{zz} = \lambda(2B + E_{zz}) + 2\mu\,E_{zz} \qquad (5)$$

The continuity of the displacement at the interface fiber/matrix allows to compute B in terms of the macroscopic strains

$$B = \frac{1}{v_f}\left(E_{TT} + \left(\frac{1-v_f}{2}\right)E_{zz}\right) \quad \text{where} \quad v_f = \left(\frac{a}{b}\right)^2 \quad .$$

Integrating the radial equilibrium equation on the matrix we obtain

$$\sigma_{rr}(b) - \sigma_{rr}(a) = \int_a^b \frac{\partial\sigma_{rr}}{\partial r}\,dr = \int_a^b \frac{\sigma_{\theta\theta} - \sigma_{rr}}{r}\,dr = \int_a^b \eta\,|\dot{\varepsilon}|^{s-2}\,\frac{2\dot{A}}{r^3}\,dr \quad .$$

The continuity of σ_{rr} across the interface fiber/matrix allows to use (5) in order to compute $\sigma_{rr}(a)$, while $\sigma_{rr}(b)$ is the transverse stress T . Finally we obtain

$$\left.\begin{aligned} T &= T^R + T^{IR} \qquad \text{where (after a few computations)} \\ T^R &= \frac{1}{v_f}\left[2(\lambda+\mu)E_{TT} + (\lambda + \mu(1-v_f))E_{zz}\right] \\ T^{IR} &= \frac{\eta}{\sqrt{3}}\left(\frac{3}{2}\right)^{s/2}|\dot{E}_{zz}|^{s-2}\,\dot{E}_{zz}\int_{x_1}^{x_1/v_f}(1+x^2)^{\frac{s-2}{2}}\,dx \end{aligned}\right\} \qquad (6)$$

$$\text{where} \quad x_1 = \frac{2\dot{E}_{TT} + \dot{E}_{zz}}{\sqrt{3}\;\dot{E}_{zz}} \quad .$$

Similarly

$$S - T = \langle\sigma_{zz}\rangle - \left\langle\frac{\sigma_{xx} + \sigma_{yy}}{2}\right\rangle = \langle\sigma_{zz}\rangle - \left\langle\frac{\sigma_{rr} + \sigma_{\theta\theta}}{2}\right\rangle$$

$$= \langle\sigma^D_{zz}\rangle - \left\langle\frac{\sigma^D_{rr} + \sigma^D_{\theta\theta}}{2}\right\rangle$$

$$= v_f(2\mu(E_{zz} - B)) + \frac{2\pi}{\pi b^2}\int_a^b \eta\,|\dot{\varepsilon}|^{s-2}\,\frac{3}{2}\dot{E}_{zz}\,r\,dr \quad .$$

Therefore the axial stress S splits into two parts

$$\left.\begin{aligned} S &= S^R + S^{IR} \\ S^R &= T^R + \mu(3v_f E_{zz} - (2E_{TT} + E_{zz})) \\ S^{IR} &= T^{IR} + \frac{\eta}{\sqrt{3}}\left(\frac{3}{2}\right)^{s/2}|\dot{E}_{zz}|^{s-2}(2\dot{E}_{TT} + \dot{E}_{zz})\int_{x_1}^{x_1/v_f}(1+x^2)^{\frac{s-2}{2}}\,\frac{dx}{x^2} \end{aligned}\right\} \qquad (7)$$

Finally the total macroscopic stress Σ splits into two parts Σ^R and Σ^{IR} :

- $\Sigma^R = T^R(\vec{e}_x \otimes \vec{e}_x + \vec{e}_y \otimes \vec{e}_y) + S^R\,\vec{e}_z \otimes \vec{e}_z$

is called the reversible part of the stress since it is related to the *present value of the macroscopic strain* E .

- $\Sigma^{IR} = T^{IR}(\vec{e}_x \otimes \vec{e}_x + \vec{e}_y \otimes \vec{e}_y) + S^{IR}\,\vec{e}_z \otimes \vec{e}_z$

is called the irreversible part of the stress since it is related only to the *macroscopic strain rate* $\dot{E}$.

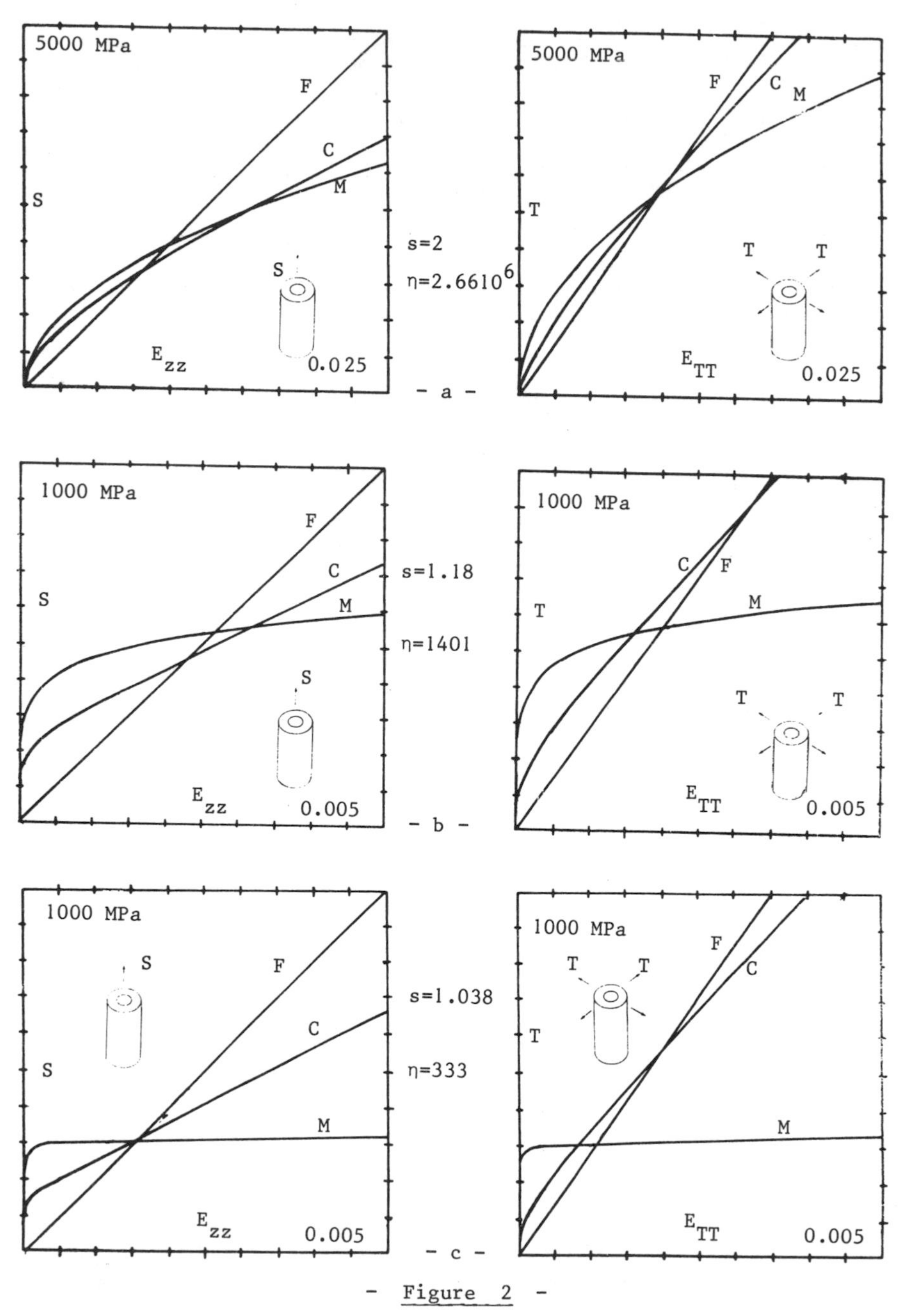

- Figure 2 -

Stress-strain curves for different values of s and η

F : fiber M : matrix C : composite

The above macroscopic constitutive law is a *generalized nonlinear Kelvin Voigt's law* where an elastic spring and a nonlinear dashpot are used in parallel. Although this is a natural result for the axial response, it is less obvious for the transverse one the constituents being usually assumed to be in series under transverse loading. The incompressibility of the matrix plays an important role in the fact that the axial and the transverse behaviors are intimately coupled.

The two thermodynamic potentials W and D can be computed according to (4)

$$W(E) = \langle w(\varepsilon)\rangle = \frac{\lambda}{2v_f}(2E_{TT} + E_{zz})^2 + \frac{\mu}{v_f}\left(2E_{TT}^2 + \frac{(1-v_f) + 2v_f^2}{2} E_{zz}^2 + 2(1-v_f)E_{TT}\, E_{zz}\right)$$

$$D(\dot{E}) = \langle d(\dot{\varepsilon})\rangle = \frac{\eta}{\sqrt{3}}\left(\frac{3}{2}\right)^{s/2} \frac{|\dot{E}_{zz}|^s}{s}\left(\frac{2\dot{E}_{TT} + \dot{E}_{zz}}{\dot{E}_{zz}}\right) \int_{x_1}^{x_1/v_f} (1 + x^2)^{s/2} \frac{dx}{x^2} .$$

It can be checked that(+)

$$2T^R = \frac{\partial W}{\partial E_{TT}} \qquad 2T^{IR} = \frac{\partial \mathcal{D}}{\partial \dot{E}_{TT}}$$

$$S^R = \frac{\partial W}{\partial E_{zz}} \qquad S^{IR} = \frac{\partial D}{\partial \dot{E}_{zz}} ,$$

which shows that the constitutive law of the composite has the standard form (2) .

On Figure 2 a,b,c are drawn the stress-strain curves of several composites submitted to an axial tensile test and to a transverse tensile test with a constant stress rate of 50 MPa h^{-1} .

a	$s = 2$	$\eta = 2.6610^6$ MPa,h	(linear viscoelasticity)
b	$s = 1.1785$	$\eta = 1401$ MPa,h	(ASTM 321 Steel [7])
c	$s = 1.038$	$\eta = 333$ MPa,h	(AU2GN [7])

It is to be noted that the matrix behavior prevails in the range of rather small strains, while the fiber behavior prevails in the rest of the stress strain curve.

LAMINATES

We now consider a laminate in which the different layers are alternatively purely elastic and purely viscoplastic. The elastic reinforcing phase is called the reinforcement. For the sake of simplicity we limit our

(+) Note that 2T is the stress associated with E_{TT} since $\Sigma_{ij}\, E_{ij} = 2TE_{TT} + SE_{zz}$.

attention to macroscopic states of stress and strain in the form

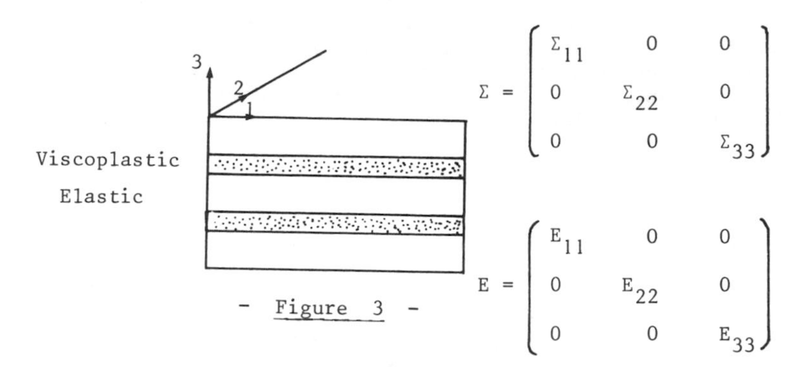

- Figure 3 -

$$\Sigma = \begin{pmatrix} \Sigma_{11} & 0 & 0 \\ 0 & \Sigma_{22} & 0 \\ 0 & 0 & \Sigma_{33} \end{pmatrix}$$

$$E = \begin{pmatrix} E_{11} & 0 & 0 \\ 0 & E_{22} & 0 \\ 0 & 0 & E_{33} \end{pmatrix}$$

As it is usual in laminated structures [8] the perfect bonding of the layers and the equilibrium equations yield

$$\varepsilon_{11} = E_{11} \quad , \quad \varepsilon_{22} = E_{22} \quad , \quad \sigma_{33} = \Sigma_{33} \quad .$$

In both phases the constitutive law shows that ε_{33} is constant

$$\varepsilon_{33} = \frac{1}{\lambda+2\mu}\left[\Sigma_{33} - \lambda(E_{11} + E_{22})\right] \quad \text{in the reinforcement}$$

$$\varepsilon_{33} = -\ (E_{11} + E_{22}) \quad \text{in the matrix (incompressibility).}$$

The macroscopic volume change is the average of the microscopic volume change

$$E_{11} + E_{22} + E_{33} = v_f(E_{11} + E_{22} + \varepsilon_{33}) \ , \ \text{(no volume change in the matrix)}$$

and thus

$$\varepsilon_{33} = \frac{1}{v_f}(E_{11} + E_{22} + E_{33}) - (E_{11} + E_{22})$$

$$\Sigma_{33} = \frac{\lambda + 2\mu}{v_f}(E_{11} + E_{22} + E_{33}) - 2\mu(E_{11} + E_{22}) \ .$$

In the matrix the pressure p is constant

$$p = -\ \Sigma_{33} - \eta\ |\dot{\varepsilon}|^{s-2}(\dot{E}_{11} + \dot{E}_{22}) \quad \text{where}$$

$$|\dot{\varepsilon}| = 2^{1/2}(\dot{E}_{11}^2 + \dot{E}_{22}^2 + \dot{E}_{11}\dot{E}_{22})^{1/2}$$

This relation allows to compute the plane components of the macroscopic stress

$$\Sigma_{11} = \langle\sigma_{11}\rangle = (E_{11} + E_{22} + E_{33})(\lambda + (\frac{1 - v_f}{v_f})(\lambda + 2\mu)) + 2\mu\ v_f\ E_{11}$$
$$+ (1 - v_f)\eta\ |\dot{\varepsilon}|^{s-2}(2\dot{E}_{11} + \dot{E}_{22}) \ .$$

Finally the macroscopic constitutive law reads as :

$$\Sigma_{11} = c^{comp}_{1111}\, E_{11} + c^{comp}_{1122}\, E_{22} + c^{comp}_{1133}\, E_{33} + (1-v_f)\eta|\dot{\varepsilon}|^{s-2}(2\dot{E}_{11} + \dot{E}_{22})$$

$$\Sigma_{22} = c^{comp}_{1122}\, E_{11} + c^{comp}_{1111}\, E_{22} + c^{comp}_{1133}\, E_{33} + (1-v_f)\eta|\dot{\varepsilon}|^{s-2}(\dot{E}_{11} + 2\dot{E}_{22})$$

$$\Sigma_{33} = \underbrace{c^{comp}_{1133}(E_{11} + E_{22}) + c^{comp}_{3333}\, E_{33}}_{\Sigma^R} \quad + \quad \underbrace{\qquad\qquad\qquad}_{\Sigma^{IR}}$$

where

$$c^{comp}_{1111} = \frac{\lambda}{v_f} + 2\mu\left(\frac{(1-v_f)^2 + v_f^2}{v_f}\right) \quad , \quad c^{comp}_{1122} = \frac{\lambda}{v_f} + 2\mu\,\frac{(1-v_f)^2}{v_f}$$

$$c^{comp}_{1133} = \frac{\lambda}{v_f} + 2\mu\left(\frac{1-v_f}{v_f}\right) \quad , \quad c^{comp}_{3333} = \frac{\lambda + 2\mu}{v_f} \quad .$$

Once more the macroscopic stress splits into a reversible part and an irreversible part

$$\Sigma^R = \frac{\partial W}{\partial E}(E) \qquad\qquad \Sigma^{IR} = \frac{\partial \mathcal{D}}{\partial \dot{E}}(\dot{E})$$

where the macroscopic potentials W and E are the averages of the macroscopic potentials

$$W(E) = \langle w(\varepsilon)\rangle = \frac{\lambda}{2v_f}(\mathrm{Tr}\ E)^2 + v_f\mu(E_{11}^2 + E_{22}^2 + \frac{1}{v_f^2}[(\mathrm{Tr}E) - v_f(E_{11} + E_{22})]^2)$$

$$\mathcal{D}(\dot{E}) = \langle d(\dot{\varepsilon})\rangle = \frac{(1-v_f)\eta}{s}\,|2|^s\,|\dot{E}_{11}^2 + \dot{E}_{22}^2 + \dot{E}_{11}\,\dot{E}_{22}|^s \quad .$$

Remark. In this specific case of laminates the elasticity of the matrix can be taken into account and more general (3 dim) loadings can be considered.

CONCLUSION

A simple model for the behavior of unidirectional and laminated elastic viscoplastic composites under axisymmetric loading has been proposed. It has been shown that the overall behavior amounts to a generalized nonlinear Kelvin Voigt model. In the general case this will not hold true but the present model could propose a reasonable approximation to the actual behavior of the composite.

REFERENCES

[1] HILL R. : "The essential structure of constitutive laws for metal composites and polycristals", J. Mech. Phys. Solids, 15, (1967), 79-95.

[2] SUQUET P. : "Elements of homogenization for inelastic solid Mechanics" in *Homogenization Techniques for Composite Materials*, C.I.S.M. Udine, 1985. To be published.

[3] DUVA J.M., HUTCHINSON J.W. : "Constitutive potentials for dilutely voided nonlinear materials", Mech. Materials, 3 , (1984), 41-54.

[4] ABOUDI J. : "A continuum theory for fiber reinforced elastic-viscoplastic composites", Int. J. Eng. Sc., 20, (1982), 605-621.

[5] RICE J.R. : "On the structure of stress-strain relations for time-dependent plastic deformation in metals", J. Appl. Mech., 37, (1970), 728-737.

[6] MANDEL J. : *Plasticité classique et viscoplasticité*, CISM Lecture Notes n° 97, Springer Verlag, Wien. 1972.

[7] LEMAITRE J., CHABOCHE J.L. : *Mécanique des Matériaux Solides*, Dunod, Paris 1985.

[8] DUMONTET H. : "Homogénéisation d'un matériau périodique stratifié de comportement linéaire et non linéaire et viscoélastique", C.R. Acad. Sc. Paris, II, 295, (1982), 633-636.

THE DETERMINATION OF THE INTERACTION COEFFICIENT IN MIXTURE EQUATIONS FOR THERMAL DIFFUSION IN UNIDIRECTIONAL COMPOSITES

JAN A. KOŁODZIEJ
Institute of Applied Mechanics, Technical University of Poznań, Piotrowo 3, 60965 Poznań, Poland

ABSTRACT

The mixture theory for thermal diffusion in a composite, proposed by Murakami, Hegemier and Meawal, involves the solution of the microstructure boundary value problem (MBVP) in order to determine the thermal interaction coefficient between the constituents of the mixture. Murakami and his co-workers solve this MBVP using the finite element method. Here the same problem is solved using the Boundary Collocation Method. This involves expressing the solution in terms of a set of trial functions which satisfy the governing differential equations and some of the boundary conditions exactly whilst the remaining boundary conditions are satisfied at the collocation points. Results are presented for composites formed of a matrix containing unidirectional cylindrical fibres arranged in square, hexagonal and triangular arrays.

INTRODUCTION

In literature we find two basic theoretical models in research of unidirectional composites with respect to their thermal properties. In the first model the composite is treated as some substitute continuous medium / model of single continuum / for which are introduced the effective thermal properties such as effective thermal conductivity, effective thermal expansion coefficient, ect., e. g. [1-3]. In the second model the composite is treated as a mixture of particular constituents which form muntually penetrating continuous media / model of many continuum / e. g. [4-6]. In the last cited paper the case for which conduction occurs primarily in the direction of the fiber axis and fibers are arranged in two dimensional periodic array, is considered. Equations of thermal diffusion in two continuous constituens are proposed and all coefficients in these equations can be determined for given properties of the constituents, their volume fractions and the geometric structure of the composite. The thermal interaction coefficient between phase of fibers and phase of matrix in mixture equations is determined from solution the so-called microstructure boundary value problem / MBVP /. In paper [6] the MBVP for the chosen cases was solved by the finite element method.

In the last years in the world literature appears a tendency to find and to apply the most optimal numerical methods for the given boundary velue problem. Therefore the boundary integral equation method was developed since it is more economical than the finite element method and the finite differences method. Another method, which has been known for at least fifty years, and which has become more and more popular recently, is the boundary collocation method / BCM /.

The purpose of this paper is a proposition of application of BCM to the solution of MBVP which was formulated in paper [6]. The selection of trial functions is given special attention, so that these functions not only satisfied exactly the differential equations, which is the essence of this method, but also part of boundary conditions. The advantages of BCM in the solution of MBVP as compared with the finite element method are pointed out.

MIXTURE EQUATIONS FOR THERMAL DIFFUSION IN UNIDIRECTIONAL COMPOSITES ACCORDING TO [6]

Let us consider a periodic, two-dimensional array of unidirectional cylindrical fibers of arbitrary cross section embedded in a matrix. With each fiber may be associated a "unit cell" as depicted in Fig. 1. Each such cell consists of regions Ω^I and Ω^{II} occupied by the fiber and matrix, respectively.

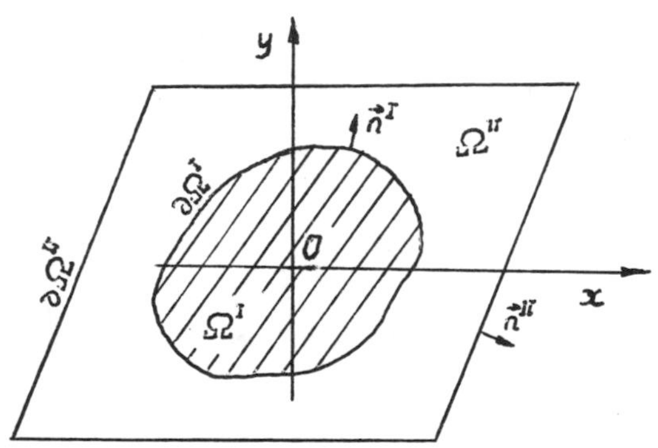

Fig . 1. Unit cell

In the considered composite two characteristic lengths are differentiated: macrodimension L , which is dimension of composite in z axis direction, and microdimension b , which characteristic dimension of the separated unit cell. The quantities L and b may be defined characteristic thermal diffusion times in longitudinal and transverse directions according to

$$\tau_L = \frac{\mu_{(m)} L^2}{\lambda_{(m)}} , \qquad \tau_b = \frac{\mu_{(m)} b^2}{\lambda_{(m)}} , \tag{1}$$

where $\mu_{(m)}$, $\lambda_{(m)}$ denote mixture heat capacity and thermal conductivity defined as follows

$$\mu_{(m)} = \varphi \mu_f + (1-\varphi)\mu_m , \qquad \lambda_{(m)} = \varphi \lambda_f + (1-\varphi)\lambda_m , \tag{2}$$

where μ_f and μ_m are heat capacity for fibers and matrix, respectively; λ_f and λ_m - thermal conductivity for fibers and matrix, respectively; φ - volume fraction of fibers.

In paper [6] , after introducing assumption that

$$\varepsilon = \frac{b}{L} = \left(\frac{\tau_b}{\tau_L}\right)^{\frac{1}{2}} \ll 1 , \tag{3}$$

the mixture equations have the form

$$\varphi\left[\bar{\lambda}_f \frac{\partial^2 \vartheta^I(Z,\tau)}{\partial Z^2} - \bar{\mu}_f \frac{\partial \vartheta^I(Z,\tau)}{\partial \tau}\right] = \zeta\left[\vartheta^I(Z,\tau) - \vartheta^{II}(Z,\tau)\right]/\varepsilon^2 , \qquad (4)$$

$$(1-\varphi)\left[\bar{\lambda}_m \frac{\partial^2 \vartheta^{II}(Z,\tau)}{\partial Z^2} - \bar{\mu}_m \frac{\partial \vartheta^{II}(Z,\tau)}{\partial \tau}\right] = -\zeta\left[\vartheta^I(Z,\tau) - \vartheta^{II}(Z,\tau)\right]/\varepsilon^2 . \qquad (5)$$

The nondimensional quantity which appear in equations (4-5) are defined as follows

$$Z = \frac{z}{L}, \quad \tau = \frac{t}{\tau_L}, \quad \vartheta^I = \frac{T^I}{T_0}, \quad \vartheta^{II} = \frac{T^{II}}{T_0}, \quad \bar{\lambda}_f = \frac{\lambda_f}{\lambda_{(m)}},$$

$$\bar{\lambda}_m = \frac{\lambda_m}{\lambda_{(m)}}, \quad \bar{\mu}_f = \frac{\mu_f}{\mu_{(m)}}, \quad \bar{\mu}_m = \frac{\mu_m}{\mu_{(m)}} , \qquad (6)$$

where T^α / $\alpha = I, II$ / - average temperature in α constituent, T_0 - reference temperarure, t - time.
The nondimensional interaction coefficient ζ in mixture equations is determined from the solution of the following boundary value problem on the microstructure level

$$\frac{\partial^2 T_*^I}{\partial X^2} + \frac{\partial^2 T_*^I}{\partial Y^2} = -\frac{1}{\varphi \bar{\lambda}_f} \qquad \text{on} \quad \bar{\Omega}^I , \qquad (7)$$

$$\frac{\partial^2 T_*^{II}}{\partial X^2} + \frac{\partial^2 T_*^{II}}{\partial Y^2} = \frac{1}{(1-\varphi)\bar{\lambda}_m} \qquad \text{on} \quad \bar{\Omega}^{II} , \qquad (8)$$

$$\vec{n}^{II} \cdot \operatorname{grad} T_*^{II} = 0 \qquad \text{on} \quad \partial\bar{\Omega}^{II} , \qquad (9)$$

$$\left.\begin{aligned} \bar{\lambda}_f \vec{n}^I \cdot \operatorname{grad} T_*^I &= \bar{\lambda}_m \vec{n}^I \cdot \operatorname{grad} T_*^{II} \qquad (10) \\ T_*^I &= T_*^{II} \qquad (11) \end{aligned}\right\} \text{on} \quad \partial\bar{\Omega}^I ,$$

$$T_*^I = 0 \qquad \text{at point} \quad O \in \bar{\Omega}^I , \qquad (12)$$

where T_*^I and T_*^{II} are nondimensional functions of two variables $X = x/b$, $Y = y/b$. After solution of the MBVP (7-12), interaction coefficient is defined by the way

$$\zeta = \frac{1}{T_*^{Ia} - T_*^{IIa}} , \qquad (13)$$

where

$$T_*^{Ia} = \frac{1}{\bar{\Omega}^I} \iint_{\bar{\Omega}^I} T_*^I(X,Y)\, d\bar{\Omega}^I , \qquad (14)$$

$$T_*^{IIa} = \frac{1}{\bar{\Omega}^{II}} \iint_{\bar{\Omega}^{II}} T_*^{II}(X,Y)\, d\bar{\Omega}^{II} . \qquad (15)$$

In paper [6] a variational principle-based finite element method was proposed for the solution of this MBVP. In aur paper we used the formulation (7-12) and the BCM is applied.

THE APPLICATION OF BCM TO SOLUTION OF MBVP

The BCM is based on exact satisfaction of differential equation / or equations / when the boundary condition / or conditions / is satisfied at finite discrete points along the boundary. The problem of choosing trial functions is one the crucial steps in application of BCM. In application of finite element method in its conventional formulation the class of used trial functions is not too large / very often these are algebraic polynomials of not too high degree / whereas in application of BCM the trial functions are varied. There are two reasons for that. Firstly, in BCM the trial functions must satisfy differential equation whereas in many other approximated methods,e. g. in finite element method, the same trial functions can be used to solve different equations. Secondly, when applying BCM our aim is that apart from differential equation also boundary condition on possibly big part of boundary was exactly satisfied and only the remaining part of boundary the condion is satisfied collocationally. In this way in applications of this method the form of trial functions is connected with the shape of the considered domain.

Many authors using the BCM for solution of boundary value problems with plane Laplace or Poisson equation applied the general solution in series form in polar coordinate system in order to determine the trial functions. The necessary condition of success here was not too complicated geometrically plane region in this sense that its maximum and minimum diameter was not too different. Since unit cell in regular reinforced composites satisfies this condition, the trial functions are determined here in the same way.

After introducing a polar coordinate system in the initial point at $O \in \Omega^I$, assuming general solution as given in [7] and taking into account boundary condition (12), after truncating infinite series the solutions of equations (7) and (8) assume the forms

$$T_*^I = -\frac{1}{\varphi \bar{\lambda}_f} \frac{R^2}{4} + \sum_{k=1}^{2N} X_k^I \varphi_k^I(R, \theta) , \tag{16}$$

$$T_*^{II} = \frac{1}{(1-\varphi) \bar{\lambda}_m} \frac{R^2}{4} + \sum_{k=1}^{4N+2} X_k^{II} \varphi_k^{II}(R, \theta) , \tag{17}$$

where the trial functions $\varphi_k^I(R,\theta)$ and $\varphi_k^{II}(R,\theta)$ are given in Tab. 1.

The directional derivatives appearing in boundary conditions (10-11) get the forms

$$\vec{n}^I \cdot \nabla T_*^I = -\frac{1}{\varphi \bar{\lambda}_f} \frac{R}{2} \cos \xi^I + \sum_{k=1}^{2N} X_k^I k R^{k-1} \cos(k\theta + \xi^I) + X_{N+k}^I k R^{k-1} \sin(k\theta + \xi^I) \tag{18}$$

$$\vec{n}^{I}\cdot\nabla T_{*}^{II}=\frac{1}{(1-\varphi)\bar{\lambda}_m}\frac{R}{2}\cos\xi^{I}+\sum_{K=1}^{4N+2}X_K^{II}\left[KR^{K-1}\cos(K\Theta+\xi^{I})-KR^{-(K+1)}\cos(K\Theta-\xi^{I})\right] \quad (19)$$

$$\vec{n}^{II}\cdot\nabla T_{*}^{II}=\frac{1}{(1-\varphi)\bar{\lambda}_m}\frac{R}{2}\cos\xi^{II}+\sum_{K=1}^{4N+2}X_K^{II}\left[KR^{K-1}\cos(K\Theta+\xi^{II})-KR^{-(K+1)}\cos(K\Theta-\xi^{II})\right] \quad (20)$$

where ξ^{I} and ξ^{II} are the angles, which are formed between radius vector and $\vec{n}^{I}$, $\vec{n}^{II}$, respectively, while $\vec{n}^{I}$ and $\vec{n}^{II}$ are the outward drawn normals to $\partial\bar{\Omega}^{I}$ and $\partial\bar{\Omega}^{II}$, respectively. The directional derivatives of trial functions can be calculated on base Tab. 1.

Assuming N_1 collocation point on boundary $\partial\bar{\Omega}^{I}$ and N_2 collocation points on boundary $\partial\bar{\Omega}^{II}$ and satisfying boundary conditions (9-11) in these points in exact way we get $2N_1+N_2$ simultaneous linear algebraic equations for unknown coefficients X_K^{I} and X_K^{II}. The number of collocation points N_1 and N_2 must be chosen in this way that the number of equations is equal to the number of unknowns.

Generally in regular unidirectional fiber-reinforced composites, the unit cell has l / l = 1, 2, 3, 4, 6 / symmetry lines crossing the point $O\in\bar{\Omega}^{I}$. Then the repeated region can be seperated from the unit cell which is embodied between two neighbouring symmetry lines. So the solution of MBVP can be sought in that region. In this case the boundary conditions on symmetry lines are satisfied exactly which simplify the form of trial functions. The trial functions for the case when / above mentioned / symmetry lines exist are given in Tab. 2.

Very often reinforced fibers in composite are of cylindrical shape. Then the boundary in unit cell cell is a circle and applying BCM the boundary conditions (10-11) can be satisfied exactly. It simplifies considerably the form of trial functions. These functions for that assuming moreover that unit cell has symmetry lines are given in Tab. 3.

Let us consider, for example, unidirectional composite with cylindrical fibers arranged in a triangular, square and hexagonal array. The radii of fibers are equal $2a$, whereas the distance between the neighbouring fibers - $2b$ / Fig. 2/. For each of the three mentioned cases the unit cells in which MBVP is formulated are given in Fig. 2b. Then unit cells have certain number of symmetry lines, namely for the square array - four / l = 4 /, for the triangular array - six / l = 6 / and for the hexagonal array - three / l = 3 /. In each case the repeated region which is seperated, as bounded by neighbouring symmetry lines, from the unit cell is right-angled triangle / the region marked by dots on Fig. 2b /. This triangle has one side equal unity in nondimensional coordinates.

Using the trial functions from Tab. 2 the solution of MBVP in repeated regions gets the form

$$T_*^{I}=-\frac{1}{\varphi\bar{\lambda}_f}\frac{R^2}{4}+\sum_{K=1}^{N}X_K\frac{2\bar{\lambda}_m}{(\bar{\lambda}_f+\bar{\lambda}_m)}R^{lK}\cos(lK\Theta), \quad (21)$$

Table 1. Trial functions for general case of unit cell

$$\varphi_k^{I}(R,\theta) = R^k \cos k\theta\,, \quad \varphi_{N+k}^{I}(R,\theta) = R^k \sin k\theta\,, \quad k = 1, 2, ..N$$

$$\varphi_1^{II}(R,\theta) = 1$$

$$\varphi_{k+1}^{II}(R,\theta) = R^k \cos k\theta\,, \quad \varphi_{N+k+1}^{II}(R,\theta) = R^{-k} \cos k\theta\,, \quad k = 1, 2, ..N$$

$$\varphi_{2N+2}^{II}(R,\theta) = \ln R$$

$$\varphi_{2N+k+2}^{II}(R,\theta) = R^k \sin k\theta\,, \quad \varphi_{3N+k+2}^{II}(R,\theta) = R^{-k} \sin k\theta\,, \quad k = 1, 2, ..N$$

Table 2. Trial functions for case when unit cell has l of symmetry lines

$$\varphi_k^{I}(R,\theta) = R^{kl} \cos(kl\theta)\,, \qquad k = 1, 2, ..N$$

$$\varphi_1^{II}(R,\theta) = 1\,, \qquad \varphi_{N+2}^{II}(R,\theta) = \ln R$$

$$\varphi_{k+1}^{II}(R,\theta) = R^{lk} \cos(kl\theta)\,, \quad \varphi_{N+k+2}^{II}(R,\theta) = R^{-kl} \cos(kl\theta)$$

$$k = 1, 2, ..N$$

Table 3. Trial functions for case when unit cell has l of symmetry lines and fibers have cylindrical shape

$$\varphi_k^{I}(R,\theta) = \frac{2\bar{\lambda}_m}{\bar{\lambda}_f + \bar{\lambda}_m} R^{lk} \cos(lk\theta)\,,$$

$$\varphi_k^{II}(R,\theta) = \left\{ R^{lk} + \frac{(\bar{\lambda}_m - \bar{\lambda}_f)\, E^{2lk}}{(\bar{\lambda}_f + \bar{\lambda}_m)\, R^{lk}} \right\} \cos(lk\theta)\,,$$

$$E = a/b \qquad k = 1, 2, ..N$$

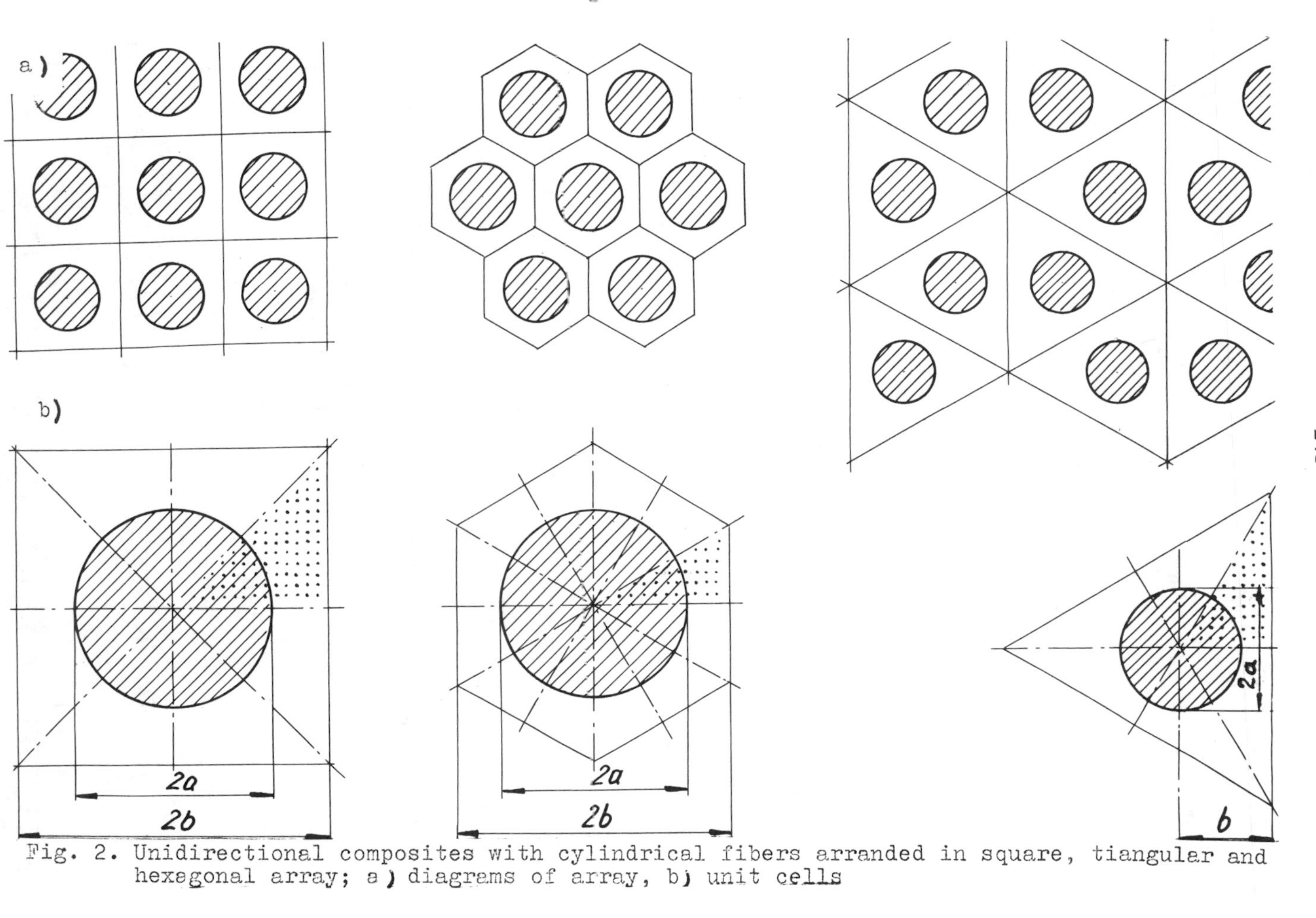

Fig. 2. Unidirectional composites with cylindrical fibers arranded in square, tiangular and hexagonal array; a) diagrams of array, b) unit cells

$$T_*^{II} = \frac{1}{(1-\varphi)\bar{\lambda}_m}\frac{R^2}{4} + \frac{E^2 \ln(E/R)}{2\bar{\lambda}_m \varphi(1-\varphi)} - \frac{1}{4}E^2\left(\frac{1}{\varphi\bar{\lambda}_f} + \frac{1}{(1-\varphi)\bar{\lambda}_m}\right) + \sum_{k=1}^{N} X_k \left[R^{lk} + \frac{(\bar{\lambda}_f - \bar{\lambda}_m)E^{2lk}}{(\bar{\lambda}_f + \bar{\lambda}_m)R^{lk}}\right]\cos(lk\Theta) \quad , \tag{22}$$

where parameter l determined the type of array. This solution exactly satisfies the equations and all boundary conditions except the condition

$$\frac{\partial T_*^{II}}{\partial X} = 0 \qquad \text{for} \quad X = R\cos\Theta = 1. \tag{23}$$

Assuming N collocation points on boundary $X = R\cos\Theta = 1$ according to the formula

$$\Theta_i = \frac{\pi(i-1)}{l(N-1)} \qquad i = 1, 2, \ldots, N \tag{24}$$

and satisfying exactly the condition (23) in these points we obtain the system of linear equations for unknown constants X_1, $X_2, \ldots, X_N$ in the form

$$\sum_{k=1}^{N}\left\{lk\left\{\frac{\cos[(lk-1)\Theta_i]}{(\cos\Theta_i)^{lk-1}} - \frac{(\bar{\lambda}_m - \bar{\lambda}_f)}{(\bar{\lambda}_f + \bar{\lambda}_m)} E^{2lk}(\cos\Theta_i)^{lk+1} \cdot \cos[(lk+1)\Theta_i]\right\}\right\} X_k = \frac{0.5}{(1-\varphi)\bar{\lambda}_m} - \frac{0.5\, E^2 \cos^2\Theta_i}{\bar{\lambda}_m \varphi(1-\varphi)} , \quad i = 1, 2, \ldots, N. \tag{25}$$

For given φ, l, $\bar{\lambda}_f$, $\bar{\lambda}_m$, $E = \frac{a}{b}$ and assumed N, after numerical solution of the system (25) the functions T_*^{I} and T_*^{II} are determined by explicit formulae (21-22).

The sought thermal interaction coefficient which is defined by formula (13) in considered case has the form

$$\zeta = \frac{1}{\frac{2l}{\pi E^2}\int_0^E\int_0^{\pi/l} T_*^{I} R\,dR\,d\Theta + \frac{2l}{l\,tg\frac{\pi}{l} - \pi E^2}\int_E^{1/\cos\Theta}\int_0^{\pi/l} T_*^{II} R\,dR\,d\Theta} . \tag{26}$$

After substitution of (21) and (22) into (26), the following integrals can be calculated analytically, mainly

$$\int_0^E\int_0^{\pi/l} T_*^{I} R\,dR\,d\Theta = -\frac{\pi E^4}{16\,l\,\varphi\,\bar{\lambda}_f} \tag{27}$$

$$\int_E^{1/\cos\Theta}\int_0^{\pi/l} T_*^{II} R\,d\Theta\,dR = \frac{1}{16(1-\varphi)\bar{\lambda}_m}\left[\frac{\sin(\pi/l)}{3\cos^3\frac{\pi}{l}} + \frac{2}{3}tg\frac{\pi}{l} - \frac{E^2\pi}{l}\right] + \frac{1}{2}\left[\frac{E^2\ln(E)}{\bar{\lambda}_m\varphi(1-\varphi)} - \frac{1}{2}E^2\left(\frac{1}{\varphi\bar{\lambda}_f} + \frac{1}{(1-\varphi)\bar{\lambda}_m}\right)\right]\left(tg\frac{\pi}{l} - \frac{\pi E^2}{l}\right) + \tag{28}$$

$$+\frac{1}{4}\frac{E^2}{\bar{\lambda}_f\varphi(1-\varphi)}\left\{\frac{\pi E^2}{l}\ln E+\frac{1}{2}tg\frac{\pi}{l}-\frac{\pi E^2}{2l}-\left[\ln(\cos\frac{\pi}{l})\,tg\frac{\pi}{l}+\right.\right.$$

$$\left.\left.+tg\frac{\pi}{l}-\frac{\pi}{l}\right]\right\}+\sum_{k=1}^{N}X_k\left\{\frac{(-1)^k\sin\frac{\pi}{l}}{(lk+2)(lk+1)(\cos\frac{\pi}{l})^{lk+1}}-\right.$$

$$\left.-\frac{(\bar{\lambda}_m-\bar{\lambda}_f)E^{2lk}}{(\bar{\lambda}_m+\bar{\lambda}_f)}\cdot\frac{}{(lk-2)(lk-1)(\cos\frac{\pi}{l})^{(1-kl)}}\right\}$$

As a result after determining the coefficients $X_1, X_2, \ldots, X_N$ by the BCM we get the explicit formula for thermal interaction coefficient identical for three kinds of arrays / the kinds of the array is determined by parametr /. The values of interaction coefficient in function of volume fraction and of kind of array are given in Fig. 3-5.

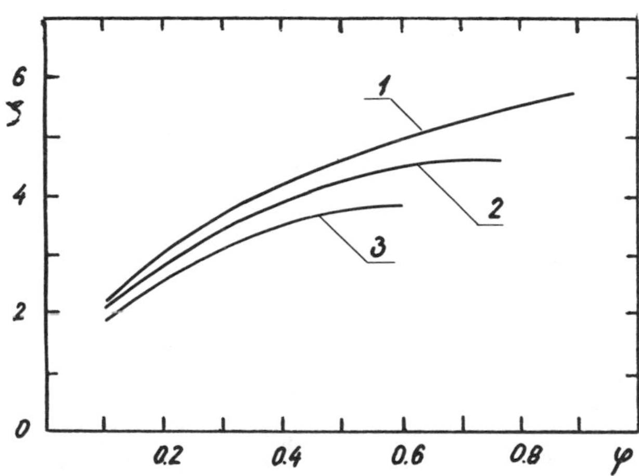

Fig. 3. Interaction coefficient ζ in function of volume fraction φ for $\lambda_f/\lambda_m = 2$; 1 - triangular array, 2 - square array, 3- hexagonal array.

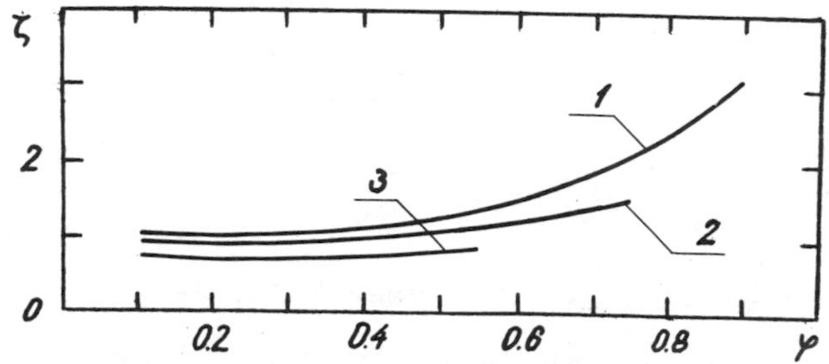

Fig. 4. Interaction coefficient ζ in function of volume fraction φ for $\lambda_f/\lambda_m = 20$; 1 - triangular array, 2 - square array, 3 - hexagonal array.

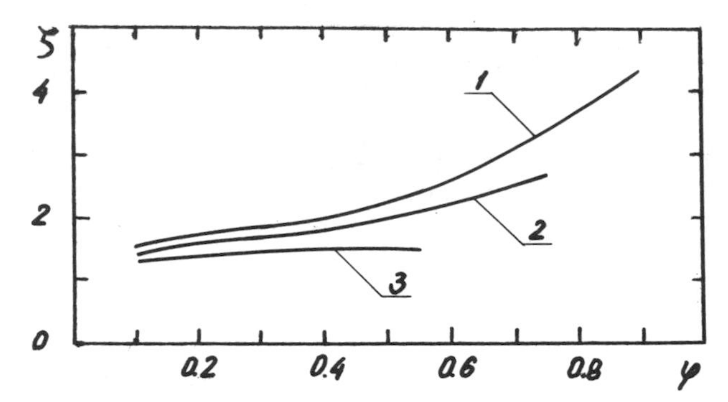

Fig. 5. Interaction coefficient ζ in function of volume fraction φ for λ_f/λ_m = 10 : 1 - triangular array, 2 - square array, 3 - hexagonal array.

CONCLUSIONS

The suggested method of solution of MBVP in this paper is the analytical-numerical method. Differentional equations and some boundary conditions are satisfied exactly, but only the rest part of boundary conditions are satisfied approximately. The application of such method is possible because the equations and boundary conditions appearing in MBVP are not too complicated.

In the considered case there is a possibility of analytical integration when calculating the interaction coefficient thanks to the analytical form of the solution of MBVP. In this way the only approximation exists in satisfying some of the boundary conditions by BCM. This approximation refers only to the part of the boundary, but not to the whole region as in the finite element method which was applied in the paper [6]. Then the dimension of the system of linear equations considered here is considerably small. Moreover, in contrast to BCM, finite element method need some quadrature formulae to compute coefficients of the resulting system of linear equations. Likewise, the appearance of large gradients of fields in the case $\lambda_f/\lambda_m \gg 1$ in the finite element method, requires the concentration of grids. In the applied method such gradients have the analytical form and dont require the concentration of collocation points on the boundary.

REFERENCES

[1] Beran M. J., Silmutzer N. R.: J. Com. Mat., 5, p. 246.
[2] Han L. S., Cosner A. A.: J. Heat Transfer, 103, p. 387.
[3] Kołodziej J. A.: Mech. Teor. Stosowana, 26, no. 4.
[4] Nayfeh A.: J. Appl. Mech., 42, p. 399.
[5] Maewal A., Bache T. C., Hegemier G.: J. Heat Transfer, 98, p. 133.
[6] Murakami H., Hegemier G., Maewal A.: Int. J. Sol. Struc., 14, p. 723.
[7] Ojalvo I. U., Linzer F. D.: Quart. J. Mech. Appl. Math., 18, p. 41.

TITLE: STRESS FIELD AROUND AN N-LAYERED SPHERICAL INCLUSION.

Author: S. Mazzullo, HIMONT ITALIA S.p.A.-Centro Ricerche "G.Natta" 44100 FERRARA (Italy).

ABSTRACT: The stress field induced by an N-layered spherical inclusion, embedded in an infinite matrix subjected to uniaxial tension, may be easily obtained by extension of a classical result by J.N.Goodier (1933). The actual determination of the field requires the solution of a linear system generated by the boundary conditions. The dimension of this system is (6N+6), where N is the number of the shells, and its solution, up to now, has been achieved numerically only, by the standard software for linear systems. For N greater than one, this procedure becomes rapidly impractical.
Taking advantage of the symmetries of the equations ,the algorithm described here, shows how the full system may be partitioned into two subsystems of lower order. Moreover, it shows that, by the use of matrix calculus and a simple recurrence relation, the first subsystem reduces to a system of two equations and the second to a single equation with one unknown. This occurs for every N. In particular, for the case of a single shell (the so-called core-shell structure), application of the formal algorithm now provides the explicit analytical form,assumed by the 12 constants solutions of the linear system.
In general, when N is greater than one this algorithm permits the numerical calculation of the (6N+6) constants more easily and more efficiently than using the standard linear software.

1. SPHERICAL INCLUSION WITH CORE-N-SHELL STRUCTURE (ONION STRUCTURE)

For convenience, the same symbols used by J.N. Goodier,/ 1 /to indicate the undetermined constants A,B,C,F,G,H, have been maintained. The elastic constants (ν,μ) of the single phases have constant given values.
The solution, with undetermined constants, both for the core (internal region to a spherical surface) and for the matrix (external region to a spherical surface) are formally identical to those derived by Goodier.
The solution for each shell is constructed, instead, as a linear combination of an internal and external solution /2/, /3/; /4/. The equations of the linear system we wish to solve, are obtained by equating displacements and radial tractions at the various interfaces. These boundary conditions generate a set of 6 equations for each interface; however,for the sake of simplicity, only equations (1) to (6), regarding the shell-shell interface $r = a_1$ have been included.
The equations for every other interface $r = a_i$ (i=0,1,... N where N is the number of shells) can be obtained by adding (i-1) to indexes "1" and "2" appearing in equations (1) to (6). There are only two exceptions to this rule and they regard the first and last interface. In the case of the first interface, $r = a_o$, the relation $A_o=B_o=C_o=0$ must be used.
This causes the displacement to vanish at the origin.

./.

The equationsat the interface $r = a_1$ are of the form

$$\underline{K}_1\underline{Z}_1 = \underline{K}_2\underline{Z}_2$$

where $\underline{Z}_1 = (A_1B_1C_1F_1G_1H_1)^T$, $\underline{Z}_2 = (A_2B_2C_2F_2G_2H_2)^T$, T denotes the transpose and the elements of the 6X6 matrices K_1 and K_2 are given in table 1.

TABLE 1 : Elements of the matrices $\underline{K}_1$ and $\underline{K}_2$.

For $\underline{K}_1$, $\bar{\nu} = \nu_1$, $\bar{\eta} = \mu_1/\mu_2$

For $\underline{K}_2$, $\bar{\nu} = \nu_2$, $\bar{\eta} = 1$.

$$\begin{array}{cccccc}
-\dfrac{1}{a_1^3} & -\dfrac{3}{a_1^5} & 0 & 1 & 2\bar{\nu}a_1^2 & 1 \\
0 & -\dfrac{9}{a_1^5} & \dfrac{5-4\bar{\nu}}{(1-2\bar{\nu})a_1^3} & 3 & 6\bar{\nu}a_1^2 & 0 \\
0 & -\dfrac{6}{a_1^5} & -\dfrac{2}{a_1^3} & -3 & -(7-4\bar{\nu})a_1^2 & 0 \\
\bar{\eta}\dfrac{2}{a_1^3} & \bar{\eta}\dfrac{12}{a_1^5} & -\bar{\eta}\dfrac{2\bar{\nu}}{(1-2\bar{\nu})a_1^3} & \bar{\eta} & -\bar{\eta}\bar{\nu}a_1^2 & \bar{\eta}\dfrac{(1+\bar{\nu})}{1-2\bar{\nu}} \\
0 & \bar{\eta}\dfrac{36}{a_1^5} & -\bar{\eta}\dfrac{2(5-\bar{\nu})}{(1-2\bar{\nu})a_1^3} & 3\bar{\eta} & -3\bar{\eta}\bar{\nu}a_1^2 & 0 \\
0 & \bar{\eta}\dfrac{24}{a_1^5} & -\bar{\eta}\dfrac{2(1+\bar{\nu})}{(1-2\bar{\nu})a_1^3} & -3\bar{\eta} & -\bar{\eta}(7+2\bar{\nu})a_1^2 & 0
\end{array}$$

In the case of the last interface, $r = a_n$ the variables F_{n+1} ; G_{n+1} ; H_{n+1} describe the "far field" conditions and assume the constant values, /1/:

$$F_{n+1} = \frac{1}{3} \frac{T}{4 \mu_{n+1}}$$

$$G_{n+1} = 0.$$

$$H_{n+1} = \frac{2}{3} \frac{1-2\nu_{n+1}}{1+\nu_{n+1}} \frac{T}{4 \mu_{n+1}},$$

where T is the applied uniaxial tension.

If the equations are considered in the order (2)(3)(5)(6)(1)(4), then the associated occurrence matrix readily shows that the full system may be partitioned into two subsystems.

	F1	G1	B1	C1	F2	G2	B2	C2	H1	A1	H2	A2
(2)	1	1	1	1	1	1	1	1				
(3)	1	1	1	1	1	1	1	1			0	
(5)	1	1	1	1	1	1	1	1				
(6)	1	1	1	1	1	1	1	1				
(1)	1	1	1	0	1	1	1	0	1	1	1	1
(4)	1	1	1	1	1	1	1	1	1	1	1	1

The first subsystem is called "system of the internal constants" and the second "system of the external constants".

Formal solution of the system of (4N+4) equations of the internal constants

With the occurrence matrix as a guide, and with reference to eqs.(2)(3)(5)(6), the matrix of coefficients to the left of the equality sign are indicated by $[L1]$, those to the right by $[R1]$ and the variables by the column vectors $\underline{x}_i$:

$$\underline{x}_i = [F_i \; G_i \mid B_i \; C_i] = [\underline{U}_i \mid \underline{V}_i]$$

If this is done for all interfaces $r = a_i$, the full system of the internal constants may be brought together into the form:

$[L_o] \underline{x}_o = [R_1] \underline{x}_1$	1st interface	
$[L_1] \underline{x}_1 = [R_2] \underline{x}_2$	2nd interface	
$[L_{n-1}] \underline{x}_{n-1} = [R_n] \underline{x}_n$	nth interface	
$[L_n] \underline{x}_n = [R_{n+1}] \underline{x}_{n+1}$	(N+1)th interface	./.

Therefore, the system has a block-diagonal form, which can be better seen from the full occurrence matrix. Note that the system is rectangular and has dimension (4N+4) x (4N+6).
If the square matrices $[R_i]$ admit the inverse matrix $[R_i]^{-1}$, then the system assumes the form:

$$
\begin{aligned}
\underline{\mathcal{X}}_1 &= [R_1^{-1} L_0] \, \underline{\mathcal{X}}_0 \equiv \begin{bmatrix} A_1 \\ C_1 \end{bmatrix} \underline{U}_0 \\
\underline{\mathcal{X}}_2 &= [R_2^{-1} L_1] \, \underline{\mathcal{X}}_1 \equiv \left[\begin{array}{c|c} A_2 & B_2 \\ \hline C_2 & D_2 \end{array}\right] \begin{bmatrix} \underline{U}_1 \\ \underline{V}_1 \end{bmatrix} \\
&\vdots \\
\underline{\mathcal{X}}_{n+1} &= [R_{n+1}^{-1} L_n] \, \underline{\mathcal{X}}_n \equiv \left[\begin{array}{c|c} A_{n+1} & B_{n+1} \\ \hline C_{n+1} & D_{n+1} \end{array}\right] \begin{bmatrix} \underline{U}_n \\ \underline{V}_n \end{bmatrix}
\end{aligned}
$$

Solving the equations, by using the partitioned form of matrices and vectors one finally obtains:

$$
\begin{cases}
\underline{U}_1 = [A_1] \, \underline{U}_0 \equiv [A_1^*] \, \underline{U}_0 \\
\underline{V}_1 = [C_1] \, \underline{U}_0 \equiv [C_1^*] \, \underline{U}_0
\end{cases}
$$

$$
\begin{cases}
\underline{U}_2 = [A_2] \, \underline{U}_1 + [B_2] \, \underline{V}_1 = [A_2 A_1^* + B_2 C_1^*] \, \underline{U}_0 \equiv [A_2^*] \, \underline{U}_0 \\
\underline{V}_2 = [C_2] \, \underline{U}_1 + [D_2] \, \underline{V}_1 = [C_2 A_1^* + D_2 C_1^*] \, \underline{U}_0 \equiv [C_2^*] \, \underline{U}_0
\end{cases}
$$

$$\vdots$$

$$
\begin{cases}
\underline{U}_{n+1} = [A_{n+1}] \underline{U}_n + [B_{n+1}] \, \underline{V}_n = [A_{n+1} A_n^* + B_{n+1} C_n^*] \, \underline{U}_0 \equiv [A_{n+1}^*] \, \underline{U}_0 \\
\underline{V}_{n+1} = [C_{n+1}] \underline{U}_n + [D_{n+1}] \, \underline{V}_n = [C_{n+1} A_n^* + D_{n+1} C_n^*] \, \underline{U}_0 \equiv [C_{n+1}^*] \, \underline{U}_0
\end{cases}
$$

Now observe that the column vector $\underline{U}_{n+1}$ describes the "far field" conditions and assumes the constant value

$$U_{n+1} = [F_{n+1}; G_{n+1}]$$

Therefore the resulting formal solution of the problem is:

$$\underline{U}_0 = [A_{n+1}^*]^{-1} \, \underline{U}_{n+1}$$

In conclusion, in order to solve the system of the internal constants, the only two burdensome operations to perform are:

1) construction of the matrices $[R_{i+1}^{-1} L_i]$
2) construction of and inversion of the 2X2 matrix $[A_{n+1}^*]$

./.

In general, this second operation can be performed numerically, while the first can easily be performed analytically. In fact, the matrix $[R_{i+1}^{-1} L_i]$ can be obtained by summing (i-1) to the indexes "1" and "2" of the matrix $[R_2^{-1} L_1]$ associated to the interface $r = a_1$.

Formal solution of the system of (2N+2) equations of the external constants:

The equations (1)(4) are initially solved to obtain the variables $(H_2 A_2)$ as a function of the variables $(H_1 A_1)$. The following is obtained:

$$(1)\quad \frac{3(1-\nu_2)}{1-2\nu_2} H_2 = (2+\eta_1 \frac{1+\nu_1}{1-2\nu_1}) H_1 - 2(1-\eta_1)\left(\frac{a_o}{a_1}\right)^3 \left[\frac{A_1}{a_o^3} + \frac{1}{3}\frac{5-4\nu_1}{1-2\nu_1}\frac{C_1}{a_o^3}\right]$$

$$(4)\quad \left[\frac{A_2}{a_1^3} + \frac{1}{3}\frac{5-4\nu_2}{1-2\nu_2}\frac{C_2}{a_1^3}\right] = H_2 - H_1 + \left(\frac{a_o}{a_1}\right)^3 \left[\frac{A_1}{a_o^3} + \frac{1}{3}\frac{5-4\nu_1}{1-2\nu_1}\frac{C_1}{a_o^3}\right]$$

Addition of (i-1) to the indexes "0", "1" and "2" in the above equations leads to the general system for the interfaces $r = a_i$.
Therefore, if

$$\alpha_{i+1} = \left(2 + \eta_i \frac{1+\nu_i}{1-2\nu_i}\right) \frac{1-2\nu_{i+1}}{3(1-\nu_{i+1})}$$

$$\beta_{i+1} = 2(1-\eta_i) \frac{1-2\nu_{i+1}}{3(1-\nu_{i+1})}$$

$$K_{i+1} = \left[\frac{A_{i+1}}{a_i^3} + \frac{1}{3}\frac{5-4\nu_{i+1}}{1-2\nu_{i+1}}\frac{C_{i+1}}{a_i^3}\right] \quad ; \; i = 0,1,2, \ldots N$$

the full system of the external constants assumes the form:

$$\begin{cases} H_1 = \alpha_1 H_o \equiv \alpha_1^* H_o \\ K_1 = H_1 - H_o \equiv \gamma_1^* H_o \end{cases}$$

$$\begin{cases} H_2 = \alpha_2 H_1 - \beta_2 \left(\frac{a_o}{a_1}\right)^3 K_1 = [\alpha_2 \alpha_1^* - \beta_2 \left(\frac{a_o}{a_1}\right)^3 \gamma_1^*] H_o \equiv \alpha_2^* H_o \\ K_2 = H_2 - H_1 + \left(\frac{a_o}{a_1}\right)^3 K_1 = [\alpha_2^* - \alpha_1^* + \left(\frac{a_o}{a_1}\right)^3 \gamma_1^*] H_o \equiv \gamma_2^* H_o \end{cases}$$

.

.

.

.

.

./.

$$\begin{cases} H_{n+1} = \alpha_{n+1} H_n - \beta_{n+1} \left(\frac{a_{n-1}}{a_n} \right)^3 K_n = \left[\alpha_{n+1} \alpha_n^* - \beta_{n+1} \left(\frac{a_{n-1}}{a_n} \right)^3 \gamma_n^* \right] H_o = \alpha_{n+1}^* H_o \\ K_{n+1} = H_{n+1} - H_n + \left(\frac{a_{n-1}}{a_n} \right)^3 K_n = \left[\alpha_{n+1}^* - \alpha_n^* + \left(\frac{a_{n-1}}{a_n} \right)^3 \gamma_n^* \right] H_o = \gamma_{n+1}^* H_o \end{cases}$$

Once again, the variable H_{n+1} describes the "far field" condition, and has a given constant value. Thus, the formal solution of the problem is:

$$H_o = H_{n+1} / \alpha_{n+1}^*$$

2. SPHERICAL INCLUSION WITH CORE-SHELL STRUCTURE

The particular case of a single shell (fig.1) has been analysed in detail by T.Riccò et al.(1977) and previously by V.A.Matonis et al. (1969) and T.T.Wang et al. (1969).

Mixed analytical-numerical techniques were used in their work, the numerical part regarding the solution of the linear system generated by the boundary conditions.

It would be highly profitable, for the purpose of research, to have a completely analytical solution; and this can be obtained with the help of the previous two formal solutions and a minimal amount of effort.

In the case of a single shell, the boundary conditions generate a system of 12 equations in 12 unknowns, whose analytical expression is the following

Internal constants

$$F_1 = + \frac{d}{\Delta} ;$$

$$G_1 a_1^2 = - \frac{c}{\Delta} ;$$

$$G_o a_o^2 = \frac{35(1-\nu_1)}{4(7-10\nu_o) + \eta_o (7+5\nu_o)} G_1 a_o^2$$

$$\left[F_o + \frac{7}{5} G_o a_o^2 \right] = \frac{15(1-\nu_1)}{(7-5\nu_1) + \eta_o (8-10\nu_1)} \left[F_1 + \frac{7}{5} G_1 a_o^2 \right]$$

./.

$$\frac{C_1}{a_o^3} = + \frac{1}{2}\ \frac{1-2\nu_1}{1-\nu_1}\ (1-\eta_o) \left[F_o + \frac{7}{5}\ G_o a_o^2 \right] ;$$

$$\frac{B_1}{a_o^5} = - \frac{1}{15} \left[-(7-10\nu_o)\ G_o a_o^2 + (7-10\nu_1) G_1 a_o^2 - \frac{3}{1-2\nu_1}\ \frac{C_1}{a_o^3} \right] ;$$

$$\frac{C_2}{a_1^3} = \frac{1}{15}\ \frac{1-2\nu_2}{1-\nu_2}\ \frac{1}{1-2\nu_1}\ \frac{C_1}{a_1^3} \left[(8-10\nu_1) + \eta_1 (7-5\nu_1) \right] +$$

$$+ \frac{1}{2}\ \frac{1-2\nu_2}{1-\nu_2} (1-\eta_1) \left[F_1 + \frac{7}{5}\ G_1 a_1^2 \right] ;$$

$$\frac{B_2}{a_1^5} = - \frac{1}{15} \left[- \frac{3}{1-2\nu_2}\ \frac{C_2}{a_1^3} + \frac{3}{1-2\nu_1}\ \frac{C_1}{a_1^3} - 15\ \frac{B_1}{a_1^5} - (7-10\nu_1)\ G_1 a_1^2 \right] .$$

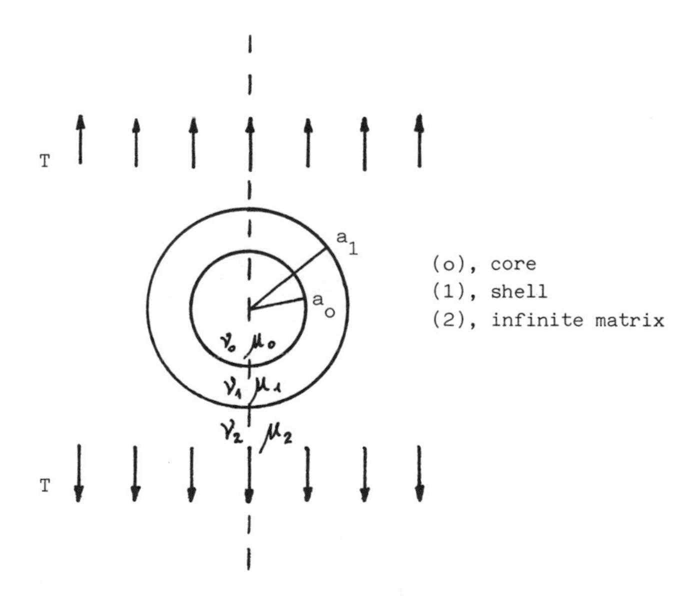

Fig. 1 Planar meridian section for the core-shell model

Where:

$$\Delta = \frac{12\mu_2}{T}\left(-(1-\rho^2)\, m_1 m_2 + \left(m_2 + \frac{m_1 m_3}{(1-\eta_1)(1-\rho^2)}\right)(m_4 - m_5)\right),$$

$$d = m_1 \rho^2 + m_4 - m_5 \; ,$$

$$c = \frac{5}{7}\, m_1,$$

and

$$m_1 = 126\, \frac{(1-\eta_1)(1-\rho^2)\,(1-\eta_0)\,\rho^3}{(7-5\,\nu_1) + \eta_0\,(8-10\,\nu_1)} \; ,$$

$$m_2 = \frac{(7-5\,\nu_2) + \eta_1\,(8-10\,\nu_2)}{15\,(1-\nu_2)},$$

$$m_3 = \frac{1}{126}\,(8-10\,\nu_1)\frac{2}{15}\;\frac{4-5\,\nu_2}{1-\nu_2}\;((8-10\,\nu_1) + \eta_1\,(7-5\,\nu_1)),$$

$$m_4 = 4\,(1-\eta_1)\,\rho^7\left((7-10\,\nu_1) - \frac{35(1-\nu_1)\,(7-10\,\nu_0)}{4(7-10\,\nu_0) + \eta_0\,(7+5\nu_0)}\right),$$

$$m_5 = 4\,(7-10\,\nu_1) + \eta_1\,(7+5\,\nu_1),$$

$$\rho = \frac{a_o}{a_1} \; .$$

External constants

$$H_o = \frac{T}{4\mu_2} \frac{\frac{3(1-\nu_1)}{1-2\nu_1} \quad \frac{2(1-\nu_2)}{1+\nu_2}}{\left(2+\eta_1 \frac{1+\nu_1}{1-2\nu_1}\right)\left(2+\eta_o \frac{1+\nu_o}{1-2\nu_o}\right) + 2\rho^3(1-\eta_1)\left(\frac{1+\nu_1}{1-2\nu_1} - \eta_o \frac{1+\nu_o}{1-2\nu_o}\right)}$$

$$H_1 = \frac{1-2\nu_1}{3(1-\nu_1)} \left(2 + \eta_o \frac{1+\nu_o}{1-2\nu_o}\right) H_o$$

$$\frac{A_1}{a_o^3} = H_1 - H_o - \frac{1}{3} \frac{5-4\nu_1}{1-2\nu_1} \frac{C_1}{a_o^3}$$

$$\frac{A_2}{a_1^3} = \frac{2}{3} \frac{T}{4\mu_2} \frac{1-2\nu_2}{1+\nu_2} - H_1 + \rho^3 (H_1 - H_o) - \frac{1}{3} \frac{5-4\nu_2}{1-2\nu_2} \frac{C_2}{a_1^3} .$$

In the limiting cases ρ=o and ρ=1, Goodier's solution results.

BIBLIOGRAPHY

/¯1¯7 - J.N. Goodier; "Concentration of stress around spherical and cylindrical inclusions and flaws"; J. Appl. Mech. 55, 39 (1933).

/¯2¯7 - V.A. Matonis, N.C. Small; "A macroscopic analysis of composites containing layered spherical inclusions", Polym. Eng. Sci. 9, 90 (1969).

/¯3¯7 - T. Riccò, A.Pavan, G.Danusso; "Analisi micromeccanica di un modello elementare di materiale composito con particelle eterogenee"; La Meccanica Italiana 108, 33 (1977).

/¯4¯7 - T.T. Wang, H.Schonhorn; "Tensile properties of polymer-filler composites", J. Appl. Phys. 40, 5131 (1969).

A FINITE ELEMENT FOR COMPOSITE SHELLS

K. Dorninger, Institute of Light Weight Structures
and Aerospace Engineering, Technical University of Vienna

Abstract

A plate/shell finite element with the capability for anisotropic layered material is introduced. The element contains the following features: geometric nonlinearity, arbitrary lamination scheme and anisotropic lamina properties including transvers shear.

Introduction

To execute a calculation efficiently it is useful to consider all requirements occuring during the analysis. In this paper the requirements in composite analysis are determined as follows:

- Most composite structures are composed of plates or shells of arbitrary shape.
- In many composite applications geometrically nonlinear behaviour occur:
  - large displacements but small strains (e.g. plate-springs)
  - buckling with a nonlinear pre- and postbuckling path (e.g. snap through of a shallow shell)
- If the lay-up is unsymmetric about the middle-plane coupling phenomena occure (e.g. bending-stretching coupling).

To fulfill all the listed requirements the following procedures and special assumptions are used:

An adequate tool to model arbitrary shaped constructions is the finite elemente method. With this a wide range of problems can be solved, e.g. connections between composites and other materials are possible.

To derive the stiffness expressions for a plate/shell element the degeneration principle [1,2] is applied , where starting from the 3-dimensional theory some special assumptions are invoked to meet a shell like behaviour. To include the geometric nonlinearities the element formulation is based on the principle of virtual work in updated Lagrangian description.

Consideration of an unsymmetric lay-up is achieved by formulating the constitutive relations for a fully anisotropic material and by carefully integrating over the shell thickness.

Theory

This chapter deals with the major steps of the derivation of the FE-procedure used. For detailed information see [1,2].

Derivation of the equations of equilibrium

The starting point is the description of a continuous body in a cartesian coordinate system at incremental load steps. In fig.1 three configurations are depicted. Configuration "Ø" denotes the unloaded structure, "m" denotes an already calculated state of equilibrium at load level "m" and "m+1" denotes the unknown configuration after the incremental load step "m" to "m+1".

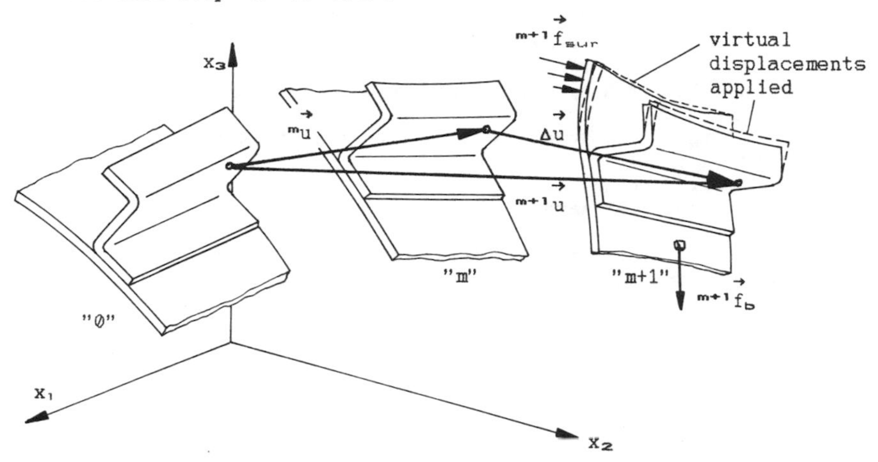

Fig.1: Schematic representation of the incremental equilibrium configurations

$^{m+1}\vec{f}_b$, $^{m+1}\vec{f}_{sur}$ ...body and surface force vector at config. "m+1"

$^{m+1}\vec{u}$, $^{m}\vec{u}$ .... total displacement vector at config. "m+1" and "m"

$\Delta\vec{u}$ .... vektor of displacement increments (unknown)

To derive the governing equations of equilibrium the principle of virtual work is applied in configuration "m+1":

$$\delta W = \delta W^{(i)} + \delta W^{(e)} = 0 \tag{1}$$

In updated Lagrangian description (U.L) these expressions follow to:

$$^{m+1}\delta W = \underbrace{-\int_{^mV} {}^{m+1}_{m}S_{ij}\,\delta\,{}^{m+1}_{m}\epsilon_{ij}\,d^mV}_{^{m+1}\delta W^{(i)}} + \underbrace{\int_{^mV} {}^{m+1}\vec{f}_b\,d^mV + \int_{^mSur} {}^{m+1}\vec{f}_{sur}\,d^mSur}_{^{m+1}\delta W^{(e)}} = 0 \tag{2}$$

Incremental decomposition of stresses and strains:

$$ {}^{m+1}_{m}S_{ij} = {}^{m}\tau_{ij} + \Delta S_{ij} \tag{3}$$

$$ {}^{m+1}_{m}\epsilon_{ij} = {}^{m}_{m}\epsilon_{ij} + \Delta\epsilon_{ij} \tag{4}$$

$$ \Delta\epsilon_{ij} = \tfrac{1}{2}\,(\underbrace{\Delta u_{i,j} + \Delta u_{j,i}}_{\text{linear part } = \Delta e_{ij}} + \underbrace{\Delta u_{k,j}\Delta u_{k,i}}_{\text{nonlinear part } = \Delta\eta_{ij}}) \tag{5}$$

${}^{m+1}_{m}S_{ij}$ ... 2nd Piola-Kirchhoff stress tensor at config. "m+1" referred referred to config. "m"
ΔS_{ij} ... incremental 2nd Piola-Kirchhoff stress tensor
${}^{m}\tau_{ij}$ ... Euler-Cauchy stress tensor at config. "m"
${}^{m+1}_{m}\epsilon_{ij}$ ... Green-Lagrange strain tensor at config. "m+1" referred to config. "m"
$\Delta\epsilon_{ij}$ ... incremental Green-Lagrange strain tensor

Incremental constitutiv relation for linear elastic material:

$$ \Delta S_{ij} = C_{ijrs}\,\Delta\epsilon_{rs} \approx C_{ijrs}\,\Delta e_{ijrs} \tag{6}$$

C_{ijrs}...constant material tensor

With the above listed expressions the principle of virtual work leads to the linearized incremental equation of equilibrium:

$$ \int_{{}^{m}V} C_{ijrs}\Delta e_{rs}\delta\Delta e_{ij}d^{m}V + \int_{{}^{m}V} {}^{m}\tau_{ij}\delta\Delta\eta_{ij}d^{m}V = \delta W^{(a)} - \int_{{}^{m}V} {}^{m}\tau_{ij}\delta\Delta e_{ij}d^{m}V \tag{7}$$

The next steps are the discretization and algebraization (in the sense of FE-procedures) of this equation. This leads to the typically incremental finite element matrix expression:

$$ ({}^{m}\underset{\approx}{\mathbf{K}}_{e} + {}^{m}\underset{\approx}{\mathbf{K}}_{g})\,\Delta\underset{\sim}{U} = {}^{m+1}\underset{\sim}{R} - {}^{m}\underset{\sim}{F} \tag{8}$$

table 1: correspondence table

virtual work expression	finite element expression	
$\int_{{}^{m}V} C_{ijrs}\Delta e_{rs}\delta\Delta e_{ij}d^{m}V$	$\left(\int_{{}^{m}V} {}^{m}\underset{\approx}{B}^{T}\,\underset{\approx}{C}\,{}^{m}\underset{\approx}{B}\,d^{m}V\right)\Delta\underset{\sim}{U}$	${}^{m}\underset{\approx}{\mathbf{K}}_{e}\,\Delta\underset{\sim}{U}$
$\int_{{}^{m}V} {}^{m}\tau_{ij}\delta\Delta\eta_{ij}d^{m}V$	$\left(\int_{{}^{m}V} {}^{m}\underset{\approx}{\bar{B}}^{T}\,{}^{m}\underset{\approx}{\bar{\tau}}\,{}^{m}\underset{\approx}{\bar{B}}\,d^{m}V\right)\Delta\underset{\sim}{U}$	${}^{m}\underset{\approx}{\mathbf{K}}_{g}\,\Delta\underset{\sim}{U}$
$\int_{{}^{m}V} {}^{m}\tau_{ij}\delta\Delta e_{ij}d^{m}V$	$\int_{{}^{m}V} {}^{m}\underset{\approx}{B}^{T}\,{}^{m}\underset{\sim}{\tau}\,d^{m}V$	${}^{m}\underset{\sim}{F}$

$^{m}\underset{\approx}{K}_{L}, {^{m}\underset{\approx}{K}_{G}}$...incremental linear and geometric stiffness matrix at config. "m", respectively

$^{m}\underset{\sim}{F}$ ......vector of the internal nodal point forces equivalent to the element stresses at config. "m"

$^{m+1}\underset{\sim}{R}$......vector of the external nodal point forces equivalent to the externally applied loads at config. "m+1"

$^{m}\underset{\approx}{\bar{B}}$........matrix relating the displacement-increments-derivatives with the node-displacement-increments

$^{m}\underset{\approx}{B}$........matrix relating the linear strain-increments with the node-displacement-increments

$^{m}\underset{\sim}{\tau}, {^{m}\underset{\approx}{\bar{\tau}}}$.....Euler-Cauchy stresses arranged as vector and as matrix, respectively

$\Delta\underset{\sim}{U}$........vector of nodal point displacement-increments

$\underset{\approx}{C}$........material matrix

Equ. (8) leads to a first approximation of the incremental displacement vector $\Delta\underset{\sim}{U}$ which can be improved by applying (8) iteratively (pure Newton-Raphson iteration, modified Newton-Raphson iteration, Riks-Wempner constant arc length iteration....).

The matrices $^{m}\underset{\approx}{B}$ and $^{m}\underset{\approx}{\bar{B}}$ depend on the element type. In the following chapter the derivation of the stiffness expressions for the degenerated 3-d plate/shell element is described in correspondence with [2].

Degenerated 3-d plate/shell element

Starting from the 3-d theory two assumptions are introduced:

I. The "normal" of the shell remains straight and inextensional during deformation.

II. The strain energy of the stresses perpendicular to the middle surface is ignored by using a modified (i.e. plain stress) material law.

To achieve the first assumption the geometry and the displacement-increments are decomposed into a part which describes the middle surface of the shell and a part which describes the variation along the shell-"normal", in isoparametric formulation :

$$^{m}x_i(r,s,t) = \underbrace{\sum_{k=1}^{M} \phi^k(r,s)\,{^{m}x_i^k}}_{\text{shell middle part}} + \underbrace{t\cdot\tfrac{1}{2}\sum_{k=1}^{M} h^k\phi^k(r,s)\cos{^{m}\Phi_i^k}}_{\text{shell normal part}} \qquad i=1,2,3 \tag{9}$$

$$\Delta u_i(r,s,t) = {^{m+1}x_i} - {^{m}x_i} \tag{10}$$

r,s,t ....natural (shell-fixed) coordinate system

$^{m}x_i$ ....cartesian coordinate of an arbitrary point of the shell

$^{m}x_i^k$ ....cartesian coordinate of the nodal point k

ϕ^k ....lagrangian shape functions of the middle surface for node k

h^k ....shell thickness at node k

$\cos{^{m}\Phi_i^k}$ ...direction cosines of the shell-"normal" at node k with respect to the global coordinate axis.

The angles ${}^m\bar{\Phi}^k_1$, ${}^m\bar{\Phi}^k_2$, ${}^m\bar{\Phi}^k_3$ are not independent, we determine:

$$\begin{aligned} \cos{}^m\bar{\Phi}^k_1 &= \cos{}^m\bar{\Phi}^k \\ \cos{}^m\bar{\Phi}^k_2 &= \cos{}^m\Omega^k \sin{}^m\bar{\Phi}^k \\ \cos{}^m\bar{\Phi}^k_3 &= \sin{}^m\Omega^k \sin{}^m\bar{\Phi}^k \end{aligned} \tag{11}$$

With the incremental decomposition

$$ {}^{m+1}\bar{\Phi}^k = {}^m\bar{\Phi}^k + \beta^k, \quad {}^{m+1}\Omega^k = {}^m\Omega^k + \alpha^k \tag{12}$$

and the linearization of the direction cosines in α^k and β^k, the displacement-increments follow to

$$\Delta u_i(r,s,t) = \sum_{k=1}^{M} \phi^k(r,s)\Delta U^k_i + t\tfrac{1}{2}\sum_{k=1}^{M} h^k \phi^k(r,s)({}^mF^k_i\alpha^k + {}^mG^k_i\beta^k) \quad i=1,2,3 \tag{13}$$

$$\begin{aligned} {}^mF^k_1 &= 0 & {}^mG^k_1 &= -\sin{}^m\bar{\Phi}^k \\ {}^mF^k_2 &= -\sin{}^m\Omega^k \sin{}^m\bar{\Phi}^k & {}^mG^k_2 &= \cos{}^m\Omega^k \cos{}^m\bar{\Phi}^k \\ {}^mF^k_3 &= \cos{}^m\Omega^k \sin{}^m\bar{\Phi}^k & {}^mG^k_3 &= \sin{}^m\Omega^k \cos{}^m\bar{\Phi}^k \end{aligned} \tag{14}$$

To calculate the displacement-functions as described above 5 DOF's per node are required: the three displacement-increments of the middle surface ΔU^k_i and additionally the two angle-increments α^k and β^k to fix the the position of the "normal". Due to this description of the "normal" it does not have to be exactly normal to the middle surface and thereby shear-deformations are included.

With the displacement-functions the ${}^m\mathbf{B}$ and ${}^m\bar{\mathbf{B}}$ matrices (see table 1) can be derived. The individual block of these matrices for one nodal point k is given in the next table:

table 2:

${}^m\underset{\approx}{B}$	...	node k					...
$\underset{\sim}{\Delta e}$	...	ΔU^k_1	ΔU^k_2	ΔU^k_3	α^k	β^k	...
Δe_{11}		ϕ^k_1	0	0	$\bar{\phi}^k_1 F^k_1$	$\bar{\phi}^k_1 G^k_1$	
Δe_{22}		0	ϕ^k_2	0	$\bar{\phi}^k_2 F^k_2$	$\bar{\phi}^k_2 G^k_2$	
Δe_{33}		0	0	ϕ^k_3	$\bar{\phi}^k_3 F^k_3$	$\bar{\phi}^k_3 G^k_3$	
$2\Delta e_{12}$		ϕ^k_2	ϕ^k_1	0	$\bar{\phi}^k_2 F^k_1 + \bar{\phi}^k_1 F^k_2$	$\bar{\phi}^k_2 G^k_1 + \bar{\phi}^k_1 G^k_2$	
$2\Delta e_{13}$		ϕ^k_3	0	ϕ^k_1	$\bar{\phi}^k_3 F^k_1 + \bar{\phi}^k_1 F^k_3$	$\bar{\phi}^k_3 G^k_1 + \bar{\phi}^k_1 G^k_3$	
$2\Delta e_{23}$		0	ϕ^k_3	ϕ^k_2	$\bar{\phi}^k_3 F^k_2 + \bar{\phi}^k_2 F^k_3$	$\bar{\phi}^k_3 G^k_2 + \bar{\phi}^k_2 G^k_3$	...

$$\begin{aligned} \phi^k_j &= J^{-1}_{j1}(\delta\phi^k/\delta r) + J^{-1}_{j2}(\delta\phi^k/\delta s) \\ \bar{\phi}^k_j &= \tfrac{1}{2}h^k(J^{-1}_{j3}\phi^k + t\cdot\phi^k_j) \end{aligned} \tag{15}$$

J^{-1}_{jk} ...Elements of the invers Jacobian matrix ${}^m\underset{\approx}{J}^{-1}(r,s,t)$

${}^{m}\bar{\underset{\approx}{B}}$	...	node k					...
$\underset{\sim}{\Delta d}$	...	ΔU_1^k	ΔU_2^k	ΔU_3^k	α^k	β^k	...
$\Delta u_{1,1}$		ϕ_1^k	0	0	$\bar{\phi}_1^k\ F_1^k$	$\bar{\phi}_1^k\ G_1^k$	
$\Delta u_{1,2}$		ϕ_2^k	0	0	$\bar{\phi}_2^k\ F_1^k$	$\bar{\phi}_2^k\ G_1^k$	
$\Delta u_{1,3}$		ϕ_3^k	0	0	$\bar{\phi}_3^k\ F_1^k$	$\bar{\phi}_3^k\ G_1^k$	
$\Delta u_{2,1}$		0	ϕ_1^k	0	$\bar{\phi}_1^k\ F_2^k$	$\bar{\phi}_1^k\ G_2^k$	
$\Delta u_{2,2}$		0	ϕ_2^k	0	$\bar{\phi}_2^k\ F_2^k$	$\bar{\phi}_2^k\ G_2^k$	
$\Delta u_{2,3}$		0	ϕ_3^k	0	$\bar{\phi}_3^k\ F_2^k$	$\bar{\phi}_3^k\ G_2^k$	
$\Delta u_{3,1}$		0	0	ϕ_1^k	$\bar{\phi}_1^k\ F_3^k$	$\bar{\phi}_1^k\ G_3^k$	
$\Delta u_{3,2}$		0	0	ϕ_2^k	$\bar{\phi}_2^k\ F_3^k$	$\bar{\phi}_2^k\ G_3^k$	
$\Delta u_{3,3}$		0	0	ϕ_3^k	$\bar{\phi}_3^k\ F_3^k$	$\bar{\phi}_3^k\ G_3^k$	

The modified material law (assumption II) for isotropic linear elastic material defined in a coordinate system which is fixed to the shell-middle-surface has the following form:

$$\begin{bmatrix} S_{11} \\ S_{22} \\ S_{33} \\ S_{12} \\ S_{13} \\ S_{23} \end{bmatrix} = \underbrace{\begin{bmatrix} C_{11} & C_{12} & 0 & 0 & 0 & 0 \\ & C_{22} & 0 & 0 & 0 & 0 \\ & & 0 & 0 & 0 & 0 \\ & & & C_{44} & 0 & 0 \\ & \text{sym.} & & & C_{55} & 0 \\ & & & & & C_{66} \end{bmatrix}}_{\bar{\underset{\approx}{C}}} \cdot \begin{bmatrix} \epsilon_{12} \\ \epsilon_{22} \\ \epsilon_{33} \\ \epsilon_{12} \\ \epsilon_{13} \\ \epsilon_{23} \end{bmatrix} \qquad \begin{aligned} C_{11} &= E/(1-\mu^2) \\ C_{22} &= C_{11} \\ C_{12} &= \mu C_{11} \\ C_{44} &= G \\ C_{55} &= G\cdot k \\ C_{66} &= G\cdot k \end{aligned} \tag{16}$$

E ... Young's modulus
μ ... Poisson number
G ... Shear modulus
k ... Shear-correction-factor

In each configuration "m" the matrix $\bar{\underset{\approx}{C}}$ is transformed to the global system to get the appropriate $\underset{\approx}{C}$ matrix for the calculation of the stiffness expressions shown in table 1.

Laminated Fibre Composite

In this section the necessary procedures are described to include the layered material correctly in the derivation of the stiffness expressions.

Fig. 2 shows the nomenclature for the laminated fibre composite element (LFC) and the definition of a local fibre-fixed coordinate system.

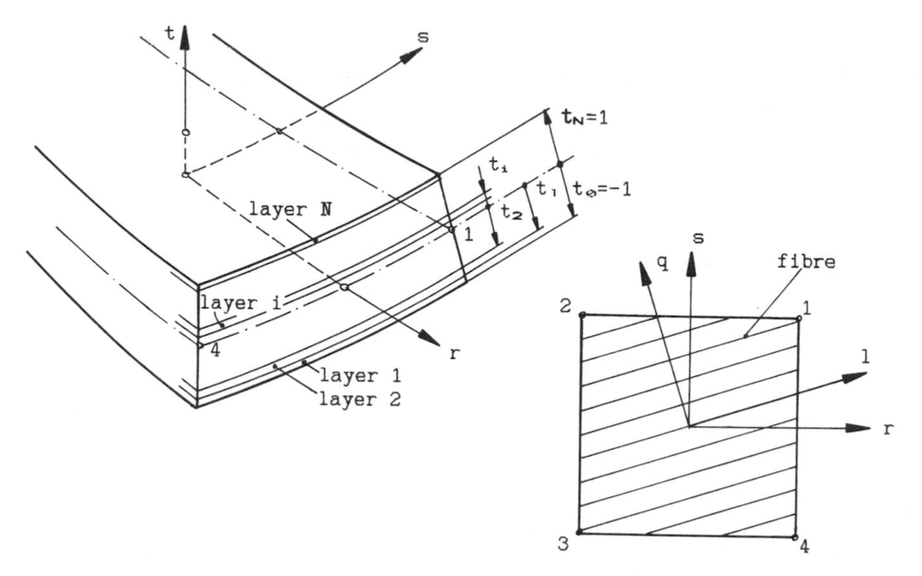

Fig.2 Nomenclature for the LFC-Element

Each layer is assumed to show orthotropic material behaviour with respect to its individual local fibre-fixed coordinate system.

For the layer i the material matrix ${}^L\bar{\underset{\approx}{C}}_i$ has the following form referred to the local coordinate system:

$$
{}^L\bar{\underset{\approx}{C}}_i = \begin{bmatrix} C_{11} & C_{12} & 0 & 0 & 0 & 0 \\ & C_{22} & 0 & 0 & 0 & 0 \\ & & 0 & 0 & 0 & 0 \\ & & & C_{44} & 0 & 0 \\ & \text{sym.} & & & C_{55} & 0 \\ & & & & & C_{66} \end{bmatrix}
\qquad
\begin{aligned}
C_{11} &= E_l/(1-\mu_{ql}^2 E_q/E_l) \\
C_{22} &= E_q/(1-\mu_{ql}^2 E_q/E_l) \\
C_{12} &= \mu_{ql} C_{22} \\
C_{44} &= G_{lq} \\
C_{55} &= G_{lt} \\
C_{66} &= G_{qt}
\end{aligned}
\qquad (17)
$$

subscript l ....longitudinal or fibre direction
subscript q ....transverse direction
subscript t ....thickness direction
subscript i ....layer i
superscript L....local coordinate system

The elements of the matrix ${}^L\bar{\underset{\approx}{C}}_i$ are derived from the engineering elastic constants of the layer as shown above or can be calculated from the material-constants of fibre and matrix [3].

Because of the orthotropy all ${}^L\bar{\underset{\approx}{C}}_i$-matrices depend on the coordinate system. Because of the layered set-up the global $\underset{\approx}{C}$-matrix depends on the thickness coordinate **t**. These two facts have an important effect on the derivation of the stiffness expressions.

In the LFC-element introduced a quasi analytical thickness integration was chosen to consider all the effects mentioned previously.

One additional assumption was necessary to perform the explicite thickness integration:

The Jacobian matrix ${}^{m}\underset{\approx}{J}$ is independent of the thickness coordinate t.

This is valid for shells with a small thickness to curvature ratio.

With this assumption one can separate the ${}^{m}\underset{\approx}{B}$ and ${}^{m}\underset{\approx}{\bar{B}}$ matrices:

$$ {}^{m}\underset{\approx}{B} = {}^{m}\underset{\approx}{B1}(r,s) + t\ {}^{m}\underset{\approx}{B2}(r,s) \qquad {}^{m}\underset{\approx}{\bar{B}} = {}^{m}\underset{\approx}{\bar{B}1}(r,s) + t\ {}^{m}\underset{\approx}{\bar{B}2}(r,s) \tag{18} $$

$$ \underset{\approx}{C}(t) = \begin{cases} \underset{\approx}{\bar{C}_1} & -1 \le t < t_1 \\ \underset{\approx}{\bar{C}_2} & t_1 \le t < t_2 \\ \dots & \\ \underset{\approx}{\bar{C}_N} & t_{N-1} \le t < t_N \end{cases} \qquad (\underset{\approx}{\bar{C}_i} = {}^{L}\underset{\approx}{\bar{C}_i} \text{ rotated into the global coordinate system}) \tag{19} $$

With the $\underset{\approx}{C}(t)$ matrix the stiffness expressions can be rewritten:

$$ {}^{m}\underset{\approx}{K_e} = \int_{-1}^{+1}\int_{-1}^{+1}\int_{-1}^{+1} ({}^{m}\underset{\approx}{B1} + t\ {}^{m}\underset{\approx}{B2})^T\ \underset{\approx}{C}(t)\ ({}^{m}\underset{\approx}{B1} + t\ {}^{m}\underset{\approx}{B2}) \det {}^{m}\underset{\approx}{J}(r,s)\, dr\, ds\, dt \tag{20} $$

$$ {}^{m}\underset{\approx}{K_g} = \int_{-1}^{+1}\int_{-1}^{+1}\int_{-1}^{+1} ({}^{m}\underset{\approx}{\bar{B}1} + t\ {}^{m}\underset{\approx}{\bar{B}2})^T\ {}^{m}\underset{\approx}{\bar{\tau}}(r,s,t) ({}^{m}\underset{\approx}{\bar{B}1} + t\ {}^{m}\underset{\approx}{\bar{B}2}) \det {}^{m}\underset{\approx}{J}(r,s)\, dr\, ds\, dt \tag{21} $$

$$ {}^{m}\underset{\approx}{F} = \int_{-1}^{+1}\int_{-1}^{+1}\int_{-1}^{+1} ({}^{m}\underset{\approx}{B1} + t\ {}^{m}\underset{\approx}{B2})^T\ {}^{m}\underset{\sim}{\tau}(r,s,t) \det {}^{m}\underset{\approx}{J}(r,s)\, dr\, ds\, dt \tag{22} $$

Performing the matrix multiplications one can separate t-dependent terms of the following form:

$$ \int_{-1}^{+1} \underset{\approx}{C}(t)\ t^p\ dt \qquad p=0,1,2,3,4 \tag{23} $$

This expressions can be integrated quasi analytically this leads to:

$$ \sum_{i=1}^{N} \underset{\approx}{C_i}(t_i^p - t_{i-1}^p)/p \qquad p=1,2,3,4,5 \tag{24} $$

The integration in the r,s plane is performed numerically using Gauss-quadrature and the described procedure was implemented in the nonlinear-FE program system NISA [4].

Numerical examples

To show the usefulness of the LFC-element two characteristic sample problems were computed and the results are compared with other once:

1) Cross-ply laminated square plate with simply supported edges under uniformly distributed load, linear computation

reference-solution: analytically obtained by NOOR and MATHERS [5]

lay-up: [0/90/0/90/0/90/0/90/0]
layerthickness = 2.0

material: $E_1=30.0\cdot10^6$,
$E_q=0.75\cdot10^6$,
$\mu_{q1}=0.25$,
$G_{1q}=G_{1t}=0.25\cdot10^6$,
$G_{qt}=0.375\cdot10^6$

model: one quarter of the plate was modelled by 2x2 16-node elements

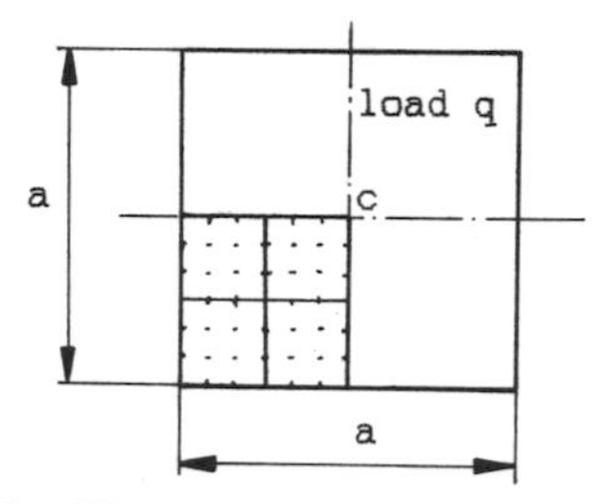

Fig.3: FE model of a square-plate

table 3:

a	h/a	load	centre deflection w_c (exact)	w_c	error [%]
180.	0.1	7.124577	0.01	0.010159	1.59
18000.	0.001	$9.31765\cdot10^{-5}$	10.0	9.9968	-0.032

The table shows the very good agreement of the centre deflection w_c for both moderately thick and thin plates.

2) Hinged cylindrical shell subjected to a central point load

reference solution: FE calculation by S.SAIGAL et al.[6]

lay-up: a) [90/0/90] (symm.)
layerthickness = 4.2
b) [+45/-45] (unsymm.)
layerthickness = 6.3

material: $E_1=3.3$,
$E_q=1.1$,
$\mu_{q1}=0.25$,
$G_{1q}=G_{1t}=0.66$,
$G_{qt}=0.55$

model: one quarter of the shell was modelled by 2x2 16-node elements

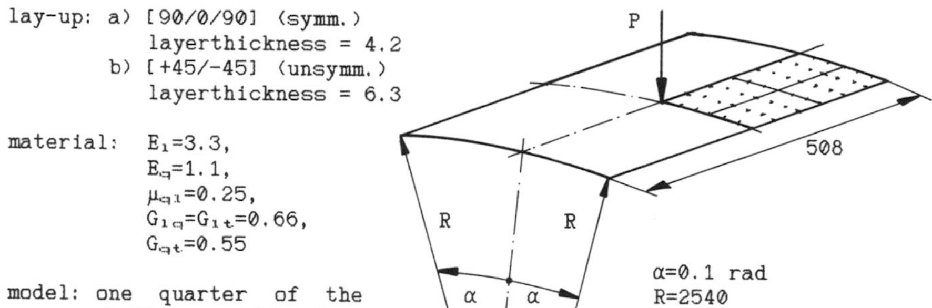

Fig.4: FE model of a cylindrical shell

In the theory used by SAIGAL et al. no transverse shear deformations are included and, hence, no shear-modulus G_{qt} was available . G_{qt} was set to $G_{1q}/1.2$.

In order to compare the results fig.5 shows the load-displacement path of the centre point a) for the cross-ply laminate and b) for the angle-ply laminate.

To calculate the unstable path of fig.5 a special equilibrium iteration method was used (Riks-Wempner constant arc-length method).

Again, reasonably good agreement is obtained. The somewhat softer response of the LFC-element is due to the shear deformations which, in contrast to the elements used by SAIGAL et al. are included in the present study.

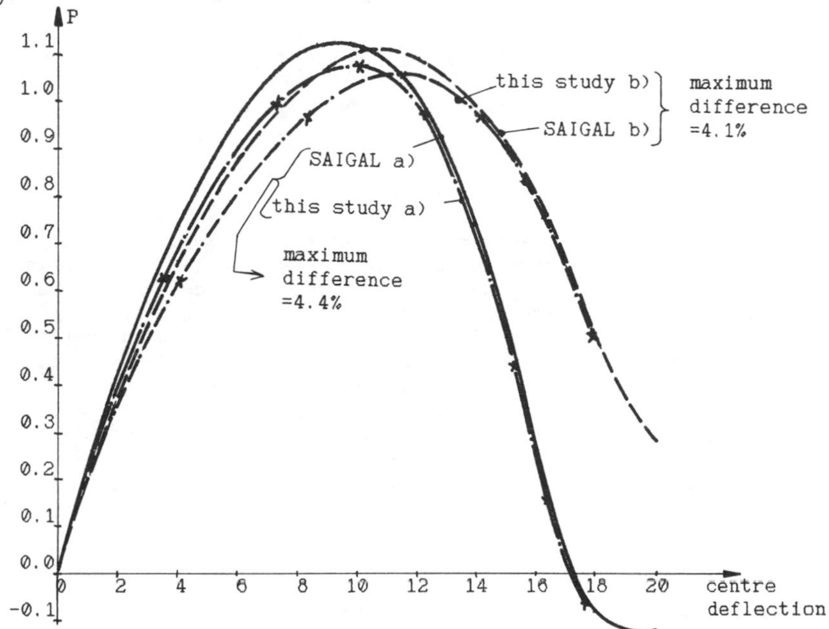

Fig.5: Load -displacement path of the centre point of a hinged cylindrical shell subjected to a central point load

References:

[1] Bathe K.J.: Finite Element Procedures in Engineering Analysis Prentice-Hall, Inc.,1982

[2] Ramm E.: A Plate/Shell Element for Large Deflection and Rotation Analysis Proc. U.S.-German Symp. MIT Cambridge,1976

[3] Anderson R.S.: The Mechanical Properties of Fibre Reinforced Composite Plates Theses at University of Aston (1975)

[4] Ramm E.,Stegmüller H.,Brendel B.,Häfner L.,Sättele J.M.: Programmsystem NISA University of Stuttgart, Institut für Baustatik

[5] Noor A.K.,Mathers M.D.: Shear-flexible Finite Element Models of Laminated Composite Plates and Shells NASA TN D-8044 (1975)

[6] Saigal S.,Kapania P.K.,Yang T.Y.: Geometrically Nonlinear Finite Element Analysis of Imperfect Laminated Shells Journal of Composite Materials, Vol.20 (1986),pp 197-214

OPTIMIZATION OF LAMINATED FIBRE COMPOSITE STRUCTURES BY FINITE ELEMENTS

Stevan Maksimović
Aeronautical institute, Niška b.b.
11133 Žarkovo-Belgrade, Yugoslavia

ABSTRACT

An efficient method is presented for the structural weight optimation of laminated fibre composite structures subject to stiffness, strength and elastic stability constraints. Approximate concepts and dual methods are combined to create an efficient procedure for minimum weight sizing of structural systems. The isoparametric finite elements are used to compute internal forces. The derivatives of finite element matrices with respect to the set of design variables are derived analytically. The conjugate gradient type maximizer is adopted to solve the approximate problem for the current set of dual variables. Using this procedure the number of actual finite element analysis is kept to a minimum. Examples are included in this paper to demonstrate the usefulness of this method.

INTRODUCTION

To desire to employ high performance light weight fiber-composite materials eficiently has stimulated the development of optimization procedures for the design laminates that will make it possible to fully exploit the design potential offered by tailoring of these new materials. Not only do composite materials provide an increase in the number of design variables, they can also cause an increase in the complexity of failure modes. For that reason, the automated structural sizing procedure must incoporate accurate structural analysis methods. Optimal design of large-scale systems is usually based on combining finite element analysis and optimization techniques into an integrated computer system. The use finite element methods in parallel with optimization techniques such as nonlinear mathematical programming or optimal criterion make it posible to attack large and complex problem. In order to efficiently and reliable solve the optimization problem, several important techniques are used here. These include design variable linking, constraint deletion, the use of reciprocal variables, and formal approximation techniques. The computational efficiency of the optimization procedure was improved by utilizing analytical sensitivity derivatives. During the early 1980s approximation concepts [1] were combined with dual method formulation to create a powerful new method for minimum weight design of structural systems [2] . The efficiency of the method is due to the fact that the dimensionality of the dual space, where most of the optimization effort is expended, is relatively low for many structural optimization problems of practical interest. Using this procedure and efficient algorithm in process optimization, the number of actual finite element analysis is kept to a minimum.

PROBLEM FORMULATION

The structural optimization problem considered in this paper consists of the weight minimization of a 3-dimensional composite structures modelled with bar, membrane and shear-panel finite elements. The design variables are the cross-sectional areas, ply thicknesses and angles of orthotropy. For higher quality linear approximation of the behavioral constraints the entire problem can therefore be expressed in terms of the reciprocal of the design variables as follows:

Find a vector x such that

$$W(x) = \sum_{i=1}^{N} \frac{m_i}{x_i} \rightarrow \min \tag{1}$$

subject to linear behavioral constraints

$$G_j(x) \equiv \bar{g}_j - g_j(x) \geqslant 0, \qquad j \varepsilon Q_R \tag{2}$$

and side constraints

$$x_i^L \leqslant x_i \leqslant x_i^U \tag{3}$$

where

$$g_j(x) = \sum_{i=1}^{N} C_{ij} x_i, \qquad j \varepsilon Q_R \tag{4}$$

and Q_R denotes the set of retained constraints. Here N is the number of independent design variables, $\bar{g}_j$ denotes an upper bound to a response quantity $g_j(x)$, the m_i are positive constants corresponding to the weight of the set of finite elements in the i-th linking group when $x_i=1$, C_{ij} are the partial derivatives of j constraints with respect to x_i, x_i^L and x_i^U are the lower and upper limits of the independent reciprocal design variables x_i. In order to establish the optimality conditions for the problem defined by aqn. (1-3) we need the associated Lagrangian

$$L(x,\lambda) = \sum_{i=1}^{N} \frac{m_i}{x_i} - \sum_{j \varepsilon Q_R} \lambda_j \left(1 - \sum_{i=1}^{N} C_{ij} x_i\right) \tag{5}$$

where λ_j are the Lagrangian multipliers. The Langrangian multipliers must be postive, hence:

$$\lambda \varepsilon \Lambda = \{ \lambda = (\lambda_1, \lambda_2, \ldots, \lambda_N)^T \mid \lambda_j \geqslant 0 \ , \ j\varepsilon Q_R \} \ . \tag{6}$$

Equations (1)-(3) define a convex optimization problem and therefore the necessary and sufficient conditions for a design x_i $(i=1,2,\ldots,N)$ to be optimal solution of the primal problem are the Kuhn-Tcker conditions [3]

$$\lambda_j^* G_j(x^*) = 0 \ , \qquad j\varepsilon Q_R \tag{7}$$

$$- \frac{m_i}{x_i^{*2}} + \sum_{j \varepsilon Q_R} \lambda_j^* \frac{\partial G_j(x^*)}{\partial x_i} = 0 \qquad i=1,2,\ldots,N \tag{8}$$

$$G_j (x^*) \geqslant 0 \tag{9}$$

$$\lambda_j^* \geqslant 0 \tag{10}$$

The foregoing approximate primal problem (1)-(4) is a convex programming problem then therefore this conditions (7)-(10) are necessary and sufficient for the solution vector x^*, λ^* to represent global optimizing point.The special case in convex programming is that of a convex separable problem, for which duality results can be readily implemented into efficient algorithms.

DUAL FORMULATION

It has been shown [4] that the explicit approximation primal problem posed by eqn. (1)-(4) is well suited for dual method formulation. The simplicity of the dual problem is due to the fact that all the function involved in the problem are explicit and separable, for which duality results can be readily implemented into efficient algorithms. In particular, the Hessian matrix of separable functions is diagonal. Without going to the development of the dual formulation, the problem can be now stated as follows:

Find λ such that explicit dual function

$$l(\lambda) = \sum_{i=1}^{N} \frac{m_i}{x_i} + \sum_{j \varepsilon Q_R} \lambda_j \left(1 - \sum_{i=1}^{N} C_{ij} x_i\right) \to \max \tag{11}$$

subject to

$$\lambda_j \geqslant 0 ; \qquad j \varepsilon \; Q_R \tag{12}$$

Here

$$l(\lambda) = \min_{x \varepsilon X} L(x,\lambda) \tag{13}$$

with

$$X = \{x : x_i^L \leqslant x_i \leqslant x_i^U ; \quad i=1,\ldots,N\} \tag{14}$$

is defined as the dual function. Here, only the main constraints (2) have to be associated with Lagrangian multipliers, or dual variables λ_j while the side constraints (3) are treated separately due to simplicity of these functions. In formulation (11)-(12) the primal variable x_i are given in terms of the dual variables λ_j by

$$x_i(\lambda) = \begin{cases} x_i^L & \text{if} \quad (x_i^L)^2 \geqslant \bar{x}_i^2 \\ \bar{x}_i & \text{if} \quad x_i^L < \bar{x}_i < x_i^U \\ x_i^U & \text{if} \quad (x_i^U)^2 \geqslant \bar{x}_i^2 \end{cases} \tag{14a}$$

where

$$\bar{x}_i = \left(\frac{m_i}{\sum_{j \varepsilon Q_R} \lambda_j C_{ij}} \right)^{1/2} \tag{15}$$

The dual problem can be solved by using different algorithms. These algorithms require the first or second derivatives of the dual function.

$$\frac{\partial l}{\partial \lambda_j} = - G_j(\lambda) = \sum_{i=1}^{N} C_{ij} x_i - \bar{g}_j \tag{16}$$

and

$$\frac{\partial^2 l}{\partial \lambda_j \partial \lambda_k} (\lambda) = \sum_{i=1}^{N} C_{ij} \frac{\partial x_i}{\partial \lambda_k} . \tag{17}$$

The second derivative of $l(\lambda)$ can be computed from equations (14a) and (15)

$$F_{ik} = \frac{\partial^2 l}{\partial \lambda_j \partial \lambda_k} (\lambda) = - \frac{1}{2} \sum_{i \varepsilon I} \frac{C_{ij} C_{ik}}{m_i} x_i^3 \tag{18}$$

where

$$\frac{\partial x_i}{\partial \lambda_k} = \begin{cases} 0 & \text{if} \quad x_i^L = x_i \\ - \dfrac{x_i^3 \; C_{ik}}{2 m_i} & \text{if} \quad x_i^L < x_i < x_i^U \\ 0 & \text{if} \quad x_i = x_i^U \end{cases} \tag{19}$$

where the summation on the index i is over the set of free primal variables

$$I = \{ i \; | x_i^L < x_i (\lambda) < x_i^U \} \tag{20}$$

From eqn.(14a) and (15) follows that discontinuites of the second derivatives exists on hyperplanes in the dual space defined by

$$\sum_{j \varepsilon Q_R} \lambda_j C_{ij} = \frac{m_i}{(x_i^L)^2} , \qquad i = 1, \ldots, N \tag{21}$$

and

$$\sum_{j \varepsilon Q_R} \lambda_j C_{ij} = \frac{m_i}{(x_i^U)^2} , \qquad i = 1, \ldots, N \tag{22}$$

OPTIMIZATION ALGORITHM

The numerical problem that must be solved in each stage is to find λ^* such that $l(\lambda) \rightarrow \max$ subject to the simple nonnegativity constraints $\lambda_j \geqslant 0$, $j \varepsilon Q_R$. For that purpose the conjugate gradient-type maximizer is used. This algorithm involves iterative modification of the dual variable vector as follows:

$$\lambda^{(k+1)} = \lambda^{(k)} + \tau \; s^{(k)} \tag{23}$$

where τ is the step length along the search direction S . This step is determined that the dual function attains its maximum along the direction S in the current dual subspace. For determination the search direction S here is used Hestenes -Stiefel formula [5] in next form

$$S^{(k+1)} = - G^{(k+1)} + \beta^{(k)} \; S^{(k)} \tag{24}$$

where
$$\beta^{(k)} = \frac{G^{(k+1)T} \, (G^{(k+1)} - G^{(k)})}{S^{(k)T} (G^{(k+1)} - G^{(k)})} \tag{25}$$

For determination τ Newthon-Raphson method is used

$$\tau^{(k)} = \tau^{(k-1)} - \frac{l'^{(k-1)}}{l''^{(k-1)}} \tag{26}$$

where l' and l'' are the first and second derivatives of dual function with respect τ

$$l' = \frac{dl}{d\tau} = \sum_{i=1}^{N} x_i \sum_{j\varepsilon Q_R} C_{ij} \, S_j - \sum_{j\varepsilon Q_R} S_j \tag{27}$$

and
$$l'' = \frac{d^2 l}{d\tau} = - \frac{1}{2} \sum_{i\varepsilon I} \frac{x_i^3}{m_i} \left[\sum_{j\varepsilon Q_R} C_{ij} \, S_j \right]^2 \tag{28}$$

where S_j are the coefficients of vector S. The critical values τ_i and $\bar{\tau}_i$ define explicitly the intersection points of the search direction with the discontinuity planes (21) and (22). These are given by

$$\underline{\tau}_i = \frac{\dfrac{m_i}{\underline{x}_i^2} - \sum_{j\varepsilon Q_R} C_{ij} \, \lambda_j}{\sum_{j\varepsilon Q_R} C_{ij} \quad S_j} \tag{29}$$

and

$$\bar{\tau}_i = \frac{\dfrac{m_i}{\bar{x}_i^2} - \sum_{j\varepsilon Q_R} C_{ij} \, \lambda_j}{\sum_{j\varepsilon Q_R} C_{ij} \quad S_j} \tag{30}$$

The procedure for computing the optimal step consists first of computing the critical values (29) and (30), then ordering them in an increasing sequence, and finally defining the internal containing the optimal step.In this internal, the function l'' is continuous and the Newthon-Raphson method (26) can be initiated without risk of divergence.

APPROXIMATE CONSTRAINTS

This study includes displacements, strength and elastic stability requirements as constraints G(x). In order to reduce the number of detailed finite element structural analysis needed to obtain optimum design, it is appropriate to construct explicit approximations for the constraints retain during the p-th stage of the optimization proces. For that purpose the first order Taylor serie is used in form

$$G_j(x) \equiv G_j(x^{(p)}) + (x-x^{(p)})^T \quad \nabla G_j \ (x^{(p)}) \qquad (31)$$

The linearized approximation of critical or potential critical constraints for membrane and shear-panel orthotropic elements based on eqn. (31) are given in follows.

1.- *Linearized approximations of MAXIMUM STRAIN CRITERION*

$$G_{s(x)} \equiv 1-Q_{ijk}(x^{(p)}) - \sum_{i=1}^{I} (x_i-x_i^{(p)}) \ \frac{\partial Q_{ijk}}{\partial x_i}(x^{(p)}) \geqslant 0; \ s\varepsilon B_s \qquad (32)$$

where

$$Q_{ijk} = A_j^i \ (\frac{\sigma_{Lik}}{E_{Li}} - \nu_{LTi} \frac{\sigma_{Tik}}{E_{Ti}}) + B_j^i \ (\frac{\sigma_{Tik}}{E_{Ti}} - \nu_{LTi} \frac{\sigma_{Lik}}{Li}) + C_j^i \ \frac{\sigma_{LTik}}{G_{LT}} \qquad (33)$$

Here σ_{Lik}, σ_{Tik}, σ_{LTik} are the components of the stress vector σ under k-th load condition, and E_L, E_T, G_{LT} ere the elastic condstants. The L,T axes are aligned respectively with the longitudinal and transverse directions of the monolayer lamina.

2.- *Linearized approximations of HILL-TSAI CRITERION*

$$G_t(x) \equiv 1-T_t(x^{(p)}) - \sum_{i=1}^{I} (x_i-x_i^{(p)}) \ \frac{\partial T_t}{\partial x_i} \ (x^{(p)}) \geqslant 0; \ t\varepsilon B_T \qquad (34)$$

where Tsai's number T_t given

$$T_t = [(\frac{\sigma_{Lik}}{F_L})^2 + (\frac{\sigma_{Tik}}{F_T})^2 - \frac{\sigma_{Lik} \ \sigma_{Tik}}{RF_L F_T} + (\frac{\sigma_{LTik}}{F_{LT}})^2]^{1/2} \qquad (35)$$

in which F_L, F_T, F_{LT}, are the stresses of failure in uniaxial tension, compression and shear, respectively.

3.- *Linearized approximation of BUCKLING CONSTRAINTS*

The linear stability of a structure is defined by the eigenvalue problem

$$[K - \lambda_j \ K_G] \ \ q_j = 0 \qquad (36)$$

where K is the total stiffness matrix of structure, K_G is the geometric stiffness matrix of the structure and q_j is the eigenvector associated with the j-th eigenvalue λ_j. The matrices K and K_G of composite stack are given in Ref [6]. The set of constraints retained can be expressed as follows

$$G_b(x) \equiv \lambda_b(x)-1 \geqslant 0 \ ; \qquad b\varepsilon B_r \qquad (37)$$

where B_r reprezents the set of buckling constraints retained. Linearized approximations of the buckling constraints retained after analyzing trial design $x^{(p)}$ are given by the following Taylor series expansion of the $G_b(x)$ (see egn. (37)) about the design $x^{(p)}$

$$G_b(x) \equiv \lambda_b(x^{(p)})-1 + \sum_{i=1}^{I} (x_i-x_i^{(p)}) \ \frac{\partial \lambda_b}{\partial x_i} \ (x_i^{(p)}) \geqslant 0 \ ; \ \ b\varepsilon B_r \qquad (38)$$

Here x_i represents the flexural rigities of the equivalent orthotropic plate D_{rs} as follows

$$D_{rs} = \frac{A_{rs}\ T^2}{12} \quad ; \qquad r,s = 1,2,6 \tag{39}$$

where

$$A_{rs} = \sum_{i=1}^{I} (C'_{rs})_i\ t_i \tag{40}$$

$$T = \sum_{i=1}^{I} t_i \tag{41}$$

The coefficients $(C'_{rs})_i$ are given in Ref [7], A_{rs} are the inplane stiffnesses of composite stack, I is the number available orientations angles in stack, t_i are the thicknesses of each ply of the composite stack for each membrane or shear-panel element.

THE SENSITIVITY ANALYSIS

The algorithm for structural optimization considered here need information about the gradients of the structural weight and constraints. The gradients are computed from finite element matrices and their derivatives. The derivatives can be computed numerically or analytically. In this paper analytical approach is applied to 4-node membrane and shear-panel isoparametric elements. Here are treated displacement constraints only but procedure is similar for all constraints. The finite element problem can be formulated as [8].

$$K\ u\ = F \tag{42}$$

The displacement behavior constraints and their derivatives with respect to design variables are given by

$$G_i(\bar{x}) = 1 - \frac{u}{\bar{u}} \geqslant 0 \tag{43}$$

$$\frac{\partial G_i}{\partial \bar{x}} = - \frac{1}{\bar{u}} \frac{\partial u_i}{\partial \bar{x}} \tag{44}$$

where $\bar{u}$ denotes the allowable displacement. The derivatives of eqn.(42) with respect to any design variables x_i will be

$$K \frac{\partial u}{\partial x_i} + \frac{\partial K}{\partial x_i} u = \frac{\partial F}{\partial x_i} \tag{45}$$

The key of sensitivity analysis is how to calculate the derivatives K and F with respect to x_i. The element stiffness matrix K_e is given by [8]

$$K_e = \int\int B^T\ D\ B\ \ |J|\ d\xi\ d\eta \tag{46}$$

The derivatives of the element stiffness matric with respect to x_i are given by

$$\left[\frac{\partial K}{\partial x_i}\right]_e = \int\int \left(\frac{B^T}{x_i}\ D\ B\ |J| + B^T D\ \frac{\partial B}{\partial x_i}\ |J| + B^T DB \frac{\partial |J|}{\partial x_i}\right) d\xi\ d\eta \tag{47}$$

or

$$\left[\frac{\partial K}{\partial x_i}\right]_e = \sum_j^{NG} \sum_k^{NG} W_j W_k \left(\frac{\partial B^T}{\partial x_i} DB|J| + B^T D \frac{\partial B}{\partial x_i}|J| + B^T DB \frac{\partial |J|}{\partial x_i} \right)_{jk} \tag{48}$$

where W_j and W_k are the weighting factors of numerical integration, NG is the number of Gauss rule adopted, and $(\ )_{jk}$ is the integrand function of a Gauss sampling point with local coordinates ξ_j and η_k. In similar manner can be derived all derivatitives.

NUMERICAL EXAMPLES

Here is considered problem weight optimization axially loaded laminate plate with a reinforced circular hole subject to strength constraints (Hill - Tsai). The formulation of this problem and finite element meshs are given in Fig 1 and 2. The iteration history and the influence of shape reinforcement for plate with fiber arangement $[0^\circ/\pm45^\circ/90^\circ]$ on total weight is given in Fig.3.

$W = 73.5$	$E_L = 14000.$	$F_{L1} = 0.12\times10^3$	$F_{LT} = 0.65\times10$
$R = 12.25$	$E_T = 500.$	$F_{T1} = 0.1\times10^3$	$R = 10$
$L = 98$	$G_{LT} = 500.$	$F_{L2} = 0.5\times10$	$\rho = 0.0018$
$\sigma_o = 46$ dN/mm	$\nu_{LT} = 0.35$	$F_{T2} = 0.12\times10^2$	$t_{min} = 0.1$

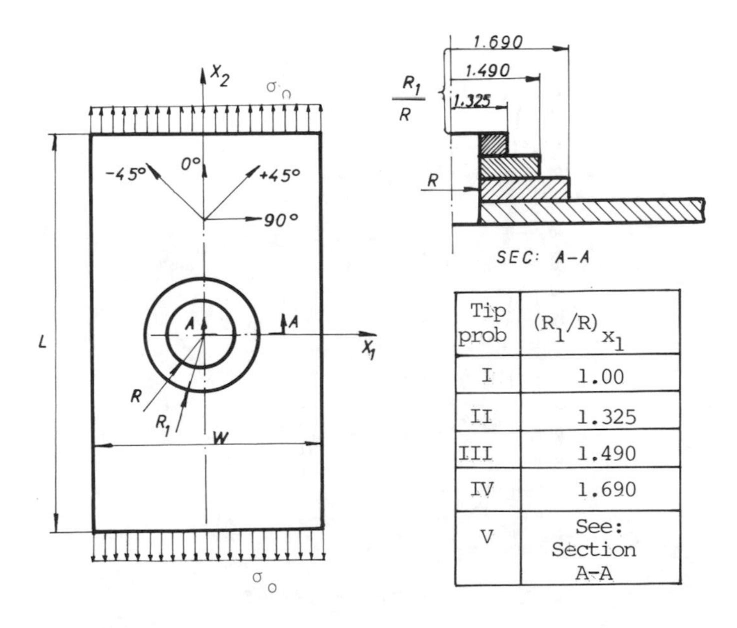

Tip prob	$(R_1/R)_{x_1}$
I	1.00
II	1.325
III	1.490
IV	1.690
V	See: Section A-A

Fig.1 The Formulation Of Problem

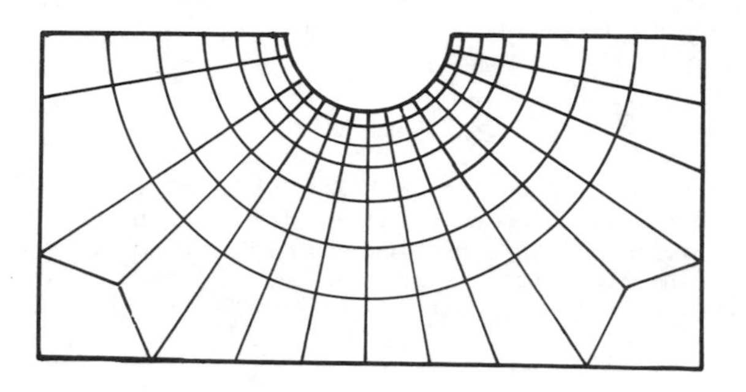

Fig. 2 The Finite Element Meshs

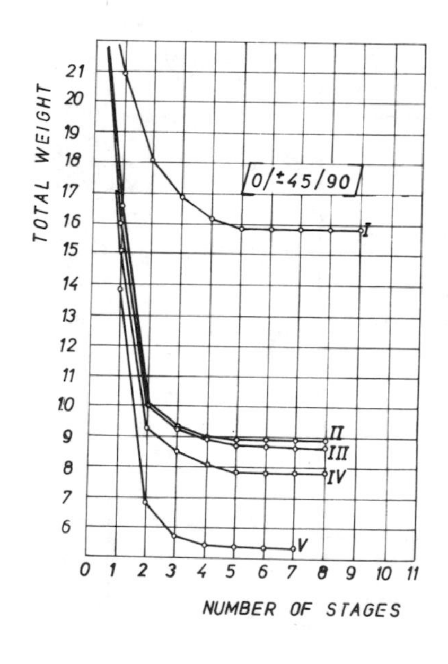

Fig. 3 Iteration Histories For Plate With Circular Hole

CONCLUSION

In this paper is presented an efficient method for the weight optimization large-scale composite structures subject to stiffness, strength and buckling constraints by using finite element method. Iteration conrvergence of results is achieved after 3-5 iterations only. The present method has had great success in the design of aircraft structures [7].

REFERENCES

1. Schmith, L.A. and Farshi, B., Some Approximation Concepts For Structural Synthesis, AIAA J, 12 (1974), 692-699.
2. Schmith, L.A. and Fleury,C., Structural Synthesis By Combining Approximation Concepts And Dual Methods, AIAAJ, 18 (1980), 1952-1260.
3. Zangwill, W.I., Nonlinear Programming: A Unified Approach, Prentice -Hall, New Jersey (1969).
4. Fleury, C. and Schmith, L.A., Dual Methods And Approximation Concepts In Structural Synthesis, NASA CR 3226 (1980).
5. Fletcher, R. and Reeves, C.M., Function Minimization By Conjugate Gradients, The Computer Journal, 7 (1964), 149-153.
6. Maksimović, S., Optimum Design Of Composite Structures, Proc. 3rd Int Conf. Composite Structures, Sept. (1985), Ed.I.H. Marshall, Appl.Sci. Publishers, London
7. Maksimović, S. and Rudić, Z., Structural Optimization Of Aircraft Structures, Raport VTI V4-3223-P (1985).
8. Zienkiewicz, O.C., The Finite Element Method In Engineering Science, McGraw-Hill, London (1977).

ON SENSITIVITY ANALYSIS AND OPTIMAL DESIGN FOR LAMINATES

PEDERSEN, PAULI

Department of Solid Mechanics
The Technical University of Denmark
Lyngby, Denmark

ABSTRACT

Sensitivity analysis with respect to fibre orientation as well as to ply thickness is performed. Well-known results from the theory of optimal design are worked out for the change in elastic energy of more general models for laminated plates.

The most simple plate model is then treated. For this model eigenfrequencies, buckling load, and displacement, are all related to the same non-dimensional functional, which is therefore taken as the objective for our optimization. A closed-form result for the optimal fibre orientation is obtained, and the problem of optimal thickness distribution is shown to be badly defined.

1. INTRODUCTION

Selection, design and optimization of materials are of vital importance for efficient use of laminated materials. The material design parameters treated are the orientations θ_k $(k = 1,2,\ldots,K)$ of the K plies of which the laminate is built up (stacked). Further design parameters are the relative positions ζ_k (indirectly the thicknesses) of the boundaries between two plies in the laminate (also indirectly the stacking sequence), and the condition of a given volume (total thickness) is thus directly accounted for.

Optimal design is a fascinating subject, and the results often give deeper insight into the structural problem itself. In a more global perspective, the main outcome is, perhaps, that we are forced to deal with sensitivity analysis. The sensitivity analysis plays an important role and should be treated as an integral part of ordinary analysis. This leads us to reconsider the basic laminate analysis, because unnecessary complications will force too early a shift to a numerical approach. A material description by non-dimensional parameters is suggested, and the method of multiple angles for rotational transformations is advanced.

Only a few of the thousands of papers on optimal design are devoted to laminates. The present paper is mainly an extension of the early work by BERT [2]. More numerically based studies are reported by RAO & SINGH [3],

SCHMIT & FARSHI [4] and, more recently, by ADALI [5]. Our goal is to minimize displacements, maximize buckling load, and maximize natural frequency, for a given laminate thickness.

An important result from the theory of design sensitivities is that the variations of certain energy functionals can be determined without displacement variations, i.e. with constant strain/curvature of the laminated plate. This is worked out for laminates with coupling terms, and a rather simple result is obtained.

Then, treating only the most simple rectangular plate model, closed-form optimization results are obtained. The optimal fibre orientation is presented as a function of the mode parameter η (defined as the ratio of the two half-wave lengths), with non-trivial solutions ($\theta_k \neq 0^0$ and 90^0) only in the range $0.55 \lesssim \eta$ or $\eta^{-1} \leq 1$. This result, which was first presented by BERT [2], is shown in this paper to be applicable in many other situations and for different materials.

The sensitivity analysis in relation to the thickness distribution gives similar simple results, and some general conclusions for specific cases can be drawn. First of all, the optimization problem will often be badly defined, and the individual plies should preferably have their maximum or minimum thickness (active side constraints). Secondly, the case of $\eta = 1$ (square displacement pattern) is a rather degenerated case, because the thickness distribution may then be chosen arbitrarily.

2. CHANGE IN ELASTIC ENERGY WITH CHANGE IN LAMINATE DESIGN

The constitutive relations for a laminated plate are traditionally written:

$$\begin{Bmatrix} \{N\} \\ \{M\} \end{Bmatrix} = \begin{bmatrix} [A] & ; & [B] \\ [B] & ; & [D] \end{bmatrix} \begin{Bmatrix} \{\varepsilon^0\} \\ \{\kappa\} \end{Bmatrix} \tag{2.1}$$

where $\{N\}$, $\{\varepsilon^0\}$ are corresponding inplane forces and midsurface strains, and $\{M\}$, $\{\kappa\}$ are corresponding bending moments and curvatures. The matrices of order three by three, $[A]$, $[B]$ and $[D]$, contain the laminate stiffnesses. Assuming linear elasticity and dead loads, the specific elastic energy $\bar{U}$ (energy per area of the laminated plate) and the specific external work $\bar{W}$ are

$$\begin{aligned} 2\bar{U} = \bar{W} &= \{\varepsilon^0\}^T\{N\} + \{\kappa\}^T\{M\} \\ &= \{\varepsilon^0\}^T[A]\{\varepsilon^0\} + \{\kappa\}^T[D]\{\kappa\} + 2\{\varepsilon^0\}^T[B]\{\kappa\} \ . \end{aligned} \tag{2.2}$$

For linear elastic models subjected to dead loads, the variation in compliance (external work) δW with respect to *design change* is obtained as: minus twice the variation of elastic energy for *fixed displacements*. In relation to (2.2), this means that midsurface strains $\{\varepsilon^0\}$ and curvatures $\{\kappa\}$ are kept (not assumed) constant during design variation, and we get

$$2\delta\bar{U} = \delta\bar{W} = - \{\varepsilon^0\}^T[\delta A]\{\varepsilon^0\} - \{\kappa\}^T[\delta D]\{\kappa\} - 2\{\varepsilon^0\}^T[\delta B]\{\kappa\} \ . \tag{2.3}$$

Thus, our attention is directed towards the variation of laminate stiffnesses.

We shall here use a *non-dimensional* description of material as well as of plies and laminate, and the method of multiple angles for rotational transformations is advanced. Referring to the book by JONES [6], we may write the laminate stiffnesses:

$$[A] = \frac{\bar{E}_L h}{8\bar{\alpha}_0}[a] = \frac{\bar{E}_L h}{8\bar{\alpha}_0}\sum_{k=1}^{K} d_k \ (\zeta_k - \zeta_{k-1})[c]_k$$

$$[B] = \frac{\bar{E}_L h^2}{8\bar{\alpha}_0}[b] = \frac{E_L h^2}{8\bar{\alpha}_0}\sum_{k=1}^{K} d_k \frac{1}{2}(\zeta_k^2 - \zeta_{k-1}^2)[c]_k \qquad (2.4)$$

$$[D] = \frac{\bar{E}_L h^3}{8\bar{\alpha}_0}[d] = \frac{\bar{E}_L h^3}{8\bar{\alpha}_0}\sum_{k=1}^{K} d_k \frac{1}{3}(\zeta_k^3 - \zeta_{k-1}^3)[c]_k$$

$$d_k := (E_L/\alpha_0)_k/(\bar{E}_L/\bar{\alpha}_0) \quad ; \quad \alpha_0 \text{ by (2.6)}$$

where the modulus in the fibre direction is E_L ($\bar{E}_L$ is some reference value, so when all plies are of the same material, we get $d_k = 1$) . In the description by TSAI & PAGANO [7], the non-dimensional constitutive matrix $[c]_k$ (omitting the ply index k) is given by

$$[c] = \begin{bmatrix} \alpha_1+\alpha_2\cos2\theta+\alpha_3\cos4\theta \ ; & \alpha_4-\alpha_3\cos4\theta & ; -\frac{1}{2}\alpha_2\sin2\theta-\alpha_3\sin4\theta \\ & \alpha_1-\alpha_2\cos2\theta+\alpha_3\cos4\theta & ; -\frac{1}{2}\alpha_2\sin2\theta+\alpha_3\sin4\theta \\ \text{symmetric} & & \frac{1}{2}(\alpha_1-\alpha_4)-\alpha_3\cos4\theta \end{bmatrix}, \qquad (2.5)$$

with the non-dimensional material parameters defined by

$$\begin{aligned} \alpha_0 &:= (1-\nu_{LT}\nu_{TL}) = (1-\nu_{LT}^2 E_T/E_L) \ ; \\ \alpha_1 &:= 3 + (E_T/E_L)(3+2\nu_{LT}) + 4(G_{LT}/E_L)\alpha_0 \ ; \\ \alpha_2 &:= 4 - (E_T/E_L)4 \ ; \\ \alpha_3 &:= 1 + (E_T/E_L)(1-2\nu_{LT}) - 4(G_{LT}/E_L)\alpha_0 \ ; \\ \alpha_4 &:= 1 + (E_T/E_L)(1+6\nu_{LT}) - 4(G_{LT}/E_L)\alpha_0 \ . \end{aligned} \qquad (2.6)$$

The engineering parameters E_L , E_T , G_{LT} and ν_{LT} for modulus in the fibre direction, the transverse direction, and in shear and Poisson's ratio, may be rather different. However, as seen in table 2.1 below, the non-dimensional parameters α_i are not too different.

Material	G Pa			ν_{LT}	α_0	α_1	α_2	α_3	α_4	$\frac{1}{2}(\alpha_1-\alpha_4)$	$\frac{\alpha_2}{4\alpha_3}$	Ref.
	E_L	E_T	G_{LT}									
Graphite /Epoxy	181.0	10.30	7.17	0.28	0.9955	3.3603	3.7724	0.8673	0.9948	1.1828	1.0874	[1]; p. 19
	138.0	8.96	7.10	0.30	0.9942	3.4383	3.7403	0.8214	0.9772	1.2306	1.1384	[1]; p. 19
	207.0	5.17	2.59	0.25	0.9984	3.1374	3.9001	0.9625	1.0125	1.0625	1.0130	[8]; p.276
Boron /Epoxy	204.0	18.50	5.59	0.23	0.9952	3.4229	3.6373	0.9399	1.1068	1.1581	0.9675	[1]; p. 19
	207.0	20.70	6.90	0.30	0.9910	3.4921	3.6000	0.9079	1.1479	1.1721	0.9913	[8]; p.276
	213.7	23.44	5.17	0.28	0.9914	3.4864	3.5613	0.9523	1.1980	1.1442	0.9349	[3]; p.103
Aramid /Epoxy	76.0	5.50	2.30	0.34	0.9916	3.3864	3.7105	0.9031	1.1000	1.1432	1.0271	[1]; p. 19
Glass /Epoxy	38.6	8.27	4.14	0.26	0.9855	4.1770	3.1430	0.6801	1.1257	1.5256	1.1555	[1]; p. 19
	53.8	17.90	8.96	0.25	0.9792	4.8168	2.6691	0.5140	1.1795	1.8187	1.2981	[8]; p.276
ISOTROPIC	E	E	$\frac{E}{2(1+\nu)}$	ν	$1-\nu^2$	8	0	0	8ν	$4(1-\nu)$	$\frac{1+\nu}{1+\nu-2\nu^2}$	-

Table 2.1: Examples of actual material constants.

With respect to *variation of fibre direction* $\delta\theta$, the stiffness variations will, by (2.4), be

$$
\begin{aligned}
[\delta A] &= \widetilde{W}\,[c]_{,\theta}\,\delta\theta_k\,(\zeta_k - \zeta_{k-1}) \\
[\delta B] &= \widetilde{W}\,[c]_{,\theta}\,\delta\theta_k\,\tfrac{1}{2}h\,(\zeta_k^2 - \zeta_{k-1}^2) \\
[\delta D] &= \widetilde{W}\,[c]_{,\theta}\,\delta\theta_k\,\tfrac{1}{3}h^2\,(\zeta_k^3 - \zeta_{k-1}^3)
\end{aligned}
\tag{2.7}
$$

where the common factor $\widetilde{W}$ of energy per area and the gradient $[c]_{,\theta}$ of the constitutive matrix are

$$
\widetilde{W} = \frac{\bar{E}_L h}{8\bar{\alpha}_0} d_k = \frac{E_L h}{8\alpha_0}
\tag{2.8}
$$

$$
[c]_{,\theta} = \begin{bmatrix} -2\alpha_2\sin2\theta-4\alpha_3\sin4\theta \; ; & 4\alpha_3\sin4\theta & ; \; -\alpha_2\cos2\theta-4\alpha_3\cos4\theta \\ & 2\alpha_2\sin2\theta-4\alpha_3\sin4\theta & ; \; -\alpha_2\cos2\theta+4\alpha_3\cos4\theta \\ \text{symmetric} & & 4\alpha_3\sin4\theta \end{bmatrix}
$$

We note the similarities of the different variations, and thus, with respect to variation of fibre direction, have in total

$$
\delta\bar{U} = -\frac{1}{2}\widetilde{W}\delta\theta_k \cdot \left(q_1 \sin 2\theta + q_2 \cos 2\theta + q_3 \sin 4\theta + q_4 \cos 4\theta\right)
\tag{2.9}
$$

The non-dimensional constants $q_1 - q_4$ in (2.9) depend on the actual strains and curvatures and on the material parameters of the actual ply k and its position in the laminate.

With respect to *variation of thickness distribution* by $\delta\zeta_k$, the stiffness variation will, by (2.4), be

$$[\delta A] = \widetilde{W}(\delta\zeta_k)\left([c]_k - \frac{d_{k+1}}{d_k}[c]_{k+1}\right)$$

$$[\delta B] = \widetilde{W}(\delta\zeta_k)\left([c]_k - \frac{d_{k+1}}{d_k}[c]_{k+1}\right)\zeta_k h \tag{2.10}$$

$$[\delta D] = \widetilde{W}(\delta\zeta_k)\left([c]_k - \frac{d_{k+1}}{d_k}[c]_{k+1}\right)\zeta_k^2 h^2$$

We note the simple dependence on ζ_k itself by ζ_k for $[\delta B]$ and ζ_k^2 for $[\delta D]$. In essence this leads to not well-defined optimization problems.

3. OPTIMAL FIBER ORIENTATION FOR THE MOST SIMPLE PLATE MODEL

The most simple plate model is a model where only part of the bending stiffnesses are involved:

$$[D] = \begin{bmatrix} D_{11} ; & D_{12} ; & 0 \\ D_{12} ; & D_{22} ; & 0 \\ 0 ; & 0 ; & D_{66} \end{bmatrix} \quad ; \quad [B] = [0] \quad ; \quad [A] = "[0]" \ , \tag{3.1}$$

as for a symmetric $([B] = [0])$, balanced $(D_{16} = D_{26} = 0)$ laminate, where in-plane displacements are not accounted for $([A] = "[0]")$. For such a model we define a non-dimensional functional ϕ by

$$\phi := \eta^4 d_{11} + d_{22} + 2\eta^2(d_{12} + 2d_{66}) \ , \tag{3.2}$$

where η is a mode parameter describing the displacement pattern. For the rectangular plate with length a, width b, and thickness h, our Cartesian coordinate system is placed so that

$$0 \le x \le a \quad ; \quad 0 \le y \le b \quad ; \quad -\frac{h}{2} \le z \le \frac{h}{2} \ , \tag{3.3}$$

and assuming a displacement field with transverse displacement w given by

$$w = w_{mn} \sin\frac{x}{\ell_x} \sin\frac{y}{\ell_y} \quad ; \quad \ell_x := \frac{a}{m\pi} \quad ; \quad \ell_y := \frac{b}{n\pi} \ , \tag{3.4}$$

the mode parameter η is the ratio of the two half-wave lengths $\ell_x ; \ell_y$, i.e.

$$\eta := \ell_y/\ell_x = (mb)/(na) \ . \tag{3.5}$$

The non-dimensional bending stiffnesses d_{ij} are defined by (2.4).

The eigenfrequency for the plate is

$$\omega_{mn} = \frac{\pi^2 n^2}{\sqrt{8}} \frac{h}{b^2} \sqrt{\frac{\bar{E}_L}{\rho\bar{\alpha}_0}} \sqrt{\phi} \ , \tag{3.6}$$

and the buckling load is,

$$(N_x)_{cr} = \frac{\pi^2}{8} \frac{h^3}{b^2} \frac{\bar{E}_L}{\bar{\alpha}_0} \frac{\phi}{\eta^2} , \qquad (3.7)$$

while the displacement amplitude corresponding to distributed load is

$$|w|_{max} = p_{mn} \frac{8}{\pi^4 n^4} \frac{b^4}{h^3} \frac{\bar{\alpha}_0}{\bar{E}_L} \frac{1}{\phi} \quad \text{for} \quad p = p_{mn} \sin\frac{x}{\ell_x} \sin\frac{y}{\ell_y} . \qquad (3.8)$$

Let us consider a laminate of total thickness $\hat{h}$ and the calculated responses $\hat{\omega}$, $(\hat{N}_x)_{cr}$ and $|\hat{w}|_{max}$. Then, if the actual constraints are $\bar{\omega}$, $(\bar{N}_x)_{cr}$, $|\bar{w}|_{max}$, we can *scale the laminate* by choosing

$$\bar{h} = \hat{h} \cdot \text{Max}\left\{ \frac{\bar{\omega}}{\hat{\omega}} \; ; \; \sqrt[3]{\frac{(\bar{N}_x)_{cr}}{(\hat{N}_x)_{cr}}} \; ; \; \sqrt[3]{\frac{|\hat{w}|_{max}}{|\bar{w}|_{max}}} \right\} , \qquad (3.9)$$

and directly obtain a feasible and at the same time "fully stressed" design.

The laminate stiffnesses depend on the orientations of the plies, as shown by (2.7) - (2.8). We can assume that the cost (weight) of the laminate is independent of the chosen orientation(s); therefore, if we want to maximize frequency/maximize buckling load/minimize displacements, the optimization problem will be an unconstrained one:

$$\underset{\text{over } \theta}{\text{Max}} \quad (\phi) , \qquad (3.10)$$

and we shall find the solution at one or more of the stationary points $\phi_{,\theta} := \partial\phi/\partial\theta = 0$. In fact, the functional ϕ is a non-dimensional elastic energy describing $\{\kappa\}^T[D]\{\kappa\}$ for our specific model.

Taking variations with respect to θ_k , we (analogously to the general result (2.9)) get

$$\phi_{,\theta_k} = (\zeta_k^3 - \zeta_{k-1}^3) \frac{2}{3}\Big(\big(\alpha_2(1-\eta^4) - 4\alpha_3(1-6\eta^2+\eta^4)\cos 2\theta\big)\sin 2\theta\Big)_k , \qquad (3.11)$$

which may also be found in the early paper by BERT [2]. The factor $(\zeta_k^3 - \zeta_{k-1}^3)$ is always positive, and as $\alpha_2/(4\alpha_3)$ is close to unity, the sign of $\phi_{,\theta_k}$ depends heavily on the function $(1-\eta^4)/(1-6\eta^2+\eta^4)$. Stationary values of ϕ , i.e. $\phi_{,\theta} = 0$ are found for

$$\theta_k = 0 \; ; \; |\theta_k| = \pi/2 \; ; \; |2\theta_k| = \text{arc}\cos\left(\frac{\alpha_2}{4\alpha_3} \frac{(1-\eta^4)}{(1-6\eta^2+\eta^4)}\right)_k , \qquad (3.12)$$

In fig. 3.1 we show the optimal orientations for the different materials (different α_2 ; α_3) .

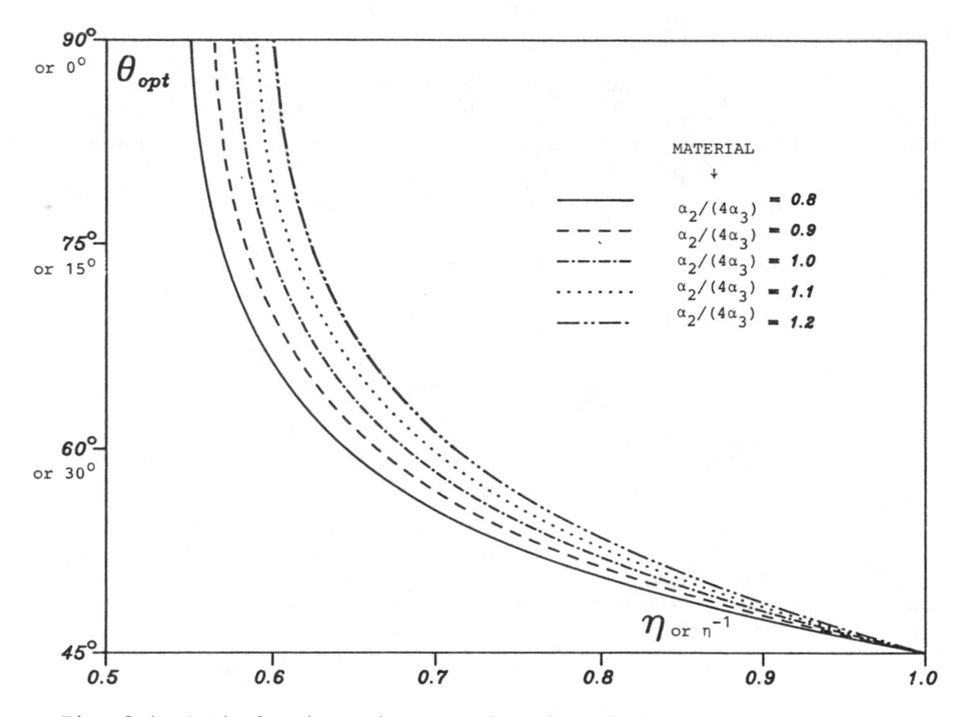

Fig. 3.1: Optimal orientation as a function of the mode parameter η .

The conclusions of this result for harmonic displacement distributions are:

- The optimal orientation depends mainly on the mode parameter η .
- Two cases for which $\eta_1 = \eta_2^{-1}$ has complementary optimal orientations: $\theta_2 = \pi/2 - \theta_1$.
- For "extreme" values of η , the optimal orientation is either 90° ($\eta \lesssim 0.6$) or 0° ($\eta \gtrsim 1.7$) .
- In the "switching" domains of η , the optimal orientation is extremely sensitive to the mode, i.e. $0.6 \lesssim \eta \lesssim 0.7$ gives $90^o \gtrsim \theta^o_{opt} \geq 60^o$, and $1.4 \lesssim \eta \lesssim 1.7$ gives $30^o \geq \theta^o_{opt} \geq 0^o$.
- The optimal orientation is rather insensitive to the material parameters.
- The optimal orientation is the same, independent of the position of the layer in the laminate, and thus the same for all layers.
- Local optima exist because several orientations return a stationary value of ϕ . For $\alpha_2/(4\alpha_3) < 1$, additional local optima exist close to η or $\eta^{-1} = 0$.

4. OPTIMAL THICKNESS DISTRIBUTION - A NOT WELL-DEFINED PROBLEM

The problem of optimal thickness distribution has also been studied, see RAO & SINGH [3] and SCHMIT & FARSHI [4], who assume the orientations to be given and all the plies to be made of the same material. We shall see that this problem is often a badly defined optimization problem when only non-conflicting constraints are included, which is the case when only the functional (3.2) is involved. The total thickness h is given, and we can thus only choose the relative ply thicknesses. In the present paper, this is described by the non-dimensional parameters ζ_k for $k = 1,2,\ldots K-1$, where K is the total number of plies.

The unconstrained problem is

$$\underset{\text{over } \zeta_k}{\text{Max}} \ (\phi) \quad ; \quad \text{where} \quad -\frac{1}{2} < \zeta_{k-1} < \zeta_k < \zeta_{k+1} < \frac{1}{2} , \quad (k = 1,2,\ldots K-1) \tag{4.1}$$

and the partial derivatives

$$\phi_{,\zeta_k} := \partial\phi/\partial\zeta_k = \eta^4 d_{11,\zeta_k} + d_{22,\zeta_k} + 2\eta^2 (d_{12,\zeta_k} + 2d_{66,\zeta_k}) , \tag{4.2}$$

are naturally of major importance. Variations $[\delta D]$ are shown by (2.10).

Plies of the same material. For this case we are in reality comparing the orientation θ_k of ply No. k with the orientation θ_{k+1} of ply No. $k+1$, when we change the boundary position ζ_k. Therefore, it is not surprising that a result similar to (3.11) is obtained:

$$\phi_{,\zeta_k} = 2\zeta_k^2 d_k \Big(-\alpha_2 (1 - \eta^4) + 2\alpha_3 (1 - 6\eta^2 + \eta^4)(\cos 2\theta_k + \cos 2\theta_{k+1}) \Big) \cdot (\cos 2\theta_k - \cos 2\theta_{k+1}) . \tag{4.3}$$

As the sign of $\phi_{,\zeta_k}$ is independent of ζ_k itself, the optimization problem is degenerated. The solution is either to choose the minimum or the maximum thickness for ply No. k as otherwise prescribed.

Some specific combinations of orientations, θ_k ; θ_{k+1}, give $\phi_{,\zeta_k} = 0$, which means that the thickness distribution (among these two plies) has no influence on the functional ϕ. These values we read directly from (4.3):

$$\phi_{,\zeta_k} = 0 \quad \text{for} \quad |\theta_k| = |\theta_{k+1}| \quad \text{and for}$$
$$\frac{1}{2}(\cos 2\theta_k + \cos 2\theta_{k+1}) = \frac{\alpha_2}{4\alpha_3} \frac{(1 - \eta^4)}{(1 - 6\eta^2 + \eta^4)} , \tag{4.4}$$

with the important specific case, which is material-independent

$$\text{for} \quad \eta = 1 \quad \text{is} \quad \phi_{,\zeta_k} = 0 \quad \text{when} \quad \theta_{k+1} = \frac{\pi}{2} - \theta_k . \tag{4.5}$$

The above results seem to conflict with the optimal designs obtained by RAO & SINGH [3], who presented detailed thickness distributions. In [3], an angle-ply laminate with $\eta = 1$ was assumed, and this case, by (4.5), always returns $\phi_{,\zeta} = 0$. The specific optimal designs presented in [3] must therefore primarily relate to the numerical procedure.

Plies of different materials. Dealing with plies of different materials, the general stiffness variations (2.10) have to be applied. We still see that the actual thickness distribution as described by ζ_k has no influence on the sign of $\phi_{,\zeta_k}$. Therefore, the optimization problem is still rather degenerated because it pays to increase or decrease ζ_k as much as possible. The sign of $\phi_{,\zeta_k}$ depends on the actual two plies and on the actual mode. The dependence on the mode parameter is by a second order polynomial in (η^2) , and the values η_0 which return $\phi_{,\zeta_k} = 0$ are thus of primary importance.

We finish this section by concluding:

- The problem of optimal thickness distribution is only a well-posed problem when conflicting constraints (multiple loads, strength constraints, etc.) are formulated.
- The parameter of main importance is the mode parameter η .

5. CONCLUSION

General results from the theory of sensitivity analysis are related to laminate design. To limit even this more general description, shear deformations are not included but would in reality just add more terms. Elastic energy is directly related to different responses and the variation of this energy with respect to design changes is determined only by variations of the laminate stiffnesses, i.e. displacement (strains) variations are not involved.

The nice and simple result of BERT [2] is rederived and shown to be valid not only in relation to fundamental frequency, but also in relation to buckling load, to higher order frequencies, and to displacements in general. The optimal fibre orientation is less dependent on the actual material but strongly dependent on the displacement mode.

Similar numerical results can be obtained for non-harmonic displacements [9], using Fourier expansions and Newton-Raphson iterations. Extensions to more advanced plate and laminate models will be based on the finite element method of analysis and the linear programming method of optimization. Then it will also be natural to treat the more practical problems with constraints on strength.

Numerical approaches are mostly used for optimizing the thickness distribution. It is shown here that some simple results can be obtained analytically for this problem, too. The problems with only constraint on a fixed total thickness are not found to be well-defined.

REFERENCES

[1] Tsai, S.W., Hahn, H.T., Introduction to composite materials, Technomic, Westport Com., 1980, 457.

[2] Bert, C.W., Optimal design of a composite-material plate to maximize its fundamental frequency, J. of Sound and Vibration, 50, 1977, 229.

[3] Rao, S.S., Singh, K., Optimum design of laminates with frequency constraints, J. of Sound and Vibration, 67, 1979, 101.

[4] Schmit, L.A., Farshi, B., Optimum design of laminated fibre composite plates, Int. J. for Num. Meth. in Eng., 11, 1977, 623.

[5] Adali, S., Design of shear-deformable antisymmetric angle-ply laminates to maximize the fundamental frequency and frequency separation, Composite Structures, 2, 1984, 349.

[6] Jones, R.M., Mechanics of composite materials, McGraw-Hill, N.Y., 1975, 355.

[7] Tsai, S.W., Pagano, N.J., Invariant properties of composite materials, in Tsai, Halpin, Pagano (eds.): "Composite Materials Workshop", Technomic, Westport Com., 1968, 233.

[8] Bert, C.W., Design of clamped composite-material plates to maximize fundamental frequency, J. of Mechanical Design ASME, 100, 1978, 274.

[9] Pedersen, P., Minimum flexibility of non-harmonic loaded laminated plates, Proc. of Symp. on Mechanical Characterisation of Fibre Composite Materials, Aalborg, Denmark, 1986.

INDEX OF CONTRIBUTORS

SUBJECT INDEX